Capillary Electrophoresis–Mass Spectrometry for Metabolomics

New Developments in Mass Spectrometry

Editor-in-chief:
Frank Sobott, *University of Leeds, UK*

Series editors:
Juan F. Garcia-Reyes, *Universidad de Jaén, Spain*
Marek Domin, *Boston College, USA*

Titles in the Series:
1: Quantitative Proteomics
2: Ambient Ionization Mass Spectrometry
3: Sector Field Mass Spectrometry for Elemental and Isotopic Analysis
4: Tandem Mass Spectrometry of Lipids: Molecular Analysis of Complex Lipids
5: Proteome Informatics
6: Capillary Electrophoresis–Mass Spectrometry for Metabolomics

How to obtain future titles on publication:
A standing order plan is available for this series. A standing order will bring delivery of each new volume immediately on publication.

For further information please contact:
Book Sales Department, Royal Society of Chemistry, Thomas Graham House, Science Park, Milton Road, Cambridge, CB4 0WF, UK
Telephone: +44 (0)1223 420066, Fax: +44 (0)1223 420247
Email: booksales@rsc.org
Visit our website at www.rsc.org/books

Capillary Electrophoresis–Mass Spectrometry for Metabolomics

Edited by

Rawi Ramautar
Leiden University, The Netherlands
Email: r.ramautar@lacdr.leidenuniv.nl

ROYAL SOCIETY
OF **CHEMISTRY**

New Developments in Mass Spectrometry No. 6

Print ISBN: 978-1-78801-104-4
PDF ISBN: 978-1-78801-273-7
EPUB ISBN: 978-1-78801-484-7
ISSN: 2044-253X

A catalogue record for this book is available from the British Library

The Royal Society of Chemistry is a charity, registered in England and Wales, Number 207890, and a company incorporated in England by Royal Charter (Registered No. RC000524), registered office: Burlington House, Piccadilly, London W1J 0BA, UK, Telephone: +44 (0) 207 4378 6556.

For further information see our web site at www.rsc.org

Printed in the United Kingdom by CPI Group (UK) Ltd, Croydon, CR0 4YY, UK

Preface

Metabolomics research makes it possible to study changes in metabolism, which may result in the identification of specific metabolites (biomarkers) or metabolic profiles that can be linked to certain states or conditions of organisms. So far, metabolomics studies have effectively shown changes in metabolism resulting from disease or environmental interactions, and provided new mechanistic insights that could not have been identified using classical biochemical approaches. A key aim of using metabolomics is to address fundamental questions in biology and biomedicine, with the ultimate goal to prevent disease and to improve health during the course of the human lifespan.

Metabolomics studies may be considered highly challenging from an analytical chemistry viewpoint; for example, the human metabolome contains more than 100 000 metabolites with different physico-chemical properties in a wide concentration range. Therefore, advanced analytical separation techniques are currently used for the global profiling of metabolites in biological samples. In this context, capillary electrophoresis–mass spectrometry (CE-MS) has emerged as a very useful analytical tool for the analysis of (highly) polar and charged metabolites, as compounds are separated on the basis of their charge-to-size ratio. In comparison to especially chromatographic-based separation techniques, the use of CE-MS in metabolomics and bioanalysis is still underrepresented, despite this approach having some unique analytical characteristics for metabolic profiling studies.

In this book, the possibilities of CE-MS for metabolomics studies are emphasized with a special focus on recent technological developments. Though not explicitly indicated, the book is roughly structured into two parts. After introducing the potential of CE and CE-MS for metabolomics from a historical perspective, the first part deals with chapters on advances in sample preparation, separation conditions, preconcentration techniques,

New Developments in Mass Spectrometry No. 6
Capillary Electrophoresis–Mass Spectrometry for Metabolomics
Edited by Rawi Ramautar

Published by the Royal Society of Chemistry, www.rsc.org

interfacing designs and to some extent data analysis. Then, as a bridge to the second part, the strengths and limitations of CE-MS for metabolomics are discussed in comparison to chromatographic-based separation techniques. The second part highlights applications for which CE-MS is a well-suited approach or may even be considered as key, such as for chiral metabolomics and single-cell metabolomics. Specific attention is also paid to quality assurance and validation strategies, and how to incorporate them in a CE-MS-based analytical workflow to obtain reliable metabolomics data. Within each chapter, the impact of developments in CE-MS methodology and procedures is demonstrated by illustrative metabolomics studies.

Overall, this book should give a contemporary representation of the main technological developments in CE-MS for metabolomics, including important application fields. I hope that the book is useful for both beginners and experts in the field of metabolomics and bioanalysis in general, and that it will stimulate further research in this area. I highly appreciate the efforts of all contributing authors and would like to thank them for their excellent chapters. This book would not have been possible without their valuable contributions. I would also like to thank the referees for their help in the evaluation of these papers and the editorial staff, especially Robin Driscoll, from the Royal Society of Chemistry for their support during the preparation of this book.

Rawi Ramautar
Leiden, The Netherlands

About the Editor

Strongly believing in the power of analytical technology to contribute to a better understanding of the biochemical mechanisms underlying diseases, Rawi Ramautar studied both Pharmacochemistry and Analytical Sciences at Vrije Universiteit of Amsterdam in order to have the right background for PhD training in this direction. In 2010, he completed his PhD on the development of capillary electrophoresis-mass spectrometry methods for metabolomics at the Utrecht University. Intrigued by metabolomics for disease prediction and diagnosis, Rawi switched to the Leiden University Medical Center to broaden his horizons on this topic. In 2013 and 2017, Rawi Ramautar received the highly prestigious Veni and Vidi personal grants, respectively, from the Netherlands Organisation for Scientific Research. Currently, he is a Principal Investigator (tenured) at the Leiden Academic Center for Drug Research of the Leiden University where his group is developing microscale analytical workflows for volume-restricted biomedical problems.

New Developments in Mass Spectrometry No. 6
Capillary Electrophoresis–Mass Spectrometry for Metabolomics
Edited by Rawi Ramautar

Published by the Royal Society of Chemistry, www.rsc.org

Contents

New Developments in Mass Spectrometry No. 6
Capillary Electrophoresis–Mass Spectrometry for Metabolomics
Edited by Rawi Ramautar

Published by the Royal Society of Chemistry, www.rsc.org

CHAPTER 1

Capillary Electrophoresis–Mass Spectrometry for Metabolomics – From Metabolite Analysis to Metabolic Profiling

RAWI RAMAUTAR

Leiden Academic Center for Drug Research, Leiden University, Einsteinweg 55, 2333 CC Leiden, The Netherlands
*E-mail: r.ramautar@lacdr.leidenuniv.nl

1.1 Introduction

One of the major goals of using a metabolomics approach is to obtain an answer to a specific biological/clinical question.[1] For this purpose, advanced analytical separation techniques are employed for the reliable profiling of (endogenous) metabolites in various biological samples. The most recent version of the Human Metabolome Database (HMDB version 4.0) contains more than 100 000 metabolite entries including both water-soluble and lipid-soluble metabolites, as well as metabolites that would be considered as either abundant (>1 μM) or low-abundant (<1 nM).[2] This number of metabolites and their concentration range clearly indicate that various analytical separation techniques with complementary separation mechanisms, like nuclear magnetic resonance (NMR) spectroscopy and mass spectrometry (MS)

New Developments in Mass Spectrometry No. 6
Capillary Electrophoresis–Mass Spectrometry for Metabolomics
Edited by Rawi Ramautar

Published by the Royal Society of Chemistry, www.rsc.org

hyphenated to gas chromatography (GC), liquid chromatography (LC) and/or capillary electrophoresis (CE), are needed in order to analyze a wide range of (endogenous) metabolites in a given biological sample.[3] For example, for the characterization of (endogenous) metabolites in human serum, a number of analytical separation techniques with different separation mechanisms have been used, and this strategy allowed the detection of more than 4000 metabolites at concentration levels from the low nM- to the high mM-range.[4]

Within the metabolomics field, two analytical strategies can be distinguished, *i.e.* the non-targeted approach, also known as global metabolic profiling, and the targeted approach.[5–8] In non-targeted metabolomics, the aim is to analyze as many (endogenous) metabolites as possible in a given biological sample without having *a priori* knowledge on the nature and identity of the measured compounds. In targeted metabolomics, the analysis is often focused on the generation of quantitative data for pre-selected metabolites. Both non-targeted and targeted approaches can be employed within a single metabolomics study, where the first approach is generally used for the screening of differential metabolites (or metabolic indicators) for diseases. In the second approach, the statistically relevant differential metabolites may then be quantified and confirmed, preferably employing standardized operating procedures.[9] Reproducibility and validation of confirmed differential metabolites for a certain disease require further assessment in large sample cohorts, preferably at multiple sites, in order to determine whether these compounds can be used as biomarkers.[10]

At present, the analytical techniques most often used for global metabolic profiling studies are ^{1}H NMR spectroscopy, reversed-phase ultra performance (RP-UP)LC-MS and GC-MS.[11–16] Notably, RP-UPLC-MS methods using columns with sub-2 µm porous particles or core–shell silica particles have received a lot of attention for the efficient and fast profiling of metabolites in complex samples.[17–19] NMR spectroscopy can still be considered the most robust and reliable analytical tool for high-throughput and reproducible metabolic profiling of body fluids (such as for example serum and urine) using minimal sample pretreatment. Typically, this approach provides quantitative data for metabolites present in the µM to mM concentration range employing a data acquisition time of 5 min per sample.[11] However, evaluation of recorded NMR data may be considered as a challenging task due to the complex spectra and superimposition of signals at certain chemical shift regions. GC-MS has a well-known track record for the profiling of (endogenous) metabolites in various body fluids in a clinical setting due to its high separation efficiency and sensitivity.[20,21] Also, the reliable identification of compounds by GC-MS is a very strong asset for metabolomics studies. However, this approach is not suitable for the analysis of non-volatile, thermolabile, and/or highly polar compounds and, therefore, derivatization is usually needed to yield volatile and thermostable compounds. Derivatization is generally a time-consuming procedure, thereby potentially limiting the high-throughput capacity of GC-MS for metabolic profiling of large sets of clinical samples.

Both LC-MS and CE-MS can be used for the analysis of polar and non-volatile metabolites in biological samples without using derivatization and laborious sample pretreatment procedures. As indicated above, RP-UPLC-MS has become a key technique for global metabolic profiling studies.[22–24] A broad array of metabolites can be analyzed by RP-LC approaches; however, the hydrophobic stationary phases generally used do not provide sufficient retention and selectivity for (highly) polar and charged metabolites. To enable the analysis of such compounds by RP-LC-MS, ion-pairing agents like tributylamine or hexylamine can be added to the mobile phase.[25–28] However, the use of ion-pair agents in RPLC-MS may result in severe ion suppression and it may contaminate the ion source and ion optics. Furthermore, stability and re-equilibration of the column prior to the next injection may also be considered an issue in ion-pair RP-LC-MS.[8] An emerging liquid chromatographic separation tool for metabolomics is hydrophilic interaction LC (HILIC), in which a polar stationary phase is used in combination with aqueous organic eluents. This approach is increasingly employed in combination with RP-LC-MS for global metabolic profiling studies.[29]

In this chapter, attention will be devoted to the potential of CE and especially CE-MS for the profiling of (endogenous) metabolites in biological samples. In comparison to chromatographic-based separation techniques, CE-MS is not as often applied in bioanalysis and metabolomics. However, this approach has some unique analytical characteristics for metabolomics studies.[30,31] In CE, referring in this context to capillary zone electrophoresis (CZE), compounds are separated on the basis of differences in their charge-to-size ratio. Therefore, the separation mechanism of CE is fundamentally different from that of chromatographic-based separation techniques. An important feature of CE is the intrinsically high separation efficiency as a result of the flat flow profile of the electro-osmotic flow (EOF), making this technique well-suited for the selective analysis of polar and charged metabolites in biological samples. Moreover, the main source contributing to band broadening is longitudinal diffusion as there is no mass transfer in CE separations. As only small sample volumes are needed (~10–50 nL), CE is highly suited for the analysis of size-limited biological samples, thereby enabling volume-restricted metabolomics studies.

Concerning the analysis of polar and charged metabolites, both CE and HILIC may be considered as useful analytical tools for this purpose. However, small injection volumes are typically used in CE, resulting in compromised concentration sensitivities for global metabolic profiling studies by CE-MS. The concentration sensitivity of CE can be improved by the use of electrokinetic- or chromatographic-based preconcentration techniques, as outlined in Chapters 5 and 6, respectively. Though both CE and HILIC are useful analytical separation techniques for the analysis of polar and ionic metabolites, recent studies have indicated a significant degree of orthogonality between HILIC and CE for metabolomics.[32,33] A comparison of CE-MS with other analytical separation techniques for metabolomics is provided in Chapter 8.

Here, CE and CE-MS approaches developed for metabolite analysis and global metabolic profiling are discussed. The chapter starts by providing a summary of the various CE separation modes used in CE and CE-MS for metabolomics. Special attention will be devoted to research work that eventually resulted in the development of CE-MS systems for global metabolic profiling of biological samples. Finally, the current situation of CE-MS in metabolomics is considered and some thoughts on where this approach may head to is provided.

1.2 Capillary Electrophoresis for Metabolite Analysis

CE can be employed in a number of separation modes, such as CZE, normally referred to as "CE" and also in this chapter, capillary electrochromatography (CEC), micellar electrokinetic chromatography (MEKC), capillary gel electrophoresis (CGE), capillary isotachophoresis (ITP), capillary iso-electric focusing (cIEF) and microemulsion electrokinetic chromatography (MEEKC). In the main CE separation mode, compounds are separated according to differences in their intrinsic electrophoretic mobility, which is mainly dependent on the size (radius) and charge of the compound. Moreover, this separation mode is also compatible with ESI-MS as volatile buffers can be used for the analysis. The CE separation mode has been used for decades now for the profiling of (endogenous) metabolites in various biological samples. Jellum *et al.* used CE for the profiling of organic acids in human body fluids in order to screen various human metabolic diseases.[34–36] It was found that human urine samples could be injected directly into the CE instrument without employing any sample pretreatment.[35] More than 50 metabolites were analyzed in 15 min and identification of a selected number of metabolites was based on migration times and characteristic diode-array spectra. This approach was used for the analysis of urine from patients with metabolic diseases, such as for the screening of pyroglutamic aciduria (Figure 1.1). This work can be considered as one of the first CE-based clinical metabolomics studies, in which the potential of CE for the analysis of small biological samples in biomedicine and clinical diagnosis was already envisioned.

Another pioneering work in the development of CE methods for the screening of metabolic disorders in body fluids emerged from Coral Barbas and co-workers.[37–39] This group developed a relatively fast and automated CE method employing UV absorbance detection for the analysis of 27 organic acids in human urine.[39] With this method, organic acids in urine samples

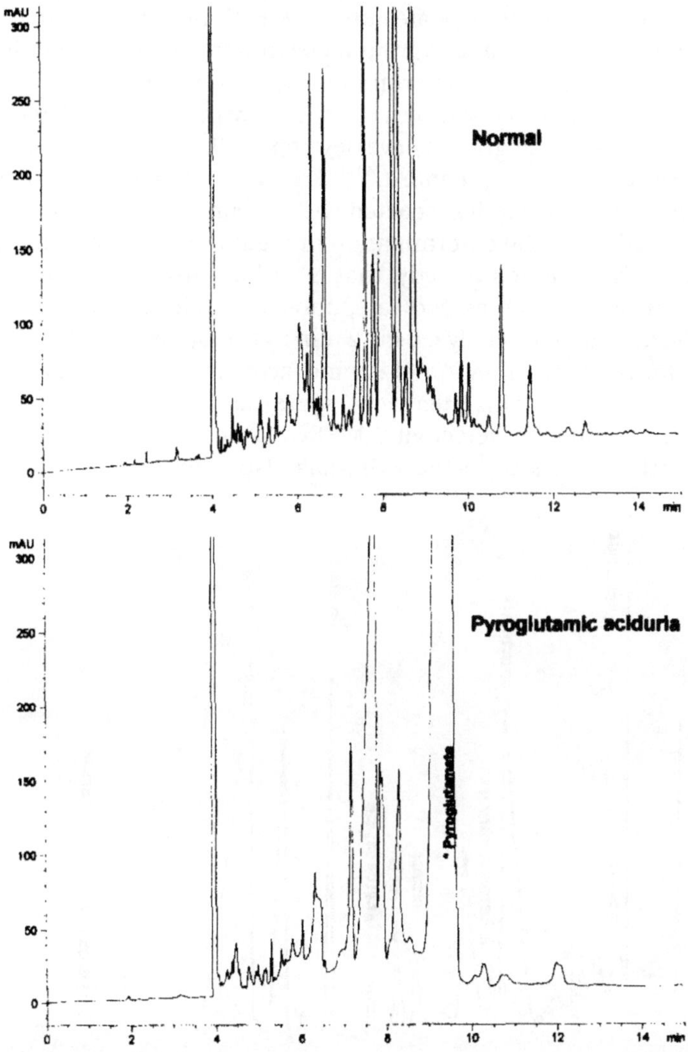

Figure 1.1 CE diode-array analysis of urine from a patient with pyroglutamic aciduria (bottom) and a control subject (top). Hydrodynamic injection was used to inject the urine (7 nL) without any sample pretreatment. Borate buffer (300 mM, pH 8.5) was used, voltage 30 kV. Capillary: effective length 56 cm, 50 µm I.D. The electropherograms were corrected for different degrees of urinary dilution as determined by the creatinine values. Reproduced from *J. Chromatogr. B: Biomed. Appl.*, **683**, E. Jellum, H. Dollekamp and C. Blessum, Capillary electrophoresis for clinical problem solving: analysis of urinary diagnostic metabolites and serum proteins, 55–65.[35] Copyright 1996, with permission from Elsevier.

could be measured directly after centrifugation and dilution. Electrophoretic separation was performed in reversed polarity mode employing a polyacrylamide-coated capillary. Phosphate buffer (pH 6.0) containing a small amount of methanol was used as a background electrolyte (BGE). The developed CE method allowed the reliable determination of methylmalonic, pyroglutamic, and glutaric acids in urine samples of patients with diseases related to these compounds. Also, other groups have developed CE methods for the analysis of organic acids in a clinical context.[40–44] For example, Dolnik *et al.* developed a CE method using a capillary coated with linear polyacrylamide and indirect UV detection for the determination of organic acids in human serum.[40] Serum could be analyzed directly, that is without using a deproteinization step. Organic acids, such as pyruvate, phosphate, citrate, malate, acetoacetate and lactate could be analyzed in 12 min by this approach. The CE method was used for the determination of organic acids in serum from a paediatric patient with respiratory insufficiency (Figure 1.2).

In order to improve selectivity, MEKC, which employs micelles as a pseudo-stationary phase in the BGE, has also been used for the analysis

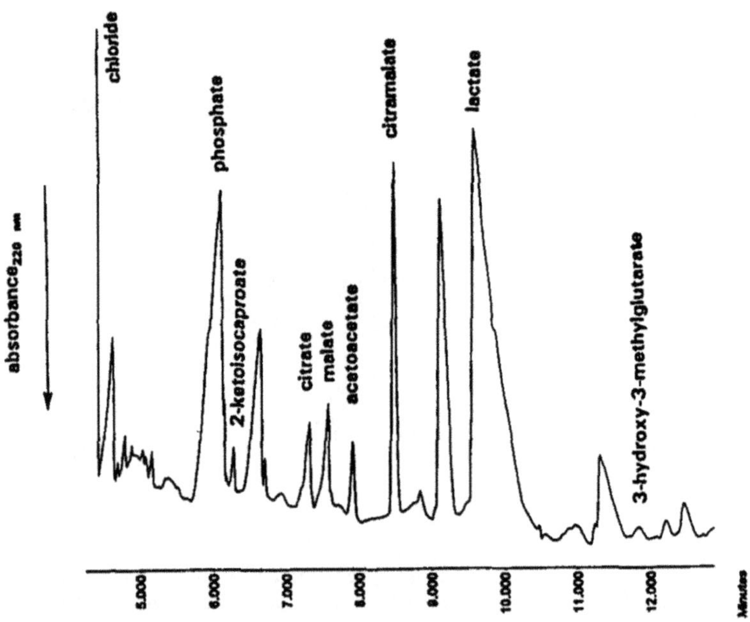

Figure 1.2 Determination of organic acids in serum from a paediatric patient with respiratory insufficiency. Experimental conditions: linear polyacrylamide-coated capillary, 75 μm I.D., 360 μm O.D., total length 51 cm, effective length 41 cm; BGE, 10 mM e-aminocaproic acid–10 mM mandelic acid; voltage, −20 kV; detection, absorption at 220 nm. Undiluted serum injected by pressure using 100 mbar for 6 s. Reproduced from *J. Chromatogr. A*, **716**, V. Dolnik and J. Dolnikova, Capillary zone electrophoresis of organic acids in serum of critically ill children, 269–277.[40] Copyright 1995, with permission from Elsevier.

of (endogenous) metabolites in biological samples.[45–48] To enhance the metabolic coverage, both CE (referring to CZE) and MEKC were used for metabolic profiling of urine samples from diabetic rats.[47] CE employing a polyacrylamide-coated capillary and reverse polarity was used for the profiling of anionic metabolites, whereas MEKC was used for the profiling of neutral and cationic metabolites. As MEKC offers extra selectivity, an extended metabolic profile of the urine sample was obtained by combining the data obtained with both CE methods. An MEKC approach was also developed for the selective analysis of nucleosides in urine samples from both healthy subjects and cancer patients.[49] Phenylboronic acid-based SPE columns have been used for the selective preconcentration of nucleosides from human urine and subsequent nucleoside profiling by MEKC. Figure 1.3 shows typical profiles of urinary nucleosides from a uterine cervical cancer patient, a myoma patient and a healthy subject, obtained by MEKC. A total of 15 nucleosides were positively identified in urine samples from eight uterine myoma patients (benign tumor group), 10 uterine cervical cancer patients (malignant tumor group), and 10 healthy females (normal group). A sulfated β-cyclodextrin-modified MEKC method has also been used for metabolic profiling of human urine, allowing the separation of 80 compounds within 25 min.[46] Although improved separation and extra selectivity can be obtained using MEKC for metabolic profiling studies, the coupling of MEKC to MS is rather problematic and often provides limited sensitivity.

The utility of CEC, an analytical separation technique that combines the high peak efficiency of CE with the stationary phase selectivity of LC, has also been assessed for metabolic profiling studies.[50,51] For example, a pressure-assisted CEC method was developed for the profiling of metabolites in rat urine and the performance of this approach was compared with chromatographic methods.[51] Figure 1.4 clearly shows that CEC was able to measure more urinary metabolites with good resolution as compared to a standard RP-LC or capillary RP-LC method. As UV absorbance was employed for detection, only eight compounds could be identified in the urine sample using comparison with standards. Though CEC can be used for the separation of a wide range of compounds and, as such, is very promising for global metabolic profiling studies, the preparation of CEC columns in a reproducible manner remains a challenging procedure.

Concerning metabolite analysis by CE in a clinical setting, detection has been mainly performed by UV absorbance detection, but contactless conductivity detection, and electrochemical and laser-induced fluorescence (LIF) detection have also been used for this purpose.[52–60] In particular, the use of CE-LIF can be considered a very powerful approach for the selective and highly sensitive analysis of a number of metabolites in small sample amounts. Moreover, CE-LIF can be used for high-speed electrophoretic separations without compromising the separation efficiency. For example, Harstad *et al.* developed a high-speed microdialysis–capillary electrophoresis (MD-CE) method for monitoring the uptake/release dynamics of branched

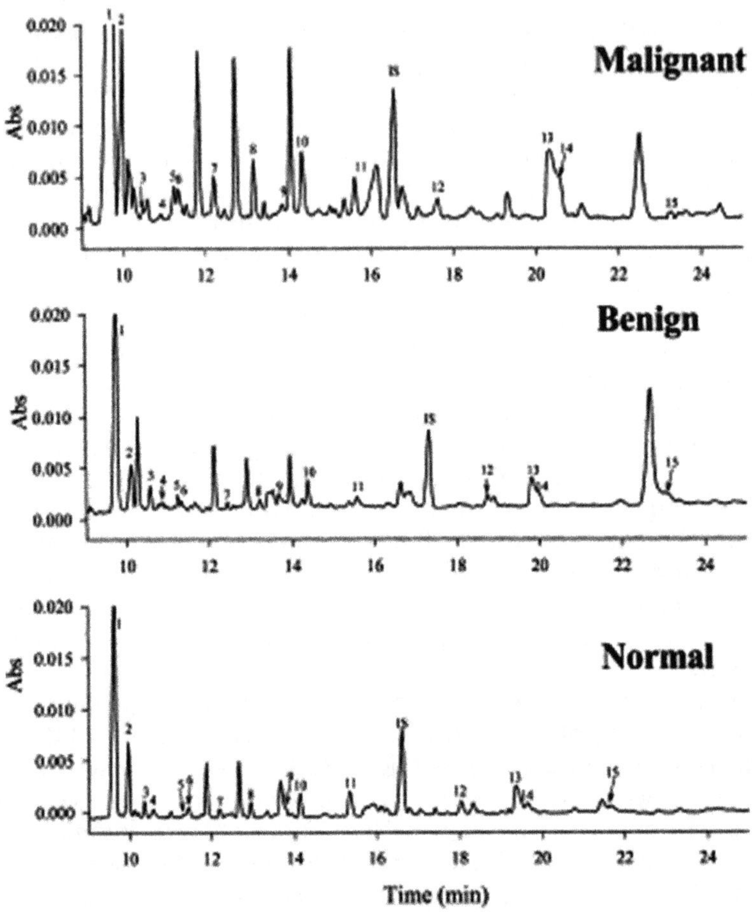

Figure 1.3 Electropherograms of urinary nucleosides from a uterine cervical cancer patient (malignant), a myoma patient (benign) and a healthy subject (normal). Experimental conditions: electrophoretic separation at 20 kV on a fused-silica capillary (57 cm × 50 µm I.D.) at 20 °C using a BGE of 200 mM SDS–25 mM borate–42.5 mM phosphate (pH 6.7) using hydrodynamic injection (4 seconds at 35 mbar), with simultaneous detection at 210 and 260 nm in dual-channel mode. Peaks: 1 = pseudouridine; 2 = dihydrouridine; 3 = uridine; 4 = cytidine; 5 = 5-methyluridine; 6 = 3-methyluridine; 7 = inosine; 8 = N4-acetylcytidine, 9 = guanosine; 10 = 1-methylguanosine; 11 = adenosine; 12 = xanthosine; 13 = N2-methylguanosine; 14 = N2,N2-dimethylguanosine; 15 = N6-methyladenosine; I.S. = 3-deazauridine. Reproduced from *J. Chromatogr. B: Biomed. Sci. Appl.*, **754**, K. R. Kim, S. La, A. Kim, J. H. Kim and H. M. Liebich, Capillary electrophoretic profiling and pattern recognition analysis of urinary nucleosides from uterine myoma and cervical cancer patients, 97–106.[49] Copyright 2001, with permission from Elsevier.

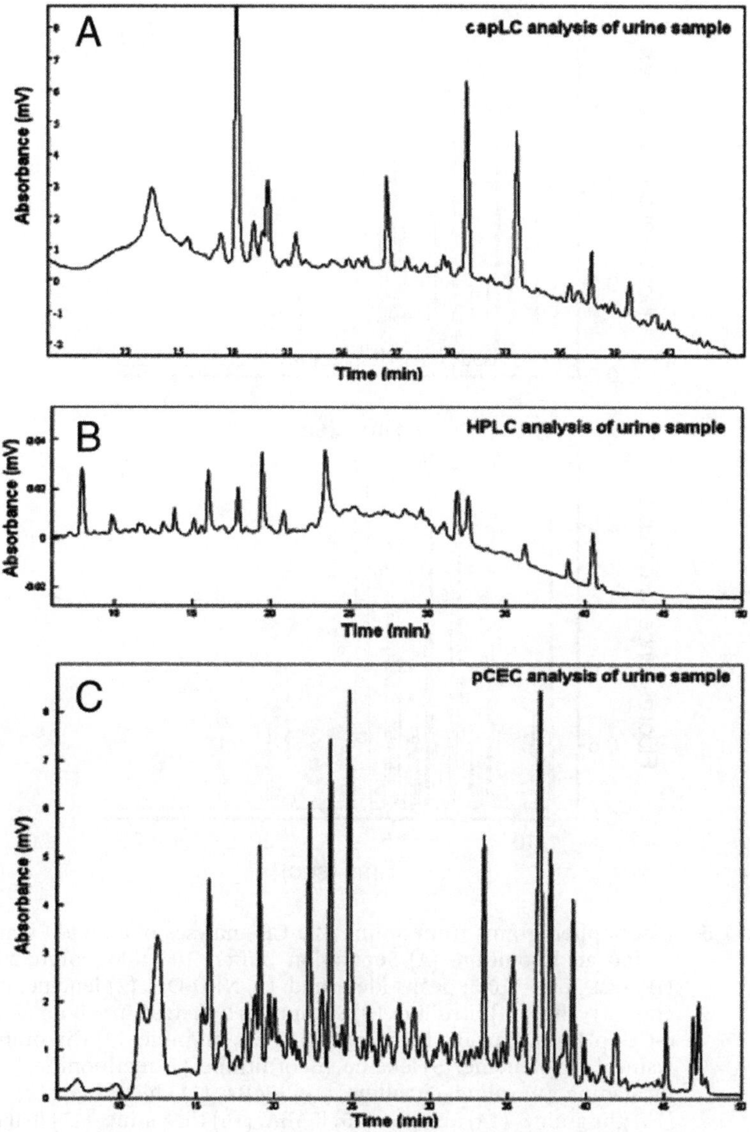

Figure 1.4 Comparison of capillary RP-LC (A), RP-LC (B), and pCEC (C) for metabolic analysis of rat urine. RP-LC conditions: column, 4.6 mm × 250 mm, packed with C18, 5 µm particle size; capillary RP-LC conditions: column, 150 µm id, 50 cm total length, packed with C18, 5 µm particle size. Gradient of water (0.02% TFA) and 95% methanol (0.02% TFA) used for LC separation. CEC conditions: column, 150 µm i.d., 50 cm total length, packed with C18, 5 µm particle size; gradient of water (0.01% TFA) and acetonitrile (0.01% TFA) used for separation. Injection volume, 5 µL; detection, UV at 214 nm. Reproduced with permission from ref. 51. Copyright © 2007 WILEY-VCH Verlag GmbH & Co. KGaA, Weinheim.

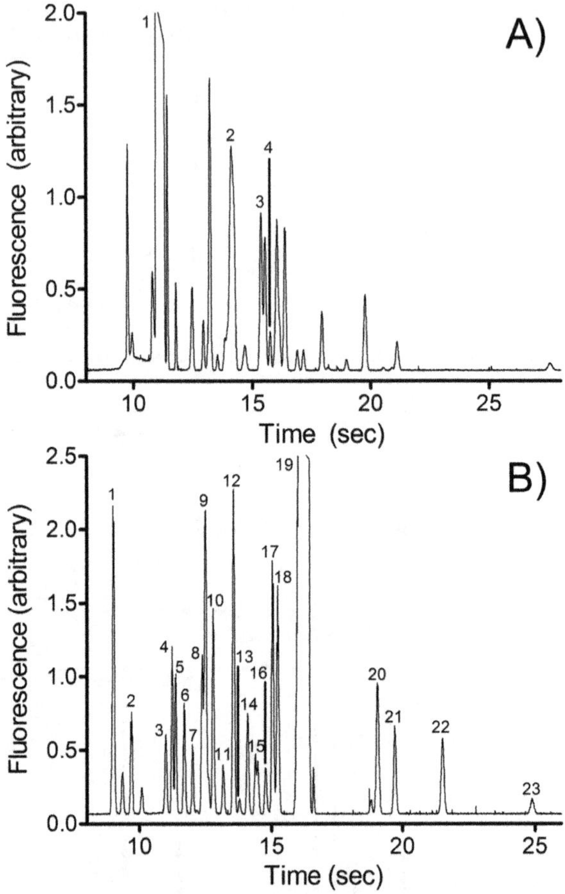

Figure 1.5 Electropherograms from online MD-CE analyses of a 20 µM standard amino acid solution. (A) Separation buffer: 100 mM borate/20 mM HP-β-CD, pH = 10.5; peaks identified: (1) NBD-OH, (2) leucine, isoleucine, (3) valine, (4) citrulline. (B) Separation buffer: 90 mM borate/35 mM α-CD, pH = 9.8; peaks identified: (1) PEA, arginine, (2) threonine, (3) lysine, (4) isoleucine, (5) leucine, (6) ornithine, (7) methionine, (8) phenylalanine, (9) valine, citrulline, (10) α-ABA, (11) histidine, (12) GABA, (13) glutamine, (14) alanine, (15) β-ABA, (16) threonine, (17) β-alanine, (18) glycine, (19) NBD-OH, (20) taurine, (21) internal standard, (22) glutamate, and (23) aspartate. Reproduced with permission from ref. 61. Copyright (2016) American Chemical Society.

chain amino acids in 3T3-L1 cells using LIF detection.[61] The separation conditions were optimized to resolve isoleucine, leucine, valine, glutamine, alanine, glutamate and many other amino acids in less than 30 s, as shown in Figure 1.5. The peak widths were around 0.2 s with plate numbers above 150 000. The high-speed MD-CE-LIF method allowed the analysis of 15 samples with 8 replicates per sample in less than 2 hours, whereas the same analysis using a traditional offline CE approach would have taken more than 20 hours.

Concerning the use of CE for comparative metabolic profiling studies, the reproducibility of the migration times and peak areas is a critical aspect.[6,62] Variability of the migration time is a particularly important issue when employing bare fused-silica capillaries. Adsorption of matrix components and/or analytes to the inner wall of the fused-silica capillaries may cause an irreproducible electro-osmotic-flow (EOF) and consequently irreproducible migration times. Moreover, when bare fused-silica capillaries are used, the separation efficiencies and repeatability of the peak area may be compromised as a result of adverse analyte–capillary wall interactions. Therefore, it is very important to incorporate, in addition to an effective sample pretreatment procedure, an effective rinsing procedure between runs when using a bare fused-silica capillary for metabolic profiling of (especially protein-rich) biological samples. Obviously, such a rinsing procedure will increase the total analysis time per sample. Another way to address this problem is to modify the inner capillary wall of the fused-silica capillaries with polymers, and this approach is further described in Chapter 3. For more details on the development of CE approaches for metabolite analysis and metabolic profiling studies from a historical perspective, the reader is referred to the following reviews.[63–68]

1.3 Capillary Electrophoresis–Mass Spectrometry for Metabolomics

In modern metabolomics, the use of a detector providing both a high selectivity and sensitivity is inevitable for the global profiling of metabolites in complex samples. Although detection in CE is still often based on on-column UV absorbance as it is easy to use and broadly applicable, it lacks sensitivity as a result of the small optical path length due to the capillary inner diameter. Moreover, the selectivity of UV detection is limited as it hardly provides any structural information on the compounds. MS has emerged as a key technology for the selective and sensitive analysis of metabolites in biological samples, providing the ability to quantify and identify metabolites. The combination of MS with a front-end separation technique is often needed in order to reduce ion suppression/enhancement, to allow the separation of isobars and isomers, and to gain additional information on the physicochemical properties of the metabolites, which may facilitate the identification process. CE-MS-based metabolomics studies generally require the use of a sample preparation step to selectively remove the highly polar and charged metabolites from proteins and lipids in a given biological sample. Sample pretreatment may result in the loss of metabolites and, therefore, proper optimization of this step in the analytical workflow designed for a metabolomics study is key for obtaining reliable metabolic profiles. In Chapter 2, attention is paid to sample preparation strategies for CE-MS-based metabolomics studies.

The coupling of CE to MS can be performed with a sheath-liquid or a sheathless interface. Until now, the sheath-liquid interface has most widely been used for CE-MS-based metabolomics studies.[31,69,70] Details about interfacing techniques for CE-MS and their utility for metabolomics can be found in Chapter 4. The ionization technique commonly used in CE-MS for metabolomics is ESI and, therefore, the selection of the separation conditions is very critical as ESI is susceptible to analyte signal suppression by relatively high buffer concentrations, non-volatile components and surfactants. Moreover, non-volatile components may cause contamination of the ion source and ion optics and as a result high background signals. Therefore, volatile BGEs, such as formic acid or ammonium acetate, are often used as BGEs for CE-MS studies.

One of the first applications reporting the use of CE-MS for metabolite analysis in a clinical setting came from the group of Egil Jellum.[71] In this work, a CE-MS/MS method was developed for the selective analysis of a set of metabolites in dried blood spots from patients with propionic aciduria, which is a metabolic disorder. CE was coupled to MS *via* a co-axial sheath-liquid interface and electrophoretic separation was performed with a bare fused-silica capillary using 20 mM ammonium acetate (pH 8.5) as a BGE. Around the same time, Soga *et al.* developed a CE-MS method for the analysis of 19 naturally occurring amino acids without using any derivatization.[72] In order to enable the simultaneous and selective analysis of all the amino acids by CE-MS, a low-pH BGE, *i.e.* 1 M formic acid (pH 1.8), was used (Figure 1.6). For eight repeated injections, the relative standard deviations for the migration times and peak areas were below 1.2% and 7%, respectively. The CE-MS method provided low µM detection limits for most amino acids (corresponding to femtomole amounts) and its potential was demonstrated by the analysis of soy sauce, beer, sake, and urine. The authors already envisioned at that time that the developed method could potentially be used for the profiling of amino acids and related compounds in body fluids and other biological samples for the screening of disorders in amino acid metabolism. Indeed, soon after the introduction of this method for the selective profiling of amino acids, CE-MS systems for global metabolic profiling studies were introduced.

The first CE-MS method used for the global profiling of metabolites in biological samples was developed by Soga *et al.*[73] Distinct CE-MS methods were used for the profiling of cationic and anionic metabolites, *i.e.* the first group was analyzed with a bare fused-silica capillary employing 1 M formic acid (pH 1.8) as the BGE (as initially developed for amino acid profiling), whereas the second group was analyzed with a cationic polymer-coated capillary employing 50 mM ammonium acetate (pH 8.5) as the BGE. A typical result for the analysis of nucleotides and CoA compounds obtained using the latter CE-MS conditions is shown in Figure 1.7. By using both CE-MS approaches, more than 1600 molecular features (a molecular feature is a detected signal that appears in the majority of the measured samples comprising a fixed interval of *m*/*z* and migration times) could be observed in an extract from *Bacillus subtilis*, of which 150 could be identified. The developed CE-MS systems allowed the selective and highly efficient analysis of diverse classes of

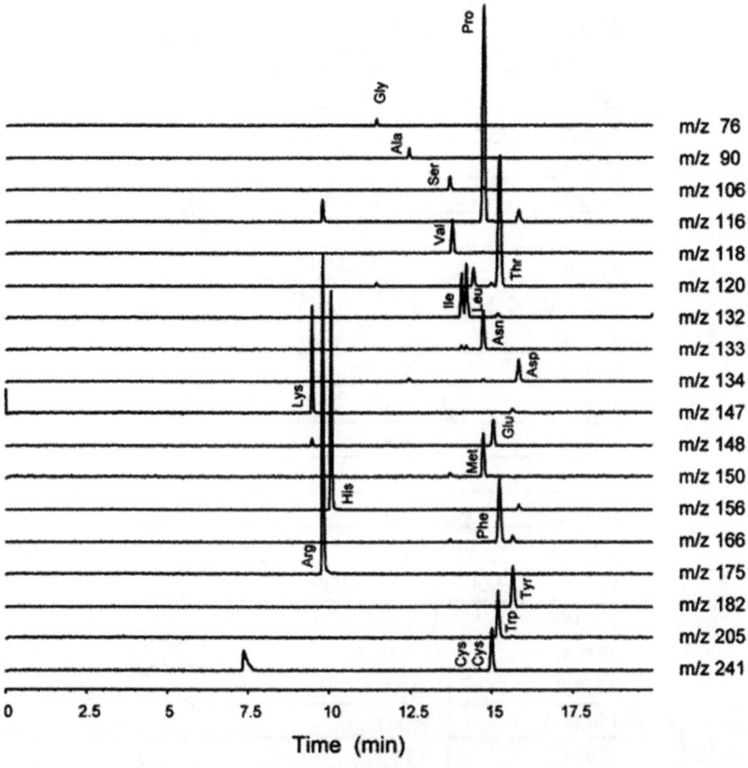

Figure 1.6 Selected ion electropherograms for a standard mixture of 19 amino acids obtained by CE-MS. Experimental conditions: sample concentration, Cys 125 μmol L^{-1} and others 250 μmol L^{-1} each; capillary, fused silica 50 μm i.d. × 100 cm; BGE, 1 M formic acid; applied potential, +30 kV; injection, 3 s at 50 mbar; sheath liquid, 10 μL min^{-1} of 5 mM ammonium acetate in 50% (v/v) methanol–water. Reproduced with permission from ref. 72. Copyright (2000) American Chemical Society.

highly polar metabolites, such as amino acids, nucleosides, small peptides, organic acids, nucleotides and sugar phosphates, which are difficult to analyze by conventional RP-LC-MS methods. More details about the design of CE-MS methods for anionic and cationic metabolic profiling can be found in Chapter 7.

For the coupling of CE to MS, volatile buffers like ammonium acetate, acetic acid or formic acid are often used as a BGE; however, these buffers may not provide optimal electrophoretic separations. In this case, organic modifiers, such as, *e.g.*, isopropanol, methanol or acetonitrile, can be added to the BGE in order to improve the separation and MS detection performance of metabolites. For example, in a CE-MS method developed for the analysis of free amino acids, a BGE of 2 M formic acid (pH 1.8) containing 20% methanol was employed in order to obtain a full baseline separation of the isomers leucine and isoleucine.[74]

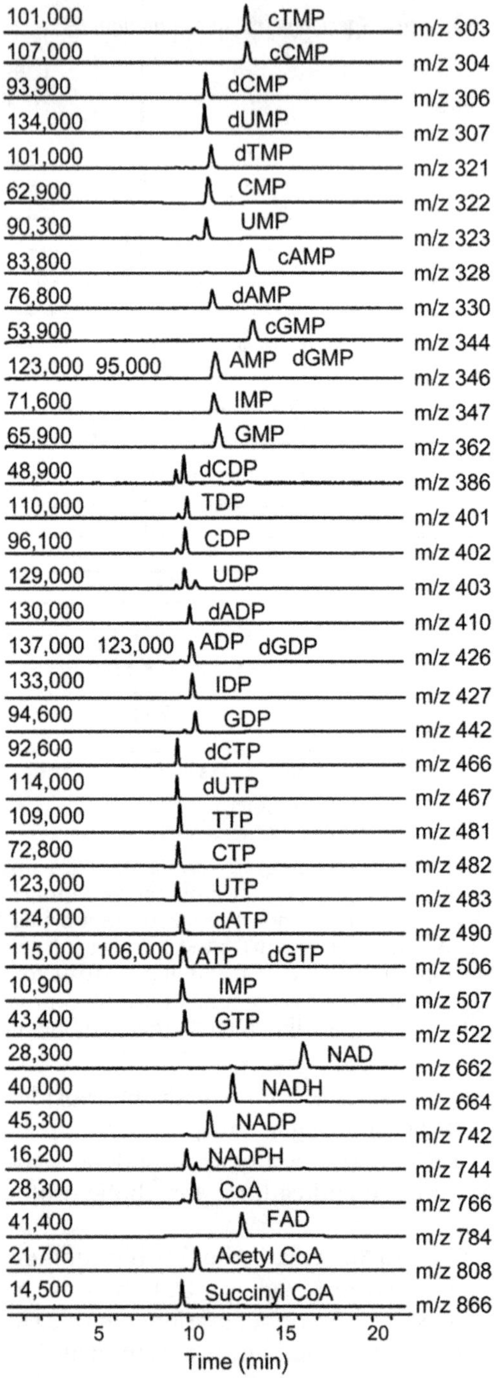

Figure 1.7 Analysis of a standard mixture (100 μM each) of nucleotides and CoA compounds by CE-MS using a cationic-coated capillary. Experimental conditions: BGE, 50 mM ammonium acetate (pH 8.5); separation at −30 kV; injection volume, 30 nL. The numbers in the upper left corner indicate the abundances associated with the tallest peak in the electropherogram, for each *m/z*. Reproduced with permission from ref. 73. Copyright (2003) American Chemical Society.

Until now, most CE-MS-based metabolomics studies have been performed with *time-of-flight* (TOF) MS instruments as they provide a high resolution and a high mass accuracy; the latter is especially important for metabolite identification purposes.[31,69] Moreover, the high spectral acquisition rate of a TOF mass analyzer is fully compatible with the relatively fast and transient signals generated during highly efficient CE separations, *i.e.*, a high number of data points can be collected across a sharp CE peak by a TOF-MS instrument. The current generation of triple quadrupole and ion trap MS instruments are also fully compatible with highly efficient CE analyses. For a complete overview of CE-MS-based metabolomics studies reported over the past decade, the reader is referred to the following reviews.[18,64,65,70,75,76]

1.4 Conclusions

In metabolomics, CE-MS has emerged as a very useful analytical tool for the profiling of highly polar and charged metabolites in biological samples, providing complementary biochemical information in comparison to chromatographic-based separation techniques. Until now, most CE-MS-based metabolomics studies have been focused on the analysis of polar and charged metabolites, including amino acids, amines, nucleosides, nucleotides, organic acids, sugar phosphates and small peptides, using an aqueous BGE system. These types of metabolites are often poorly retained by standard RP-LC-MS systems. It would be interesting to assess the utility of CE-MS also for non-polar and charged metabolites, such as charged hydrophobic lipids, in order to further expand the metabolic coverage of this approach.

Over the past few years, significant developments have been made in CE-MS methodology for metabolomics studies in order to address important analytical challenges, including sensitivity (or metabolome coverage), sample throughput, strategies for data analysis and unknown metabolite identification. All these aspects are considered in various chapters throughout this book. As metabolic profiling approaches are now more widely employed for large-scale phenotyping studies (for example, biological samples from Biobanks), the ability to measure hundreds of samples with a reliable high-throughput precision has become very important. In this context, the recently developed multi-segment injection (MSI-)CE-MS method for global metabolic profiling can be considered a very pragmatic tool (as outlined in Chapter 12). This approach can be used for serial hydrodynamic injection of seven or more discrete sample segments within a single capillary (Figure 1.8), thereby increasing sample throughput up to one order of magnitude without ion suppression and maintaining separation performance.[77]

So far, CE-MS has only been used by a limited number of groups for metabolomics with the majority of the studies reported by Soga and co-workers. Performance metrics for large-scale metabolic profiling studies by CE-MS including standard operating procedures and vendor support are urgently needed in order to expand the role of CE-MS in metabolomics and bioanalysis in general. A highly encouraging study in this context emerged very

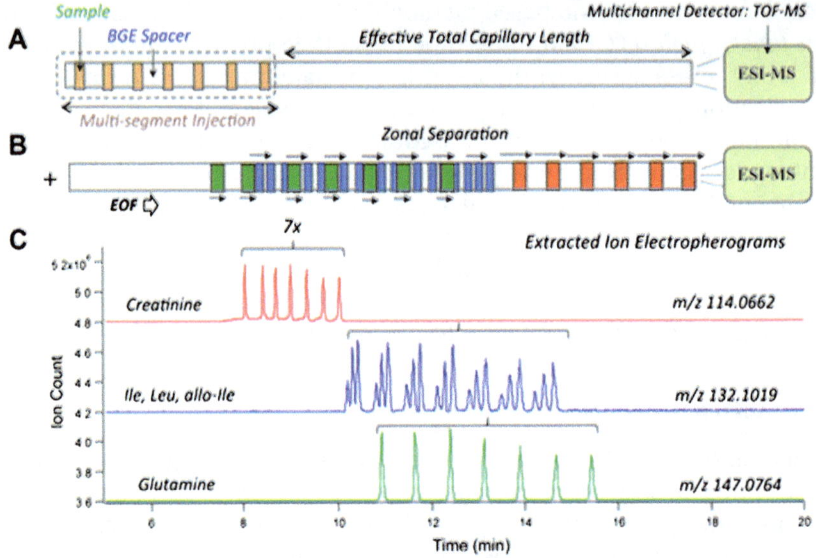

Figure 1.8 (A) Multiplexed separation based on serial injection of seven discrete sample segments within a single capillary by MSI-CE-MS; (B) ions migrate as a series of zones in free solution prior to ionization; (C) the procedure enables reliable quantification of polar metabolites and their isomers in different samples as ionization occurs within a short-time interval (≈2–6 min) under steady-state conditions when using TOF-MS. Reproduced with permission from ref. 77. Copyright (2013) American Chemical Society.

recently,[78] in which CE-MS was used for the reliable profiling and absolute quantification of polar and charged metabolites in plasma samples from the Tsuruoka Metabolomics Cohort Study (>10 000 human subjects).

Acknowledgements

Dr Rawi Ramautar would like to acknowledge the financial support of the Veni and Vidi grant scheme of the Netherlands Organisation for Scientific Research (NWO Veni 722.013.008 and Vidi 723.016.003).

References

1. R. Ramautar, R. Berger, J. van der Greef and T. Hankemeier, *Curr. Opin. Chem. Biol.*, 2013, **17**, 841–846.
2. D. S. Wishart, Y. D. Feunang, A. Marcu, A. C. Guo, K. Liang, R. Vazquez-Fresno, T. Sajed, D. Johnson, C. Li, N. Karu, Z. Sayeeda, E. Lo, N. Assempour, M. Berjanskii, S. Singhal, D. Arndt, Y. Liang, H. Badran, J. Grant, A. Serra-Cayuela, Y. Liu, R. Mandal, V. Neveu, A. Pon, C. Knox, M. Wilson, C. Manach and A. Scalbert, *Nucleic Acids Res.*, 2018, **46**, D608–D617.

3. A. Gonzalez-Dominguez, E. Duran-Guerrero, A. Fernandez-Recamales, A. Maria Lechuga-Sancho, A. Sayago, M. Schwarz, C. Segundo and R. Gonzalez-Dominguez, *Curr. Top. Med. Chem.*, 2017, **17**, 3289–3295.
4. N. Psychogios, D. D. Hau, J. Peng, A. C. Guo, R. Mandal, S. Bouatra, I. Sinelnikov, R. Krishnamurthy, R. Eisner, B. Gautam, N. Young, J. Xia, C. Knox, E. Dong, P. Huang, Z. Hollander, T. L. Pedersen, S. R. Smith, F. Bamforth, R. Greiner, B. McManus, J. W. Newman, T. Goodfriend and D. S. Wishart, *PLoS One*, 2011, **6**, e16957.
5. O. Begou, H. G. Gika, I. D. Wilson and G. Theodoridis, *Analyst*, 2017, **142**, 3079–3100.
6. R. Ramautar, *Adv. Clin. Chem.*, 2016, **74**, 1–34.
7. D. Dudzik, C. Barbas-Bernardos, A. Garcia and C. Barbas, *J. Pharm. Biomed. Anal.*, 2018, **147**, 149–173.
8. R. Ramautar, *Adv. Clin. Chem.*, 2016, **74**, 1–34.
9. S. Naz, M. Vallejo, A. Garcia and C. Barbas, *J. Chromatogr. A*, 2014, **1353**, 99–105.
10. A. Zhang, H. Sun, G. Yan, P. Wang and X. Wang, *BioMed Res. Int.*, 2015, **2015**, 354671.
11. A. C. Dona, B. Jimenez, H. Schafer, E. Humpfer, M. Spraul, M. R. Lewis, J. T. Pearce, E. Holmes, J. C. Lindon and J. K. Nicholson, *Anal. Chem.*, 2014, **86**, 9887–9894.
12. O. Beckonert, M. Coen, H. C. Keun, Y. Wang, T. M. Ebbels, E. Holmes, J. C. Lindon and J. K. Nicholson, *Nat. Protoc.*, 2010, **5**, 1019–1032.
13. A. Roux, D. Lison, C. Junot and J. F. Heilier, *Clin. Biochem.*, 2011, **44**, 119–135.
14. E. J. Want, I. D. Wilson, H. Gika, G. Theodoridis, R. S. Plumb, J. Shockcor, E. Holmes and J. K. Nicholson, *Nat. Protoc.*, 2010, **5**, 1005–1018.
15. E. C. Chan, K. K. Pasikanti and J. K. Nicholson, *Nat. Protoc.*, 2011, **6**, 1483–1499.
16. O. Fiehn, *Curr. Protoc. Mol. Biol.*, 2016, **114**, 30 34 31–30 34 32.
17. P. Yin and G. Xu, *J. Chromatogr. A*, 2014, **1374**, 1–13.
18. N. L. Kuehnbaum and P. Britz-McKibbin, *Chem. Rev.*, 2013, **113**, 2437–2468.
19. I. Kohler and M. Giera, *J. Sep. Sci.*, 2017, **40**, 93–108.
20. M. M. Koek, R. H. Jellema, J. van der Greef, A. C. Tas and T. Hankemeier, *Metabolomics*, 2011, **7**, 307–328.
21. M. M. Koek, B. Muilwijk, M. J. van der Werf and T. Hankemeier, *Anal. Chem.*, 2006, **78**, 1272–1281.
22. H. G. Gika, C. Zisi, G. Theodoridis and I. D. Wilson, *J. Chromatogr. B: Anal. Technol. Biomed. Life Sci.*, 2016, **1008**, 15–25.
23. H. G. Gika, G. A. Theodoridis, R. S. Plumb and I. D. Wilson, *J. Pharm. Biomed. Anal.*, 2014, **87**, 12–25.
24. P. A. Vorkas, G. Isaac, M. A. Anwar, A. H. Davies, E. J. Want, J. K. Nicholson and E. Holmes, *Anal. Chem.*, 2015, **87**, 4184–4193.
25. W. Lu, M. F. Clasquin, E. Melamud, D. Amador-Noguez, A. A. Caudy and J. D. Rabinowitz, *Anal. Chem.*, 2010, **82**, 3212–3221.

26. R. M. Seifar, C. Ras, A. T. Deshmukh, K. M. Bekers, C. A. Suarez-Mendez, A. L. da Cruz, W. M. van Gulik and J. J. Heijnen, *J. Chromatogr. A*, 2013, **1311**, 115–120.
27. R. M. Seifar, C. Ras, J. C. van Dam, W. M. van Gulik, J. J. Heijnen and W. A. van Winden, *Anal. Biochem.*, 2009, **388**, 213–219.
28. R. M. Seifar, Z. Zhao, J. van Dam, W. van Winden, W. van Gulik and J. J. Heijnen, *J. Chromatogr. A*, 2008, **1187**, 103–110.
29. D. Q. Tang, L. Zou, X. X. Yin and C. N. Ong, *Mass Spectrom. Rev.*, 2016, **35**, 574–600.
30. W. Zhang, T. Hankemeier and R. Ramautar, *Curr. Opin. Biotechnol.*, 2017, **43**, 1–7.
31. A. Hirayama, M. Wakayama and T. Soga, *TrAC, Trends Anal. Chem.*, 2014, **61**, 215–222.
32. M. G. Kok, G. W. Somsen and G. J. de Jong, *Talanta*, 2015, **132**, 1–7.
33. S. Naz, A. A. Calderon, A. Garcia, J. Gallafrio, R. T. Mestre, E. G. Gonzalez, C. M. de Cabo, M. C. Delgado, J. A. Balanza, A. V. Simionato, N. N. Vaeza, C. Barbas and F. J. Ruperez, *Electrophoresis*, 2015, **36**, 2303–2313.
34. E. Jellum, A. K. Thorsrud and E. Time, *J. Chromatogr.*, 1991, **559**, 455–465.
35. E. Jellum, H. Dollekamp and C. Blessum, *J. Chromatogr. B: Biomed. Sci. Appl.*, 1996, **683**, 55–65.
36. E. Jellum, H. Dollekamp, A. Brunsvig and R. Gislefoss, *J. Chromatogr. B: Biomed. Sci. Appl.*, 1997, **689**, 155–164.
37. C. Barbas, N. Adeva, R. Aguilar, M. Rosillo, T. Rubio and M. Castro, *Clin. Chem.*, 1998, **44**, 1340–1342.
38. A. Garcia, C. Barbas, R. Aguilar and M. Castro, *Clin. Chem.*, 1998, **44**, 1905–1911.
39. A. Garcia and C. Barbas, *Clin. Chem. Lab. Med.*, 2003, **41**, 755–761.
40. V. Dolnik and J. Dolnikova, *J. Chromatogr. A*, 1995, **716**, 269–277.
41. R. Ramautar, G. W. Somsen and G. J. de Jong, *Anal. Bioanal. Chem.*, 2007, **387**, 293–301.
42. M. Shirao, R. Furuta, S. Suzuki, H. Nakazawa, S. Fujita and T. Maruyama, *J. Chromatogr. A*, 1994, **680**, 247–251.
43. C. J. Petucci, H. L. Kantes, T. G. Strein and H. Veening, *J. Chromatogr. B: Biomed. Sci. Appl.*, 1995, **668**, 241–251.
44. H. Chen, Y. Xu, F. Van Lente and M. P. Ip, *J. Chromatogr. B: Biomed. Sci. Appl.*, 1996, **679**, 49–59.
45. C. Guillo, D. Perrett and M. Hanna-Brown, *Chromatographia*, 2004, **59**, S157–S164.
46. C. Guillo, D. Barlow, D. Perrett and M. Hanna-Brown, *J. Chromatogr. A*, 2004, **1027**, 203–212.
47. M. Vallejo, S. Angulo, D. Garcia-Martinez, A. Garcia and C. Barbas, *J. Chromatogr. A*, 2008, **1187**, 267–274.
48. C. Barbas, M. Vallejo, A. Garcia, D. Barlow and M. Hanna-Brown, *J. Pharm. Biomed. Anal.*, 2008, **47**, 388–398.
49. K. R. Kim, S. La, A. Kim, J. H. Kim and H. M. Liebich, *J. Chromatogr. B: Biomed. Sci. Appl.*, 2001, **754**, 97–106.

50. H. Zhang, Y. Wang, X. Gu, J. Zhou and C. Yan, *Electrophoresis*, 2011, **32**, 340–347.
51. G. Xie, M. Su, P. Li, X. Gu, C. Yan, Y. Qiu, H. Li and W. Jia, *Electrophoresis*, 2007, **28**, 4459–4468.
52. J. Hong and R. P. Baldwin, *J. Capillary Electrophor.*, 1997, **4**, 65–71.
53. J. Schneede and P. M. Ueland, *Anal. Chem.*, 1995, **67**, 812–819.
54. T. M. Phillips, *Electrophoresis*, 2018, **39**, 126–135.
55. M. P. Lorenzo, A. Villasenor, A. Ramamoorthy and A. Garcia, *Electrophoresis*, 2013, **34**, 1701–1709.
56. M. P. Lorenzo, A. Navarrete, C. Balderas and A. Garcia, *J. Pharm. Biomed. Anal.*, 2013, **73**, 116–124.
57. J. Lacna, F. Foret and P. Kuban, *Electrophoresis*, 2017, **38**, 203–222.
58. T. Lapainis, C. Scanlan, S. S. Rubakhin and J. V. Sweedler, *Anal. Bioanal. Chem.*, 2007, **387**, 97–105.
59. S. A. Sheeley, H. Miao, M. A. Ewing, S. S. Rubakhin and J. V. Sweedler, *Analyst*, 2005, **130**, 1198–1203.
60. C. A. Dailey, N. Garnier, S. S. Rubakhin and J. V. Sweedler, *Anal. Bioanal. Chem.*, 2013, **405**, 2451–2459.
61. R. K. Harstad and M. T. Bowser, *Anal. Chem.*, 2016, **88**, 8115–8122.
62. C. Huhn, R. Ramautar, M. Wuhrer and G. W. Somsen, *Anal. Bioanal. Chem.*, 2010, **396**, 297–314.
63. R. Ramautar, A. Demirci and G. J. de Jong, *TrAC, Trends Anal. Chem.*, 2006, **25**, 455–466.
64. R. Ramautar, G. W. Somsen and G. J. de Jong, *Electrophoresis*, 2009, **30**, 276–291.
65. R. Ramautar, O. A. Mayboroda, G. W. Somsen and G. J. de Jong, *Electrophoresis*, 2011, **32**, 52–65.
66. M. R. Monton and T. Soga, *J. Chromatogr. A*, 2007, **1168**, 237–246; discussion 236.
67. I. Garcia-Perez, M. Vallejo, A. Garcia, C. Legido-Quigley and C. Barbas, *J. Chromatogr. A*, 2008, **1204**, 130–139.
68. C. Barbas, E. P. Moraes and A. Villasenor, *J. Pharm. Biomed. Anal.*, 2011, **55**, 823–831.
69. R. Ramautar, G. W. Somsen and G. J. de Jong, in *Metabolomics in Practice*, Wiley-VCH Verlag GmbH & Co. KGaA, 2013, pp. 177–208.
70. R. Ramautar, G. W. Somsen and G. J. de Jong, *Electrophoresis*, 2013, **34**, 86–98.
71. K. B. Elgstoen, J. Y. Zhao, J. F. Anacleto and E. Jellum, *J. Chromatogr. A*, 2001, **914**, 265–275.
72. T. Soga and D. N. Heiger, *Anal. Chem.*, 2000, **72**, 1236–1241.
73. T. Soga, Y. Ohashi, Y. Ueno, H. Naraoka, M. Tomita and T. Nishioka, *J. Proteome Res.*, 2003, **2**, 488–494.
74. O. A. Mayboroda, C. Neususs, M. Pelzing, G. Zurek, R. Derks, I. Meulenbelt, M. Kloppenburg, E. P. Slagboom and A. M. Deelder, *J. Chromatogr. A*, 2007, **1159**, 149–153.
75. R. Ramautar, G. W. Somsen and G. J. de Jong, *Electrophoresis*, 2015, **36**, 212–224.

76. R. Ramautar, G. W. Somsen and G. J. de Jong, *Electrophoresis*, 2017, **38**, 190–202.
77. N. L. Kuehnbaum, A. Kormendi and P. Britz-McKibbin, *Anal. Chem.*, 2013, **85**, 10664–10669.
78. S. Harada, A. Hirayama, Q. Chan, A. Kurihara, K. Fukai, M. Iida, S. Kato, D. Sugiyama, K. Kuwabara, A. Takeuchi, M. Akiyama, T. Okamura, T. M. D. Ebbels, P. Elliott, M. Tomita, A. Sato, C. Suzuki, M. Sugimoto, T. Soga and T. Takebayashi, *PLoS One*, 2018, **13**, e0191230.

CHAPTER 2

CE-MS Workflows for Metabolomics: A Focus on Sample Preparation

V. GARCÍA-CAÑAS AND C. SIMÓ*

Molecular Nutrition and Metabolism, Institute of Food Science Research (CIAL, CSIC), Nicolas Cabrera 9, 28049 Madrid, Spain
*E-mail: c.simo@csic.es

2.1 Introduction

Metabolomics is the comprehensive study of the metabolome, the total small-molecular weight molecules, products, intermediates or substrates of reactions, within biological systems (microorganisms, plants, animals, humans). The high complexity, variety of physicochemical properties (from small polar or volatile compounds to large hydrophobic lipids) and dynamic range of metabolite concentrations (from femtomolar to millimolar) are some of the main challenges in metabolomics. NMR and MS-based technologies are the key techniques to measure hundreds or even thousands of metabolites all at once. Unlike NMR spectroscopy, in which no prior metabolite separation or chemical modification is usually required, MS is rarely used alone in metabolomics applications. As the ion suppression caused by matrix effects results in inaccurate quantification of the metabolites by MS, analytical platforms based on chromatographic and electrophoretic techniques are

New Developments in Mass Spectrometry No. 6
Capillary Electrophoresis–Mass Spectrometry for Metabolomics
Edited by Rawi Ramautar

Published by the Royal Society of Chemistry, www.rsc.org

coupled to MS to enhance the selectivity and sensitivity in metabolome analysis, namely, LC, GC and CE.

CE-MS coupling is of special interest in metabolomics for the analysis of highly polar and ionic metabolites in a wide range of biological samples. Hyphenation of CE to MS is generally performed through electrospray ionization (ESI) *via* an orthogonal coaxial sheath–flow interface. Using this configuration, post-capillary dilution of the eluted metabolites reduces sensitivity, and somewhat limits the metabolome coverage. The sheathless porous tip interface developed by Moini[1] has been successfully used in metabolomics applications[2] to increase metabolome coverage by increasing sensitivity; however, the classical sheath–liquid interface continues to be the most predominantly used for metabolomics by CE-MS due to its greater robustness. Other sample preconcentration procedures are utilized in order to improve the overall sensitivity of CE-MS in metabolomics, such as sample stacking, as has been recently reviewed.[3]

Cationic metabolite profiling is typically performed in uncoated fused-silica capillaries with low pH aqueous solutions containing relatively low concentrations of volatile electrolytes (*e.g.* formic acid and acetic acid), and sometimes organic modifiers (*e.g.* methanol, acetonitrile), for compatibility with MS analysis. Under these conditions, small electroosmotic flow (EOF) toward the cathode is generated, due to little ionization of the capillary wall silanol groups; the CE system is operated under normal polarity so that the EOF moves towards the cathode at the column outlet (where the MS detector is located) and so the CE electrical current is stabilized. As a result, rapid analysis with good resolution is obtained since both the EOF and the electrophoretic mobility of positively charged metabolites are toward the ESI-MS, which is conducted in positive ion mode. In some protocols, cationic metabolite profiling is performed following the same strategy aforementioned, but also with an external pressure (~25 mbar) together with the separation voltage to maintain a conductive liquid junction at the capillary outlet and hence for a greater stability of the CE-MS system.

For anionic metabolite profiling in uncoated fused-silica capillaries and the CE system operating under normal polarity, a volatile BGE with a high pH is needed to create EOF high enough to conduct the anionic compounds, which migrate counter-electroosmotically toward the inlet capillary tip (where the anode is located). In this regard, it is also important to mention that it has been reported that the typically used ammonia/ammonium salt-containing background electrolytes give rise to less efficient ionization and to ion suppression effects. In addition, alkaline ammonia-based buffers ($pH > 9$) have been shown to cause aminolysis of the outer polyimide coating of fused-silica capillaries causing a higher incidence of capillary breakages and current instability events.[4] Kok *et al.*[5,6] demonstrated the better performance of triethylamine (TEA) rather than ammonia/ammonium salt-containing BGEs, for global anionic metabolic profiling; however, it should be taken into account that TEA strongly adsorbs onto surfaces and ion suppression may occur when using the MS instrument under positive ion detection. Another solution for the CE-MS analysis of anionic metabolites was developed by Soga and co-workers in 2002 using reversed CE polarity and a static cationic coating to reverse the EOF toward the anode (in the direction of the MS detector).[7] When using this configuration the stainless steel ESI spray needle is prone to oxidation and corrosion due to electrolysis and, therefore, the same group developed a platinum ESI spray needle in order to prevent most of these corrosion effects and metal–anion complex formation under these conditions.[8] The COSMO(+) polymer capillary coating[8] is currently one of the most employed cationic coatings for anionic metabolite profiling by CE-MS; however, other cationic polymers have also been used for anionic metabolic profiling with reverse CE polarity and negative ESI-MS ion mode, such as PolyE-323,[9] FunCap-Ce Type S (with sulfonyl groups)[10] and more recently, TEDETAMA-*co*-HPMA (50 : 50) copolymer.[11] When using cationic coatings, multivalent ions (*e.g.*, nucleotides and CoA compounds) tend to adsorb onto the cationic-coated capillary inner wall. To prevent adsorption of these types of metabolites, the use of a non-charged polymer (poly(dimethylsiloxane))-coated capillary has been proposed.[12] However, one major disadvantage is that excessively long analysis times are required, and thus, anionic metabolites are often screened in this format by pressure-assisted CE-MS.

For non-targeted metabolome analysis, high resolution and highly accurate mass techniques are compulsory. Time-of-flight (TOF) is the predominant MS instrument used in CE-MS-based metabolomics. It compiles the high spectral acquisition rates required to fully exploit the narrow peaks obtained by CE and a high level of mass accuracy. As CE-MS generates large amounts of information, data processing is also a critical step in the metabolomics field. The existing and emerging computational algorithms are a key component in most of the preprocessing software tools. They typically include noise filtering, peak picking, deisotoping, alignment of peaks across multiple datasets, normalization, *etc.* However most of these tools have been developed for GC/LC-MS-based metabolomics, and it has to be considered that the peak shapes in CE-MS show more diversity, and the migration time variations are larger in CE-MS; that is why this has been a specific topic of study.[13,14]

Thousands of chemically distinct metabolites measured in a specific experiment are obtained from CE-MS analysis, and thus, data analysis applying appropriate statistical methodologies is essential for the identification of similarities and differences among experimental groups to facilitate interpretation of the experiment. Some previous studies have been focused on this topic.[15,16] After data analysis, the next step in a metabolomics experiment requires the identification of statistically significantly altered metabolites. This process is crucial to derive biological knowledge from samples. Tentative identification and peak annotation of metabolites is typically performed through the use of mass spectral libraries.[17]

2.2 Sample Preparation: General Aspects

To obtain a representative and accurate metabolome, data sample handling and processing is a crucial step in every metabolomics study. The choice of the type of specimen to be collected will be defined by the biological question and other ethical and cost considerations. Sample collection and handling procedures are important variability factors in metabolomics and sometimes they are often underestimated in metabolomics studies. Weaknesses and limitations of sample preparation methods should be understood and considered before designing a sample preparation protocol in a metabolomics study. Apart from the variability between subjects, some biological samples (*e.g.*, urine, saliva, feces) are more likely to widely vary intra-individually, depending on hydration, food intake, and other environmental conditions. For this reason, in untargeted metabolomics, the samples should ideally be subjected to pre-acquisition normalization, generally applying a dilution factor, followed by post-acquisition normalization to make the quantitative results comparable between samples. The main purpose of pre-acquisition normalization is to avoid huge concentration variations, which would result in nonlinear intensities in MS detection. On the other hand, post-acquisition normalization is applied after instrumental analysis to remove both biological and experimental variation.

The choice of the most appropriate sample treatment method for a given application will depend on the aim of the study, the analytical technique and the nature of the biological sample to be analyzed. MS-based technologies suffer from ion suppression, which adversely affects the sensitivity to low-abundance metabolites. The protein content of biological samples is an important aspect to consider in CE-MS because protein adsorption to the inner capillary surface causes severe problems. Protein-rich samples, such as plasma, serum, cells and tissues, are typically subjected to separation of metabolites from proteins prior to CE-MS. Adsorption phenomena in CE result in poor reproducibility with the migration times increasing after each sample injection. The adjustment of the pH of the running buffer to control such adsorption effects, and using capillary coatings compatible with MS detection are strategies that can be employed to minimize adsorption to the inner silica wall of the capillary. Metabolite extraction will also have a high impact on the outcome quality of a metabolomics study. The metabolites need to be dissolved in an appropriate solvent for CE-MS analysis.

Internal standards are needed in metabolomics to assess the extraction efficiency of metabolites during sample preparation. Internal standards should be spiked into the sample or added to the extraction solvent at a known concentration. These chemical species should be similar but not identical to the metabolite species in the sample, so that they can be detected during the analysis and distinguished from those compounds in the sample. Quality control in metabolomics should also be performed to evaluate the quality and assess the analytical variance of the data throughout the entire experiment. Typically, identical aliquots of all samples are taken and homogenized in a pooled QC. As the QC sample injections are identical in composition they should cluster together on a PCA scores plot.[18] If it is not possible to make a pooled QC sample due to limited sample, a commercial cocktail including multiple metabolite representatives from each class of metabolites should be prepared under identical conditions to the studied samples.

In metabolomics studies the range of samples that can be collected are varied and involve body fluids, cells and tissues, plants, microorganisms, *etc.* In the next sections, important aspects with regard to the choice of a sample pretreatment technique for metabolomics and the subsequent analysis by CE-MS will be presented and discussed.

2.3 Sample Preparation for Body Fluids

2.3.1 Urine

Urine is a biofluid generated by the kidneys containing soluble waste products as well as an excess of water from the bloodstream. As a biological waste material, urine typically contains metabolic products from dietary components, pharmaceuticals, environmental contaminants, endogenous waste metabolites and gut microbial by-products, among others. For this reason, urine samples are a very attractive source for metabolite biomarker discovery in biomedicine and drug development, and in pharmacology and toxicology, as well as in nutritional research.

From a practical point of view, urine is an interesting biological fluid that has a well-known number of advantages over other types of samples. Firstly, it is non-invasive to obtain, which aids the recruitment process in clinical trials. Moreover, it can be collected over time, in large quantities, and it is also characterized by its ease of collection. On the other hand, since the patients themselves often collect sample urine specimens, there is a greater risk of contamination and degradation compared with blood, which is usually collected in clinics. During sample collection, microbial contamination may occur and bacterial overgrowth might have an impact on the metabolome by consuming and producing metabolites. Thus, sodium azide addition or 0.20 μm filtration is recommended to prevent bacterial growth during the pre-analytical stage. In addition to bacterial contamination, biogenic enzymatic reactions and oxidation reactions of metabolites may occur if the conditions for storing urine samples until the time of analysis are not appropriate. For these reasons, the collection, transportation, and sample preparation conditions must be carefully controlled and applied to all samples in the same way.[19] Some controversy exists regarding the short-term and long-term stability of metabolites in urine. Although there is a lack of comprehensive investigation on urine metabolome stability, several studies have confirmed the necessity of storing urine at −20 °C or lower temperatures. Also, as a general rule, freeze–thaw cycles should be avoided or minimized as much as possible in urine metabolome analysis.

The knowledge about urine metabolome composition is constantly increasing. Urine contains a variety of high polarity and water-soluble compounds, such as organic acids, amino acids, inorganic salts, sugars, hormones, peptides, xenobiotics and end products of metabolism. The physicochemical characteristics of urine make CE-MS a particularly suitable technique for direct metabolite profiling with minimal sample preparation. However, the use of urine as a biofluid in CE-MS-based metabolomics can lead to specific complications. For example, variations in the pH and ionic strength of urine samples are important factors affecting the migration time shifts between different samples, and also affecting peak broadening in the worst case. On the other hand, ESI is highly susceptible to adduct formation, and thus, high concentrations of non-volatile compounds in the urine, such as salts, increases the occurrence of adducts and might even negatively affect the mass spectrometer performance due to the presence of these non-volatile compounds. Dilution, followed in some cases by centrifugation or filtration, has been used as an effective procedure to avoid eventual urine matrix effects in CE-MS.

Large intra- and inter-individual variations in the metabolome of urine samples have been reported since metabolite amount may vary as a function of the individual's liquid intake and other physiological and external factors (temperature, physical activity, *etc.*). Metabolite concentrations should therefore be normalized for reliable biological interpretation. In urine, metabolites can be normalized to the endogenous metabolite creatinine. In the absence of kidney diseases, creatinine is excreted in measurable and relatively constant amounts into the urine and it has been often used to normalize urinary metabolite levels. However, many urinary metabolomics studies target disease or abnormal physiological conditions and thus creatinine can be altered.[20] Alternatively, osmolality and specific gravity measurements can be used in a pre-analysis or pre-acquisition normalization strategy for urinary metabolomics[21,22] to equalize urine concentration by diluting the samples to the lowest concentration. Additionally, post-acquisition data normalization by using MS total useful signal (MSTUS) to reduce the impact of concentration variability has also been recommended.[21,23] Other alternative normalization strategies have been developed for the exploration of the metabolite content in urine by CE-MS. For instance, Boizard *et al.*[24] selected a set of most stable endogenous metabolites observed in a range of urine samples from healthy and diseased individuals for normalization purposes. They found 267 stable endogenous "housekeeping" metabolic features spanning a CE migration time from 17 to 36 min and in the range *m*/*z* 82–650.

A significant number of proteins and peptides are present in urine,[25] and in this regard, CE-MS is increasingly used for proteomic and peptidomic analysis, and in diagnosis, therapeutic treatment, and drug development.[26]

Although the presence of proteins cannot be dismissed in urine samples, normal urine has less protein content in comparison to other biological fluids, such as serum or plasma. Therefore, simple sample preparation, such as dilution of urine by adding water or BGE (usually containing internal standards), followed by centrifugation (or filtration) for elimination of cells and non-cellular suspended particles is typically used in CE-MS.[6,27–32] Additionally, the urine sample can be ultrafiltrated using 3–30 kDa cutoff centrifugal devices to avoid eventual matrix effects from proteins or other macromolecules present.[24] Solid-phase extraction (SPE) is also an efficient method to separate metabolites from the sample matrix to analyze target metabolite biomarkers of disease in urine by CE-MS.[33]

As already mentioned CE-MS is particularly useful for urinary metabolomics as it is well suited to the polar/ionic nature of the metabolites in this kind of sample; in this sense, hydrophilic interaction chromatography (HILIC) has grown strongly in urinary metabolomics and also provides good performance for a large set of polar metabolites.[34] In comparative studies with HILIC-MS, so far CE coupled to MS *via* a co-axial sheath liquid interface has shown complementarity and added value for anionic urinary metabolic profiling.[35] However, analysis of anionic metabolites by CE-MS generally results in less sensitive methods, and thus, there is a practical need to improve the concentration sensitivity for urinary metabolic profiling. Excellent improvement in sensitivity (up to 3 orders of magnitude) was achieved for cationic metabolite profiling of human urine using the sheathless CE-MS approach with a porous tip nanospray design,[29] as illustrated in Figure 2.1. Nevertheless, unsatisfactory results were obtained for anionic compounds in negative ionization mode using the same sheathless interface but employing an anionic metabolite profiling CE-MS configuration (uncoated capillary, 25 mM ammonium acetate at pH 9 as BGE and reverse CE polarity), as already observed with the sheath–liquid interface using the same separation conditions.[36]

In a different study, profiling of anionic metabolites by CE-MS in urine samples was achieved in an uncoated fused-silica capillary, with a BGE containing 25 mM TEA (pH 11.7), under normal CE polarity and negative ionization mode.[6] Urine samples from rats were submitted to minimal sample pretreatment (only 1 : 1 dilution with BGE). The methodology was applied to rats receiving various antibiotic treatments and the methodology revealed 17 discriminatory metabolites. It must be said that very few studies have reported the potential of the anionic metabolite profiling strategy by CE-MS in urine. Its application in clinical practice has not been fully proven. In general, and not exclusive to urine samples, analysis of anionic metabolites under alkaline conditions, normal CE polarity and negative ionization mode requires further optimization studies in order to improve the sensitivity and stability of the analytical system. An alternative approach for metabolite profiling combining a positively pre-charged deuterium-labeled coding agent for carboxylic acid derivatization and CE-MS under cationic metabolite profiling conditions was developed by Yang *et al.*[37] Although the methodology was successfully applied to rat urine, it has not yet been fully

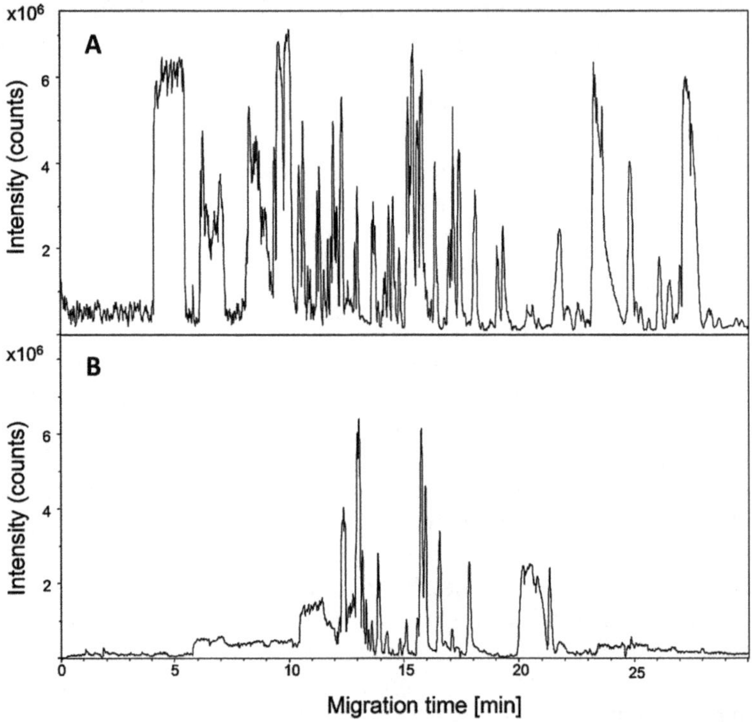

Figure 2.1 Base peak electropherogram of human urine obtained with (A) sheathless CE-MS using a porous tip sprayer and (B) sheath–liquid interface. For further information, the reader is referred to the original version of this article.[29] Reproduced with permission from ref. 29. Copyright (2012) American Chemical Society.

implemented with a large number of biological samples for profiling and biomarker discovery.

On the other hand, cationic profiling of urine by CE-MS has been demonstrated to be of value in clinical research applications. For instance, CE-MS based on a cationic profiling strategy (bare fused silica capillary, 1.2 M formic acid at pH 1.8 as BGE and normal CE polarity) was applied to search for potential tumor markers in urine samples from gastric cancer patients.[38] Urine samples were simply centrifuged and filtered through a 0.22 µm membrane prior to CE-MS analysis. In that work, the moving reaction boundary was optimized and showed potential for on-line sample stacking in CE to improve endogenous metabolite sensitivity in urine samples. In a different study, a pH-mediated stacking strategy was implemented to improve the sensitivity in urinary amino acid analysis using a cationic profiling strategy with a BGE composed of 0.8 M formic acid (pH 1.96) containing 15% methanol.[39] Prior to analysis, the urine samples were simply centrifuged and diluted with water (3 : 1 solvent-to-sample dilution ratio); also, 0.1% formic acid was added to the diluted urine to ensure that the pH-mediated stacking would be effective. CE-MS has also been applied to compare the urinary metabolome of smokers and non-smokers through the profiling of cationic metabolites under low-pH separation conditions using a bare fused-silica capillary.[40] Also, in this work, simple urine pre-treatment, which included centrifugation prior to dilution with BGE, was followed. CE-MS has also been applied for the diagnosis of ureteropelvic junction (UPJ) obstruction following a cationic metabolite profiling analytical set-up.[24] In this case, urine samples were ultrafiltrated using a 20 kDa cutoff centrifugal filter device. The urinary metabolome of 34 newborns with UPJ obstruction and 15 healthy newborns was analyzed by CE-MS and 32 clinically relevant metabolite features were detected to be differentially excreted in the urine of newborns with UPJ.

CE-MS has also been applied for metabolic profiling of volume-limited samples, such as samples from rodent models used in translational studies. For instance, only four microliters of urine was required for the profiling of cationic metabolites by CE-MS in a mouse model of accelerated aging.[28] Prior to CE-MS, the mouse urine samples were just mixed with methanol–water–BGE (4:11:1, v/v/v) and subsequently centrifuged to eliminate any sediment remaining. This strategy permitted the identification of metabolite classifiers between accelerated aging TTD mutants and wild-type animals. CE-MS has also demonstrated its usefulness for metabolic profiling of volume-limited urine samples from animal models to assess the nutraceutical properties of a variety of natural sources,[41] and the toxic effect of drugs[42] or contaminants,[43] among others.

2.3.2 Blood, Plasma and Serum

Blood derivatives like serum and plasma are the most common biofluids used in metabolomic studies by CE-MS. Blood is the circulating fluid that surrounds all tissues, and thus, changes in metabolite concentrations can be potentially associated with disease, disease progression, therapeutic intervention, *etc.* Blood sample collection commonly requires blood to be drawn by venous puncture. Although the collection of blood samples is slightly more invasive compared to urine, it is a more homogeneous sample that is less affected by daily variations than urine. Obtaining serum or plasma will depend on whether coagulation is allowed or blocked, respectively. Plasma is obtained from whole blood by addition of an appropriate anticoagulant to prevent the fibrinogen clotting cascade, before removal of blood cells by centrifugation. The most common anticoagulants are heparin, citrate and EDTA, and they are typically supplemented in the blood-collection tube. The presence of anticoagulants in plasma collection tubes may affect the resulting metabolite profile due to different mechanisms involved in anticoagulation. No consensus exists yet on the choice of anticoagulant for metabolomics studies of plasma by MS-based approaches, and thus, this is actually a specific topic of study and debate. For example, it has been reported that the anticoagulant counter cation (Na or K) may modify the ionization efficiency of analytes, making some metabolites more dominant in positive ESI mode.[44] On the other hand, serum is obtained by blood collection into tubes without the addition of anticoagulants to allow the natural process of the fibrinogen clotting cascade, followed by centrifugation to isolate the resulting supernatant fluid. The protein content of serum is somewhat lower than that of plasma since it lacks fibrinogen and other clotting proteins. Another reported difference between serum and plasma is related to a higher amount of small peptides, which are believed to emerge from the clotting cascade reaction and from protease enzymatic activities during the coagulation process.[44,45]

It has been stated that the discriminatory ability of the two biofluids is equivalent using non-targeted metabolomics by MS-based approaches, and that the differences between individuals are markedly similar for the two biofluids.[46] However, serum typically requires longer processing times and thus is subjected to an increased risk of *ex vivo* metabolism; moreover, artifact formation may occur during blood clotting. For these reasons, plasma is usually preferred over serum in metabolomics studies. Hirayama *et al.*[47] evaluated the effects of different sampling procedures and storage conditions on the stability of metabolomic profiles obtained by CE-MS in both plasma and serum. The effects of the times (60, 180 and 360 min) for clotting (serum) or incubation (plasma), the storage temperature (0 °C, RT) and the number of freeze–thaw cycles (1, 2, 5 and 10), were assessed. Cationic metabolites were separated using a fused-silica capillary, 1 M formic acid as BGE and normal CE polarity, whilst anionic metabolites were separated using a COSMO(+) coated capillary, 20 mM ammonium acetate (pH 8.5) as BGE and reversed CE polarity. An example of the effects of incubation time on the metabolomic profiles of serum and plasma (Figure 2.2A), and of serum and plasma storage temperature (0 °C, RT) and duration (0, 180, and 360 min) (Figure 2.2B) prior to metabolite extraction is given in Figure 2.2. Score plots of PCA were used to evaluate the changes in metabolomic profiles dependent on variations in metabolite features obtained by CE-MS.

It was observed that the differences in the metabolomic profiles due to differences in incubation/clotting time were smaller for plasma than for serum. Although high correlations between plasma and serum were observed, plasma showed higher stability in all three experiments (incubation/clotting time, storage temperature and freeze–thaw cycles). Overall, plasma profiles were more stable than serum profiles, and intra-individual variations were smaller than inter-individual variations in both plasma and serum samples.

Compared to urine, the higher protein content of blood requires protein removal to prevent adsorption of proteins to the inner wall of the fused-silica capillary during CE-MS analysis. Protein precipitation with cold organic solvents (typically, acetonitrile, methanol or ethanol) is the simplest treatment for the isolation and preconcentration of metabolites from plasma and

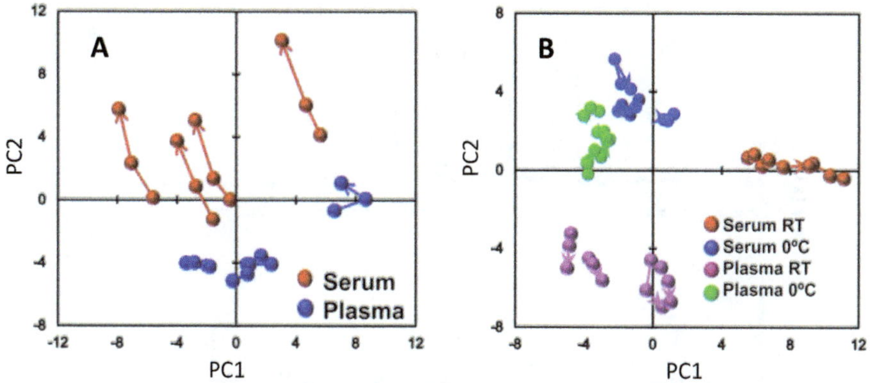

Figure 2.2 (A) Effects of incubation time of serum and plasma samples on the metabolomic profiles. Serum (red) and plasma (blue) score plots of PCA for each individual (n = 4). (B) Effects of serum and plasma storage temperature (0 °C and RT) and duration (0, 180, and 360 min) on metabolomic profile. For further information, the reader is referred to the original version of this article.[47] Reproduced with permission from ref. 47. © 2015 WILEY-VCH Verlag GmbH & Co. KGaA, Weinheim.

serum samples. It allows for simultaneous quenching and non-selective sample pre-treatment. The main disadvantage of deproteinization is the potential loss of protein-bound metabolites. In addition to proteins, endogenous phospholipids in serum and plasma samples are associated with matrix effects. This is due to the highly ionic nature of phospholipids, which makes them responsible for ion suppression in electrospray MS sources. As CE-MS-based metabolome analysis is mainly used to analyze polar metabolites, the combination of solvent deproteinization and removal of phospholipids has been revealed to be the most suitable method, in terms of improvement of metabolite coverage with the best analytical performance. Thus, liquid–liquid extraction (LLE) has become a popular approach in serum/plasma metabolomics by CE-MS. It involves the use of non-polar organic solvents immiscible with water for the removal of lipid fractions including phospholipids and other lipid species. Proteins and other macromolecules are precipitated and typically trapped in the solvent layer between the aqueous and organic phase.

Alternatively, SPE sorbents enable the removal of phospholipids from the sample and can also selectively enrich low-concentration metabolites. Several commercially available novel materials are available for removing proteins and phospholipids from serum and plasma samples in a single step (hybrid precipitation/SPE technology), such as HybridSPE™ (Sigma-Aldrich), Ostro™ (Waters), ISOLUTE® PLD (Biotage) and Phree™ (Phenomenex). The potential of on-line integration of SPE to CE-MS has been demonstrated for the profiling of cationic metabolites in plasma samples from mice.[48] This approach permitted sample clean-up and concentration

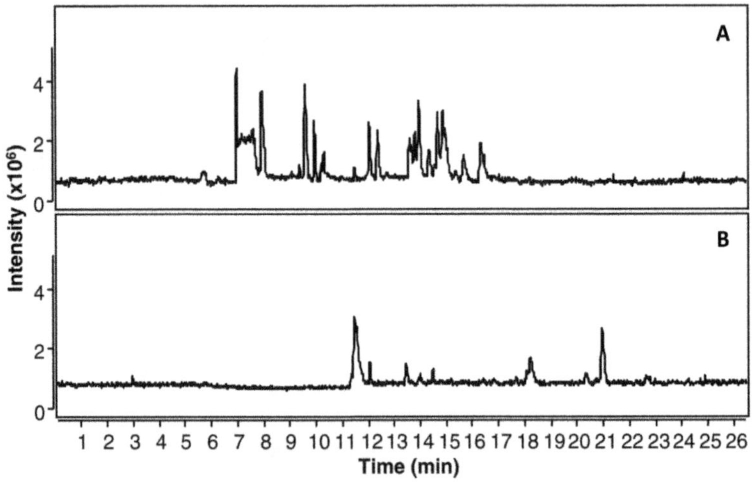

Figure 2.3 Comparison of the total ion electropherograms from rat serum samples subjected to (a) ultrafiltration and (b) SPE, prior to CE-MS analysis. For further information the reader is referred to the original version of this article.[49] Reproduced from *Analytical and Bioanalytical Chemistry*, Method development and validation for rat serum fingerprinting with CE-MS: application to ventilator-induced-lung-injury study, **405**, 2013, 4849–4858, S. Naz, A. Garcia, M. Rusak and C. Barbas, (© Springer-Verlag Berlin Heidelberg 2013) with permission from Springer.

enhancement with minimum sample handling before separation and detection. Potential metabolites useful to discriminate between control and Huntington's disease (HD) mice, as well as to follow up HD progression, were obtained.

Ultrafiltration using molecular weight cutoff filters of 3 kDa or higher has also been used for the removal of proteins from plasma and serum samples. This approach has been proved to provide a good coverage for low molecular weight compounds in global plasma/serum metabolome analysis by CE-MS. Naz *et al.*[49] compared the performance of ultrafiltration (30 kDa cutoff) with a hybrid precipitation/SPE method for global metabolomics studies on rat serum by CE-MS. They obtained better sensitivity (two-fold molecular features), lower variability, and shorter sample treatment with the ultrafiltration method. Figure 2.3 shows the CE-MS metabolic profiles of serum samples obtained from both ultrafiltration and hybrid precipitation/SPE procedures. The optimized sample treatment method was applied to real serum samples from rats with ventilator-induced lung injury, an experimental rat model for acute lung injury (VILI). Cationic metabolite profiling by CE-MS revealed distinctive metabolites that change markedly in the serum of rats with VILI compared with that of control rats.

The utility of CE-MS-based metabolomics to provide potential diagnostic serum/plasma markers for a variety of diseases is well documented and some outstanding examples are shown in Table 2.1.

Table 2.1 Summary of the performance characteristics of sample preparation methods employed for blood/serum/plasma metabolite profiling by CE-MS.[a]

Reference	Sample	Study	Metabolite extraction	CE-MS method
50	Human serum	Biomarkers of neuro-degenerative dementia	1. MeOH 2. UF (5 kDa cutoff)	1. CMP, UC, NP BGE: 1 M FA SL: MeOH–water (50 : 50 v/v) ESI MS: positive 2. AMP, SMILE(+) CC, RP BGE: 50 mM AA (pH 8.5) SL: MeOH–water (50 : 50 v/v), 5 mM AA ESI MS: negative
51	Human plasma	Biomarkers for diabetic nephropathy	Sample: 100 μL 1. 900 μL MeOH	1. CMP, UC, NP BGE: 1 M FA
52	Human serum	Biomarkers of alcohol intake and alcohol-induced liver injury	2. 1.4 mL Water–Chl (4 : 10, v/v) 3. UF (5 kDa cutoff) of aqueous layer	SL: MeOH–water (50 : 50 v/v) ESI MS: positive
53	Rat plasma	Effects of omeprazole and famotidine on gastric injury induced by aspirin	Sample: 50 μL 1. 450 μL MeOH 2. 700 μL Water–Chl (2 : 5, v/v)	2. AMP, COSMO(+) CC, RP BGE: 50 mM AA (pH 8.5) SL: MeOH–water (50 : 50 v/v),
54	Rat serum	Biomarkers of fatigue	3. UF (5 kDa cutoff) of aqueous layer	5 mM AA
55	Human plasma	Biomarkers for progression of end-stage kidney disease	Sample: 40 μL 1. 400 μL MeOH 2. 560 μL Water–Chl (3 : 4, v/v) 3. UF (5 kDa cutoff) of aqueous layer	ESI MS: negative

(continued)

Table 2.1 (*continued*)

Reference	Sample	Study	Metabolite extraction	CE-MS method
56	Human serum	Biomarkers of Alzheimer's disease	Sample: 100 µL 1. 100 µL Water 0.2 M FA 2. UF (30 kDa cutoff)	CMP, UC, NP (assisted pressure) BGE: 0.8 M FA, 10% MeOH SL: MeOH–water (60 : 40 v/v), 1 mM FA ESI MS: positive
57	Human serum	Biomarker related to acute coronary syndrome		
58	Rat plasma	Onion intake in hypercholesterolaemia		
59	Human serum	Biomarkers of obesity-associated hyperinsulinemia		
60	Human serum	Progression of type 2 diabetes mellitus	Sample: 100 µL 1. 100 µL Water 0.2 M FA 2. UF (30 kDa cutoff)	CMP, UC, NP (assisted pressure) BGE: 0.8 M FA, 10% MeOH SL: MeOH–water (50 : 50 v/v), 1 mM FA ESI MS: positive
61	Rat plasma	Impact of hyper-cholesterolemic diet		
62	Human serum	Biomarkers for hepato-cellular carcinoma	Sample: 50 µL 1. 450 µL MeOH 2. 600 µL Chl–water (5 : 1, v/v) 3. UF (5 kDa cutoff) of aqueous layer	1. CMP, UC, NP BGE: 1 M FA SL: MeOH–water (50 : 50 v/v) ESI MS: positive 2. AMP, UC, NP BGE: 50 mM AB (pH 8.5) SL: MeOH–water (50 : 50 v/v) ESI MS: negative
63	Mouse plasma	Molecular changes after cortical spreading depression (CSD)	Protein precipitation with EtOH	CMP, CC, NP (assisted pressure) BGE: 10% acetic acid Porous sheathless interface ESI MS: positive

48	Mouse plasma	Biomarkers of Huntington's disease	Sample: 60 μL On-line SPE-CE-MS (C_{18} Sep-pak)	CMP, UC, NP BGE: 50 mM AcA, 50 mM FA (pH 3.5) SL: 2-propanol–water (60 : 40, v/v), 0.05% FA ESI MS: positive
64	Human Dried blood spot	Biomarkers for inborn errors of metabolism	Sample: 3.2 mm diameter cut-out DBS 1. 100 μL MeOH and sonication 2. UF (14 kDa cutoff)	CMP, UC (MSI-CE-MS), NP BGE: 1 M FA, 15% v ACN (pH 1.8) SL: MeOH–water (60 : 40 v/v), 0.1% FA ESI MS: positive AMP, UC (MSI-CE-MS), NP BGE: 50 mM AB (pH 8.5) SL: MeOH–water (60 : 40 v/v) ESI MS: negative

[a] *AA*: ammonium acetate; *AB*: ammonium bicarbonate; *AcA*: acetic acid; *ACN*: acetonitrile; *AMP*: anionic metabolite profiling; *CC*: coated capillary; *Chl*: chloroform; *CMP*: cationic metabolite profiling; *EtOH*: ethanol; FA: formic acid; *MeOH*: methanol; *NP*: normal polarity; *MSI-CE-MS*: Multisegment injection-CE-MS; *PP*: protein precipitation; *SL*: sheath liquid; *UC*: uncoated capillary; *UF*: ultrafiltration.

According to the published information, amongst the strategies for metabolite extraction, one of the most often considered for serum/plasma metabolomics by CE-MS and followed by Soga's group[51–55] is based on a two-step protocol: first, the use of methanol followed by a water–chloroform mixture (known as the Bligh–Dyer method), and second, centrifugal filtration of the aqueous phase obtained to get an enriched fraction of polar metabolites. On the other hand, Barbas' group[56–61] follow a simpler protocol in which serum or plasma samples are diluted in 0.2 M formic acid at a 1 : 1 solvent-to-sample ratio, and afterwards the diluted samples are ultrafiltrated through a 30 kDa cutoff centrifugal device.

Whole blood can also be collected by heel or finger puncture, spotted on filter paper, and dried. Dried blood spot (DBS) testing has a long history in the clinic, particularly in newborn screening to profile for inherited diseases. However, there is a renewed interest in the use of DBSs to search for novel biomarkers in clinical metabolomics applications.[65,66] Of particular attraction is the greater ease of collection, and simplified transport and storage requirements of DBSs over conventional plasma or serum sampling. Thus, the use of DBSs may have advantages for large-scale epidemiological studies; however, despite the advantages accompanying this type of blood sampling, the effects of different storage conditions on the stability of the metabolite profiles of blood samples collected in this way have not yet been fully investigated. Nevertheless, the potential of DBSs to generate global metabolite profiles by CE-MS has been demonstrated by DiBattista *et al.*[64] Ultrasound-assisted extraction using methanol followed by filtration through a 3 kDa cutoff membrane was used for polar/ionic metabolite extraction from the DBS absorbed onto filter paper. Both CE-MS methods for cationic (bare-fused silica capillary, employing 1 M formic acid BGE and 15% acetonitrile at pH 1.8, and normal CE polarity) and anionic (bare-fused silica capillary, employing 50 mM ammonium bicarbonate BGE at pH 8.5, and normal CE polarity with hydrodynamic pressure) metabolite profiling were employed in positive and negative ion mode, respectively. Moreover, multi-segment injection was conducted to enhance sample throughput. Unambiguous identification and accurate quantification of new biomarkers for inborn errors of metabolism was achieved in this study, although larger-scale clinical and epidemiological studies are needed to validate these preliminary results.

2.3.3 Other Biofluids

Although most clinical analyses for prognosis or diagnosis of a variety of diseases are performed on biological fluids such as plasma, serum or urine, it has been demonstrated that there can be other potential biochemical markers present within other biofluids. Cerebrospinal fluid (CSF) is, for instance, one of the preferred samples to study central nervous system (CNS) pathologies mainly due to its proximity to the site of neuropathology. CSF is typically obtained by lumbar puncture; although this is a non-traumatic sample collection procedure, it is considered a more invasive method when compared to blood sampling. It has already been demonstrated that CSF is a rich source of small-molecule biomarkers for neurological and neurodegenerative disease, as it has been already reviewed.[67] On the other hand, CSF sample volumes are often limited due to the nature of the collection procedure. In this sense, CE-MS is highly suited for the analysis of metabolites in these types of samples.[68]

CE-MS has been applied to monitor metabolomic alterations in CSF related to Alzheimer's disease (AD), the most prevalent form of dementia nowadays.[69] CSF samples were taken from subjects with different cognitive statuses: patients with subjective cognitive impairment, AD patients, MCI patients that progressed to AD within 2 years, and MCI patients that remained stable after this follow-up period. Although the total protein concentration in CSF is about 200 times lower than that in blood plasma, protein removal was performed by simple ultrafiltration using 3 kDa cutoff centrifugal filters to improve reproducibility between injections. Cationic metabolite profiling (using a fused-silica capillary with a low-pH BGE and normal CE polarity) allowed the detection of 160 metabolites, and potential CSF biomarkers of disease progression were identified.

In addition to the biofluids already discussed, there is an increasing interest in alternative biofluids that are non-invasive and easily accessible. Saliva is another interesting sample for screening since it meets the requirements of being a non-invasive and accessible biofluid that can be collected in sufficient quantities for analysis. It is a clear, aqueous biofluid produced by the salivary glands to protect and lubricate the oral cavity. Since saliva is a filtrate of blood, it reflects the physiological state of the organism and therefore it is increasingly being used in metabolomics to gain a mechanistic understanding of a variety of diseases.[70]

The main difficulty in saliva metabolomics is that numerous factors, such as age, gender, diurnal rhythms, smoking status, and oral microbiota, as well as nutritional and environmental exposures, can affect its chemical composition. As with others biofluids, the conditions for collection should be standardized to eliminate confounding factors when searching for disease biomarkers, thereby providing consistent and accurate results. Typical procedures such as mouthwashes, tooth brushing, saliva stimulation, a fasting period, time of collection, devices, *etc.* have been carefully considered. For example, it has been shown that the metabolite profiles of stimulated and unstimulated saliva are different.[71] Ishikawa *et al.*[72] studied by CE-MS the effect on the salivary metabolomic profile of the duration between the previous meal and saliva collection. They revealed that the collection period after a meal affects salivary metabolites for oral cancer screening. Macromolecule removal by ultrafiltration using low molecular weight cutoff filters has been mainly used to purify metabolites in saliva samples in CE-MS-based metabolomics.[50,72–75]

2.4 Sample Preparation for Mammalian Cultured Cells

Cell culture metabolomics is an interesting research strategy to understand how metabolism and biochemical pathways are affected by pathologies and by external or internal stimuli. *In vitro* cell metabolomics analysis offers several advantages compared to classic biofluid or tissue analyses. The major advantage is that cell lines are grown in effectively controlled experimental conditions and for this reason they display lower levels of biological variability than animal models and human subjects. Moreover, *in vitro* cell cultures do not require the same level of ethics consideration. The characteristics of the cells selected for the metabolomics study determine the most suitable sample preparation. Metabolomics of cultured cells involves additional issues related to sample preparation, such as isolating the cells from media and metabolism quenching.[76] No general consensus has yet been reached regarding this issue and the optimization of sample processing of cultured cells for metabolomics is still today an active field of research.

There are two basic systems for growing cells in culture: in suspension for non-adherent cells or as monolayers (attached to an artificial substrate) for adherent cells. The cells grown in suspension are typically separated from the growth medium by filtration or centrifugation. On the other hand, adherent cells are commonly detached from the surface by direct scraping from the culture dish or by trypsinization. Both scraping and trypsinization have their own inherent complications. For instance, scraping may cause cell membrane damage and metabolite leakage, and trypsinization may also cause cell membrane damage and might introduce chemical artifacts into metabolomic data. In addition, these complications during harvesting are most likely dependent on the cell type. In this regard, some developments have been made to reduce the time consumption and physiological stress produced by conventional harvesting methods. For instance, in the MxP® CellCollect system for adherent cells, the cells are grown directly on top of a membrane that can be easily picked up with tweezers and quickly submitted to washing and quenching procedures;[77] then, simultaneous cell detachment and metabolite extraction is performed with the most appropriate solvent.

The culture medium composition is another issue to be considered in cell metabolomics. Culture media components (*i.e.* nutrients, growth factors, salts, serum, and other additives) are typically not compatible with metabolome analysis by MS-based approaches. They are considered contaminants that interfere in intracellular metabolite analysis, and might also result in shifts in the migration times by CE-MS. In a common procedure, harvested cells are typically washed with buffer under specific conditions to minimize perturbation of the metabolic state of the cells and avoid metabolite leakage.

In any metabolomics application, cell metabolism must be stopped or quenched immediately upon cell sampling. Quenching procedures aim to stop all enzymatic activities as quickly as possible. Metabolism is usually quenched by rapidly changing the temperature or pH, and/or by the immediate addition of organic solvents. Freezing in liquid nitrogen is generally considered one of the easiest ways of stopping enzyme activity. It enables cell storage before metabolite extraction and it can also be a previous step to freeze-drying.

After cell quenching, intracellular metabolite extraction is performed by efficient cell disruption by mechanical and/or non-mechanical means. Mechanical methods include ultrasonic disruption, mechanical agitation (homogenizer, blender, *etc.*) and pressurized disruption; and non-mechanical methods mainly consist of chemical permeabilization with detergents and/or organic solvents, physical disruption (osmotic shock, freezing and thawing) and enzymatic permeabilization. For comprehensive, reliable, and representative metabolic profiles, optimization of the extraction solvent composition is typically performed for each type of cell line under study. Simó *et al.*[78] investigated different metabolite purification approaches for the metabolic profiling of human colorectal adenocarcinoma HT-29 cells by CE-MS: deproteinization with organic solvents, ultrafiltration and SPE. SPE showed good extraction efficiency but with different selectivity compared to protein precipitation with methanol and ultrafiltration. On the other hand, no major differences were observed in the metabolic profiles using ultrafiltration and methanol deproteinization. Overall, the methanol deproteinization procedure brought about a higher number of metabolites associated to known metabolic pathways and, therefore, wider metabolomic information could be obtained.

Normalization is an important issue in cell culture metabolomics, especially when the cells are exposed to certain conditions (drugs, chemicals, toxicants) that reduce/induce cell proliferation and/or death, which lead to major differences in cell number among the studied groups. Thus, normalization is needed to avoid erroneous data interpretation; however, suitable normalization is still an unresolved problem in cell culture metabolomics and it has often been underestimated. Cell counting can be employed for normalization purposes in CE-MS-based cell culture metabolomics studies. Following this strategy, prior to metabolite extraction, the cells in suspension are counted and split into aliquots containing the desired cell number.[79] When working with adherent cells, direct scraping of the cells from the culture surface impedes cell counting; in such cases, cell number can be determined from experiments conducted in parallel. Total protein or DNA content, rather than cell number, should be used for normalization in cases where cells display differential increases in volume or when trypsinization is intended to be avoided. Data-based normalization methods, such as normalization to MSTUS or to the total signal of all compounds detected per sample, can also be applied as post-acquisition normalization methods. It has also been reported that there is a linear correlation between the signal intensities of some intracellular metabolites and cell amount, so they can potentially be used as an index of cell amount for the normalization of the data.[80] However, this approach has not yet been used in cell culture metabolomics by CE-MS.

Overall, the application of CE-MS-based metabolomics in the area of cell cultures is relatively less developed compared to biofluids. Nevertheless, CE-MS has been employed for metabolite profiling of a variety of cultured cell lines. In most cases, analyses of cationic and anionic metabolites are conducted according to the methods developed by Soga *et al.*[81] In Table 2.2, a summary of sample preparation procedures reported for metabolomic analysis by CE-MS of cultured cells is given.

As can be seen, often quenching and extraction procedures are combined. Cells are typically quenched by ice-cold methanol followed the addition of chloroform and water to form two-phase solvent systems.[82–89] The aqueous phase contains water soluble low-molecular-weight metabolites, while the organic phase contains non-polar metabolites. Cell debris, proteins and other macromolecules are precipitated and trapped in the middle layer. Afterwards, the aqueous phase is ultrafiltrated using low molecular weight cutoff centrifugal devices. The obtained extracts are concentrated to dryness and later reconstituted in water before their injection into the CE-MS system.

The complete cell metabolome might include the exo-metabolome, namely, all metabolites in the surrounding extracellular medium. The composition of the extracellular medium reflects the metabolic activity of cells and changes in response to environmental conditions. Technically, the measurement of extracellular metabolic profiles is simpler as it only requires filtration or centrifugation to separate the culture medium and cells. Mastrangelo *et al.*[90] characterized the cellular exo- and endo-metabolome by CE-MS from obesity- and non-obesity-derived adipose-derived stem cells (ASCs). Since the culture medium was supplemented with 10% fetal bovine serum, a precipitation method with methanol was applied at a 3 : 1 solvent-to-sample ratio to remove protein interferents. Important alterations in metabolites associated with glycolysis, the tricarboxylic acid cycle, the pentose phosphate pathway and the polyol pathway were observed in the exo-metabolome of obesity-derived human ASCs.

2.5 Sample Preparation for Solid Samples: Tissues and Plants

Human or rodent tissue metabolomics has some advantages over biofluids since it allows for the spatial description of metabolite distribution. This makes it an interesting approach for monitoring the effects of pharmaceuticals, toxicants or bioactive compounds in foods, in specific tissues, which will give us essential information about their mechanisms of action. Moreover,

Table 2.2 Overview of sample preparation methods used in global CE-MS metabolomics studies using cultured cells.

Cell line	Cell number	Washing	Harvesting/ quenching	Extraction	Concentration	Normalization[c]	Reference
Endometrial cancer (HEC-1-A) (adherent)	N.I.	5% maltose	Scraping in MeOH	1. MeOH–Chl–water (10 : 10 : 4, v/v/v)[b] 2. UF (5 kDa cutoff) of aqueous layer	Yes	Cell counting	82
Human fetal hepatocytes and primary human hepatocytes (HEP220) (adherent)	$2–3 \times 10^6$	5% mannitol	Suspended MeOH	1. MeOH–Chl–water (10 : 10 : 4, v/v/v)[b] 2. UF (5 kDa cutoff) of aqueous layer	Yes	Cell counting	83
Human lung adenocarcinoma (PC-9) (adherent)	1.8×10^6[a]	5% mannitol	Suspended in MeOH	1. MeOH–Chl–water (4 : 4 : 2, v/v/v)[b] 2. UF (5 kDa cutoff) of aqueous layer	Yes	N.I.	84
Human oral squamous cell carcinoma (HSC-2) (adherent)	6×10^5[a]	5% mannitol	Suspended in MeOH	1. MeOH–Chl–water (4 : 4: 2, v/v/v)[b] 2. UF (5 kDa cutoff) of aqueous layer	Yes	Cell counting	85
Mouse primordial and embryonic germ cells (adherent)	5×10^5	5% mannitol	Suspended MeOH	1. MeOH–Chl–water (4 : 4 : 2, v/v/v)[b] 2. UF (5 kDa cutoff) of aqueous layer	Yes	Cell volume	86
Mouse skeletal muscle cells (C2C12) (adherent)	1×10^5[a]	5% mannitol	Dispersed in MeOH	1. MeOH–Chl–water (10 : 10 : 4)[b] 2. UF (5 kDa cutoff) of aqueous layer	Yes	Cell counting	87

(continued)

Table 2.2 (*continued*)

Cell line	Cell number	Washing	Harvesting/quenching	Extraction	Concentration	Normalization[c]	Reference
Human glioma (U87) (adherent)	N.I.[c]	5% mannitol	Scraping in MeOH-water (21 : 79, v/v)	Chl is added	Yes	Total protein	88
Human colorectal adenocarcinoma (HT-29) (adherent)	10×10^6	PBS	1. Trypsinization 2. Centrifugation 3. Freezing (liquid nitrogen)	1. Water 2. UF (3 kDa cutoff)	No	Cell counting	79
Human corneal endothelial cell (HCEC) (suspension)	N.I.	5% mannitol	1. Centrifugation 2. Methanol is added	1. MeOH–water (10 : 7 v/v)[b] 2. UF (3 kDa cutoff)	Yes	N.I.	89
Murine and human adipose-derived stem cells (adherent)	1×10^6	N.I.	Scraping in MeOH	Protein precipitation with MeOH	Yes	N.I.	90

[a]Cell number at the time of seeding. The number of cells at the time of harvesting is not indicated.
[b]MeOH–Chl–water or MeOH–water ratio considering the solvent used for the harvesting/quenching procedure.
[c]N.I.: Not indicated.

direct metabolomic assessment of tumor tissues can be performed. Since tumor metabolism differs considerably from that of normal tissue, understanding the metabolic profile of tumors is important for therapeutic approaches and molecular diagnosis. In contrast, tissues are considerably less available samples than biofluids and invasive sampling techniques are needed.

The first step in tissue metabolomics is a fast rinsing of the solid sample with a buffered aqueous solution to remove any contamination, such as blood, followed by immediate quenching of metabolic activity. As mentioned above, the conventional strategies for quenching are based on quick modification of sample conditions (pH, temperature). In tissue metabolomics applications, the quenching step is frequently carried out by freezing the samples in liquid nitrogen. Prior to metabolite extraction, the tissue needs to be disrupted and homogenized to a fine powder. Tissues are more difficult to break down than cultured cells, and thus, mechanical methods using a wide variety of equipment are applied.[91] This step should be carried out preferably under liquid nitrogen to prevent tissue defrosting. Next, the powder can be lyophilized or stored at low temperatures until metabolite extraction. In some protocols, mechanical disruption and homogenization are performed with the tissue and extraction solvent combined. The solvent is selected with the aim to obtain metabolic profiles covering the maximum number of metabolites in a reproducible way.

A wide array of tissues have been studied applying metabolomic approaches by CE-MS, including mouse lungs,[92] rodent muscle,[93,94] livers from different animal models,[95–98] several tumor tissues,[99,100] and multiple tissues from mice,[101] among others. Biphasic extraction has been the preferred strategy since water-soluble metabolites can be separated from lipophilic metabolites through the use of methanol–chloroform–water to obtain two immiscible liquids and an interphase between them with the unwanted proteins, nucleic acids, and cellular debris. Although less used, monophasic extraction by using acidified water[92] or methanol–water mixtures[94,95,98] has also been employed in tissue metabolomics by CE-MS. In both approaches (biphasic and monophasic), the aqueous solutions are submitted to ultrafiltration with cutoff membranes (5–30 kDa) prior to CE-MS analysis.

Concerning plant metabolomics, proper harvesting of fresh material is a fundamental step since rapid metabolite changes may occur during this process and may therefore affect the results considerably. The time of day at which the plant is harvested is also important, since the levels of the metabolites vary throughout the day in the plant. For this reason, harvesting and separating the required material from the original plant (leaves, roots, flowers, fruits, *etc.*) must be carried out very rapidly and at the appropriate time. Once harvested, the plant material must be submitted to rapid freezing in liquid nitrogen to avoid metabolite changes caused by enzymatic or chemical reactions. Freeze-drying is recommended to eliminate the aqueous environment and to stabilize the specimens preventing further metabolic reactions and degradation. Moreover, different levels of water in the studied samples will affect the quantitation of metabolites.

A good contribution in plant metabolomics has also been made by CE-MS. It has been applied to the metabolite profiling of both the cationic and anionic metabolites[81] of a variety of leaves, roots and shoots in *Gentiana triflora*,[102,103] leaves and stems in *Rumex obtusifolius*,[104,105] shoots and roots in *Arabidopsis thaliana*,[106] tobacco leaves[107] and herbal medicines.[108] Homogenization of the plant tissue was typically conducted by using a mechanical bead-mill homogenizer with the addition of chilled methanol–water or methanol–chloroform–water, and this was followed by centrifugal filtration of the aqueous phase through a 3–5 kDa cutoff filter prior to CE-MS analysis.

2.6 Sample Preparation for Microbial Cells, Parasites and Microbiota

There is growing interest in applying microbial metabolomics in clinical research, food processing and spoiling applications, in ecological/environmental studies, and in industrial processes, among others. Sample preparation for microbial metabolomics also faces some important challenges, most of which are related to improving the quenching of cell metabolism during sampling and the metabolite extraction protocol for effective extraction to obtain accurate metabolome information without artifact formation or compound degradation for both intra- and extracellular metabolites.

Different types of cultured microorganisms have been studied by CE-MS, including *Escherichia coli*,[109] *Synechocystis* sp. PCC6803,[110] *Ralstonia eutropha* H16[111] and *Saccharomyces cerevisiae*.[112,113] The considerations regarding sample preparation of microbial cells are similar to those already described for mammalian cultured cells (harvesting, washing, quenching, extraction with solvents). In practice, filtration has been the preferred method to separate cells from the nutritional medium. After the culture medium is removed and the microbial cells are washed with water, the filter is immersed in methanol to suspend the cells completely. Immediately after that, chloroform and water are sequentially added to create two-phase separation. The polar metabolite fraction in the top methanol–water layer is subsequently centrifugally ultrafiltrated (5 kDa cutoff filter). Then, the low molecular weight fraction is evaporated to dryness and reconstituted in water before CE-MS analysis.

Metabolomics by CE-MS has also been demonstrated to be a powerful tool in understanding the mechanisms of drug action and resistance in the *Leishmania* protozoan parasite, an infectious organism which causes Leishmaniasis. Parasites were lysed and homogenized in a bead-mill, and methanol–ethanol (1 : 1, v/v) or methanol–water (1 : 1, v/v) was used for metabolite extraction. After centrifugation, the obtained metabolite extract was evaporated to dryness and resuspended in water prior to CE-MS analysis.[114–117] CE-MS-based metabolomics has also been used to understand the structure and dynamics of the metabolic network during encystation of the protozoan parasite *Entamoeba invadens*.[118] Metabolism was quenched by the addition of ice-cold methanol and intracellular metabolites were extracted by sequential addition of chloroform and methanol in order to get two immiscible liquid phases with both hydrophilic and lipophilic metabolites. The aqueous layer was ultrafiltrated and concentrated prior to CE-MS analysis, as described for the *Leishmania* parasite.

The human body coexists with a vast number of microorganisms (namely, bacteria, archaea, microbial eukaryotes and viruses) which together constitute the microbiota. It is now widely known that human phenotypes are greatly affected by microorganisms residing in their bodies. In this context, the gut microbiota is increasingly being studied since it has profound effects in physiology and the immune system; indeed, alterations in the equilibrium between microbiota and host have been observed in recent studies.[119,120] Whole community metabolomics can be considered as a function-based approach that is growing exponentially to identify key microbial functions in the gut. In this sense, metabolomic analysis of feces can provide useful insight into the relation between a microbial community and its host. There are currently no general guidelines for fecal sample collection and storage conditions as there is still a lack of comprehensive investigation of the effect of sampling procedure, and the effect of time before processing the samples, temperature of storage, *etc.* Gratton *et al.*[121] studied storage duration-, temperature-, and freeze/thaw cycle-induced metabolic changes in crude stool and fecal water by NMR-based metabolomics. They observed that the metabolic profiles of fecal water samples were more stable compared to crude samples. Crude fecal samples were remarkably unstable over time and exhibited distinct metabolic profiles at different storage temperatures. On the other hand, they observed that inter-individual variation of fecal metabolites was greater than that attributable to storage conditions, and that metabolites from fecal samples should ideally be extracted within 1 h of collection.

CE-MS together with LC-MS was employed in different metabolic profiling studies to assess the effect of gut bacterial metabolism in *Clostridium difficile*-associated diarrhea,[122] lupus erythematosus and obesity,[123] and in immune recovery in HIV-infected patients.[124] In all such studies, microbial cells were separated from the fecal matrix by mixing the fecal sample in a buffered solution followed by centrifugation. Microbial cells in the supernatants were isolated by centrifugation and immediately quenched in the presence of ice-cold methanol. A two-step extraction method was then applied to obtain the optimal extraction efficiency. First, ultrasound-assisted extraction was carried out in methanol, and immediately after the methanol extract was obtained, it was stored at −80 °C. The remaining cell pellet was re-suspended in cold water and then subjected to a second ultrasound-assisted extraction. The final metabolite extract was obtained by combining equal amounts of each of the extracts (methanol and water), and it was directly injected into the CE-MS system.

2.7 Conclusions

Over the past years, CE-MS has been demonstrated to be a useful approach in metabolomics studies for improving the metabolome coverage, since it is highly suitable for the analysis of highly polar and ionic metabolites which are difficult to analyze using other analytical platforms.

In addition to technical and methodological developments in CE-MS, appropriate sample preparation is crucial in order to obtain representative and accurate metabolome data. Depending on the type of sample and the aim of the experiment, appropriate sample preparation methods must be chosen. This can vary from simple procedures such as dilution, to more specific metabolite extraction/concentration protocols such as liquid–liquid, solid–liquid, protein precipitation with solvents, ultrafiltration, SPE, or any combination thereof. Nevertheless, there is still no consensus on the most suitable sample preparation procedure to be used for a certain sample or biological problem and thus it is still an active topic of research. Additionally, to further improve the concentration sensitivity, and thus metabolic coverage, CE-MS coupling *via* a porous sheathless interface holds great promise. In this respect, we must still face important challenges for sample preparation in CE-MS-based metabolomics, such as standardization of the sample handling, preparation and storage conditions, validation of existing sample pretreatment methods, and increase of sample throughput. As the CE-MS throughput is somewhat compromised as two systems for cationic and anionic metabolite profiling are required, the application of MSI-CE-MS to increase throughput will probably have a major impact in future metabolomics applications. Moreover, further development and improvements in robustness, as well as large-scale application of sophisticated on-line and in-line SPE-CE-MS systems for sample preparation and analysis in one single step, will also significantly increase system automation and throughput in metabolomics applications.

Acknowledgements

Support from AGL2017-89055-R (Spanish Ministry of Economy, Industry and Competitiveness) is acknowledged.

References

1. M. Moini, *Anal. Chem.*, 2007, **79**, 4241.
2. W. Zhang, T. Hankemeier and R. Ramautar, *Curr. Opin. Biotechnol.*, 2017, **43**, 1.
3. A. Šlampová, Z. Malá, P. Gebauer and P. Boček, *Electrophoresis*, 2017, **38**, 20.
4. M. Yamamoto, R. Ly, B. Gill, Y. Zhu, J. Moran-Mirabal and P. Britz-McKibbin, *Anal. Chem.*, 2016, **88**, 10710.

5. M. G. M. Kok, G. J. de Jong and G. W. Somsen, *Electrophoresis*, 2011, **32**, 3016.
6. M. G. M. Kok, M. M. A. Ruijken, J. R. Swannlan, D. Wilson, G. W. Somsen and G. J. de Jong, *Anal. Bioanal. Chem.*, 2013, **405**, 2585.
7. T. Soga, Y. Ueno, H. Naraoka, Y. Ohashi, M. Tomita and T. Nishioka, *Anal. Chem.*, 2002, **74**, 2233.
8. T. Soga, K. Igarashi, C. Ito, K. Mizobuchi, H. S. Zimmermann and M. Tomita, *Anal. Chem.*, 2009, **81**, 6165.
9. S. Ullsten, R. Danielsson, D. Bäckström, P. Sjöberg and J. Bergquist, *J. Chromatogr. A*, 2006, **1117**, 87.
10. R. Yoshida, T. Tamura, C. Takaoka, K. Harada, A. Kobayashi, Y. Mukai and E. Fukusaki, *Aging Cell*, 2010, **9**, 616.
11. T. Acunha, C. Simó, C. Ibáñez, A. Gallardo and A. Cifuentes, *J. Chromatogr. A.*, 2016, **1428**, 326.
12. T. Soga, Y. Ueno, H. Naraoka, K. Matsuda, M. Tomita and T. Nishioka, *Anal. Chem.*, 2002, **74**, 6224.
13. M. Sugimoto, A. Hirayama, T. Ishikawa, M. Robert, R. Baran, K. Uehara, K. Kawai, T. Soga and M. Tomita, *Metabolomics*, 2010, **6**, 27.
14. M. Sugimoto, M. Kawakami, M. Robert, T. Soga and M. Tomita, *Curr. Bioinf.*, 2012, 7, 96–108.
15. J. Bartel, J. Krumsiek and F. J. Theis, *Comput. Struct. Biotechnol. J.*, 2013, **4**, e201301009.
16. B. Worley and R. Powers, *Curr. Metabolomics*, 2013, **1**, 92.
17. E. Rathahao-Paris, S. Alves, C. Junot and J.-C. Tabet, *Metabolomics*, 2016, **2**, 10.
18. T. Sangster, H. Major, R. Plumb, A. J. Wilson and I. D. Wilson, *Analyst*, 2006, **131**, 1075.
19. J. Delanghe and M. Speeckaert, *Biochem. Med.*, 2014, **24**, 89.
20. D. Ryan, K. Robards, P. D. Prenzler and M. Kendall, *Anal. Chim. Acta*, 2011, **684**, 17.
21. A. J. Chetwynd, A. Abdul-Sada, S. G. Holt and E. M. Hill, *J. Chromatogr. A.*, 2016, **1431**, 103.
22. Y. Gagnebin, D. Tonoli, P. Lescuyer, B. Ponte, S. de Seigneux, P. Y. Martin, J. Schappler, J. Boccard and S. Rudaz, *Anal. Chim. Acta*, 2017, **955**, 27.
23. B. M. Warrack, S. Hnatyshyn, K.-H. Ott, M. D. Reily, M. Sanders, H. Zhang and D. M. Drexler, *J. Chromatogr. B*, 2009, **877**, 547.
24. F. Boizard, V. Brunchault, P. Moulos, B. Breuil, J. Klein, N. Lounis, C. Caubet, S. Tellier, J.-L. Bascands, S. Decramer, J. P. Schanstra and B. Buffin-Meyer, *Sci. Rep.*, 2016, **6**, 34453.
25. J. Adachi, C. Kumar, Y. Zhang, J. V. Olsen and M. Mann, *Genome Biol.*, 2006, 7, R80.
26. M. Pejchinovski, D. Hrnjez, A. Ramirez-Torres, V. Bitsika, G. Mermelekas, A. Vlahou, P. Zürbig, H. Mischak, J. Metzger and T. Koeck, *Proteomics: Clin. Appl.*, 2015, **9**, 453.
27. R. Ramautar, A. A. van der Plas, E. Nevedomskaya, R. J. Derks, G. W. Somsen, G. J. de Jong, J. J. van Hilten, A. M. Deelder and O. A. Mayboroda, *J. Proteome Res.*, 2009, **8**, 5559.

28. E. Nevedomskaya, R. Ramautar, R. Derks, I. Westbroek, G. Zondag, I. Van der Pluijm, A. M. Deelder and O. A. Mayboroda, *J. Proteome Res.*, 2010, **9**, 4869–4874.
29. R. Ramautar, J.-M. Busnel, A. M. Deelder and O. A. Mayboroda, *Anal. Chem.*, 2012, **84**, 885.
30. J. V. Alberice, A. F. Amaral, E. G. Armitage, J. A. Lorente, F. Algaba, E. Carrilho, M. Márquez, A. García, N. Malats and C. Barbas, *J. Chromatogr. A*, 2013, **1318**, 163–170.
31. C. Balderas, F. J. Rupérez, E. Ibañez, J. Señorans, J. Guerrero-Fernández, I. G. Casado, R. Gracia-Bouthelier, A. García and C. Barbas, *Electrophoresis*, 2013, **34**, 2882–2890.
32. D. Dudzik, M. Zorawski, M. Skotnicki, W. Zarzycki, G. Kozlowska, K. Bibik-Malinowska, M. Vallejo, A. García, C. Barbas and M. P. Ramos, *J. Proteomics*, 2014, **103**, 57.
33. L. C. Soliman, Y. Hui, A. K. Hewavitharana and D. D. Y. Chen, *J. Chromatogr. A*, 2012, **1267**, 162.
34. T. Zhang, D. J. Creek, M. P. Barrett, G. Blackburn and D. G. Watson, *Anal. Chem.*, 2012, **84**, 1994.
35. M. G. M. Kok, G. W. Somsen and G. J. de Jong, *Talanta*, 2015, **132**, 1.
36. R. Ramautar, J. S. Toraño, G. W. Somsen and G. J. de Jong, *Electrophoresis*, 2010, **31**, 2319.
37. W. C. Yang, F. E. Regnier and J. Adamec, *Electrophoresis*, 2008, **29**, 4549.
38. J. L. Chen, J. Fan and X. J. Lu, *Electrophoresis*, 2014, **35**, 1032.
39. K. T. Rodrigues, D. Mekahli, M. F. M. Tavares and A. Van Schepdael, *Electrophoresis*, 2016, **37**, 1039.
40. I. Garcia-Perez, J. C. Lindon and E. Minet, *Bioanalysis*, 2014, **6**, 2733–2749.
41. J. Godzien, D. García-Martínez, P. Martinez-Alcazar, F. J. Ruperez and C. Barbas, *Metabolomics*, 2013, **9**, 188.
42. N. Shima, I. Miyawaki, K. Bando, H. Horie, K. Zaitsu, M. Katagi, T. Bamba, H. Tsuchihashi and E. Fukusaki, *Toxicology*, 2011, **287**, 29.
43. J. Zeng, H. Kuang, C. Hu, X. Shi, M. Yan, L. Xu, L. Wang, C. Xu and G. Xu, *Environ. Sci. Technol.*, 2013, **47**, 7457.
44. T. Barri and L. O. Dragsted, *Anal. Chim. Acta*, 2013, **768**, 118.
45. H. Wu and F. Feng, *RSC Adv.*, 2016, **6**, 28279.
46. D. C. Wedge, J. W. Allwood, W. Dunn, A. A. Vaughan, K. Simpson, M. Brown, L. Priest, F. H. Blackhall, A. D. Whetton, C. Dive and R. Goodacre, *Anal. Chem.*, 2011, **83**, 6689.
47. A. Hirayama, M. Sugimoto, A. Suzuki, Y. Hatakeyama, A. Enomoto, S. Harada, T. Soga, M. Tomita and T. Takebayashi, *Electrophoresis*, 2015, **36**, 2148.
48. L. Pont, F. Benavente, J. Jaumot, R. Tauler, J. Alberch, S. Ginés, J. Barbosa and V. Sanz-Nebot, *Electrophoresis*, 2016, **37**, 795.
49. S. Naz, A. Garcia, M. Rusak and C. Barbas, *Anal. Bioanal. Chem.*, 2013, **405**, 4849.
50. M. Tsuruoka, J. Hara, A. Hirayama, M. Sugimoto, T. Soga, W. R. Shankle and M. Tomita, *Electrophoresis*, 2013, **34**, 2865.

51. A. Hirayama, E. Nakashima, M. Sugimoto, S. Akiyama, W. Sato, S. Maruyama, S. Matsuo, M. Tomita, Y. Yuzawa and T. Soga, *Anal. Bioanal. Chem.*, 2012, **404**, 3101.
52. S. Harada, T. Takebayashi, A. Kurihara, M. Akiyama, A. Suzuki, Y. Hatakeyama, D. Sugiyama, K. Kuwabara, A. Takeuchi, T. Okamura, Y. Nishiwaki, T. Tanaka, A. Hirayama, M. Sugimoto, T. Soga and M. Tomita, *Environ. Health Prev. Med.*, 2016, **21**, 18.
53. K. Takeuchi, M. Ohishi, K. Endo, K. Suzumura, H. Naraoka, T. Ohata, J. Seki, Y. Miyamae, M. Honma and T. Soga, *Metabolomics*, 2014, **10**, 995.
54. S. Kume, M. Yamato, Y. Tamura, G. Jin, M. Nakano, Y. Miyashige, A. Eguchi, Y. Ogata, N. Goda, K. Iwai, E. Yamano, Y. Watanabe, T. Soga and Y. Kataoka, *PLoS One*, 2015, **10**, e0120106.
55. T. Kimura, K. Yasuda, R. Yamamoto, T. Soga, H. Rakugi, T. Hayashi and Y. Isaka, *Sci. Rep.*, 2016, **6**, 26138.
56. R. González-Domínguez, A. García, T. García-Barrera, C. Barbas and J. L. Gómez-Ariza, *Electrophoresis*, 2014, **35**, 3321.
57. S. Naz, A. A. Calderón, A. García, J. Gallafrio, R. T. Mestre, E. G. González, C. M. de Cabo, M. C. Delgado, J. A. Balanza, A. V. Simionato, N. N. Vaeza, C. Barbas and F. J. Rupérez, *Electrophoresis*, 2015, **36**, 2303.
58. D. González-Peña, D. Dudzik, C. Colina-Coca, B. de Ancos, A. García, C. Barbas and C. Sánchez-Moreno, *J. Funct. Foods*, 2015, **19**, 363.
59. A. Mastrangelo, G. Á. Martos-Moreno, A. García, V. Barrios, F. J. Rupérez, J. A. Chowen, C. Barbas and J. Argente, *Int. J. Obes.*, 2016, **40**, 1494.
60. M. Ciborowski, E. Adamska, M. Rusak, J. Godzien, J. Wilk, A. Citko, W. Bauer, M. Gorska and A. Kretowski, *Electrophoresis*, 2015, **36**, 2286.
61. D. González-Peña, D. Dudzik, C. Colina-Coca, B. de Ancos, A. García, C. Barbas and C. Sánchez-Moreno, *Eur. J. Nutr.*, 2016, **55**, 997.
62. J. Zeng, P. Yin, Y. Tan, L. Dong, C. Hu, Q. Huang, X. Lu, H. Wang and G. Xu, *J. Proteome Res.*, 2014, **13**, 3420.
63. R. Shyti, I. Kohler, B. Schoenmaker, R. J. Derks, M. D. Ferrari, E. A. Tolner, O. A. Mayboroda and A. M. van den Maagdenberg, *Mol. BioSyst.*, 2015, **11**, 1462.
64. A. DiBattista, N. McIntosh, M. Lamoureux, O. Y. Al-Dirbashi, P. Chakraborty and P. Britz-McKibbin, *Anal. Chem.*, 2017, **89**, 8112.
65. I. Wilson, *Bioanalysis*, 2011, **3**, 2255.
66. M. Wagner, D. Tonoli, E. Varesio and G. Hopfgartner, *Mass Spectrom. Rev.*, 2014, **35**, 361.
67. C. Ibáñez, A. Cifuentes and C. Simó, *Int. Rev. Neurobiol.*, 2015, **122**, 95.
68. R. Ramautar, R. Shyti, B. Schoenmaker, L. de Groote, R. J. Derks, M. D. Ferrari, A. M. van den Maagdenberg, A. M. Deelder and O. A. Mayboroda, *Anal. Bioanal. Chem.*, 2012, **404**, 2895.
69. C. Ibáñez, C. Simó, P. J. Martín-Álvarez, M. Kivipelto, B. Winblad, A. Cedazo-Mínguez and A. Cifuentes, *Anal. Chem.*, 2012, **84**, 8532.
70. B. Cuevas-Córdoba and J. Santiago-García, *OMICS*, 2014, **18**, 87.
71. E. Neyraud, M. Tremblay-Franco, S. Gregoire, O. Berdeaux and C. Canlet, *Metabolomics*, 2013, **9**, 213.

72. S. Ishikawa, M. Sugimoto, K. Kitabatake, M. Tu, A. Sugano, I. Yamamori, A. Iba, K. Yusa, M. Kaneko, S. Ota, K. Hiwatari, A. Enomoto, T. Masaru and M. Iino, *Amino Acids*, 2017, **49**, 761.
73. M. Sugimoto, J. Saruta, C. Matsuki, M. To, H. Onuma, M. Kaneko, T. Soga, M. Tomita and K. Tsukinoki, *Metabolomics*, 2013, **9**, 454.
74. S. Ishikawa, M. Sugimoto, K. Kitabatake, A. Sugano, M. Nakamura, M. Kaneko, S. Ota, K. Hiwatari, A. Enomoto, T. Soga, M. Tomita and M. Iino, *Sci. Rep.*, 2016, **6**, 31520.
75. N. Okuma, M. Saita, N. Hoshi, T. Soga, M. Tomita, M. Sugimoto and K. Kimoto, *PLoS One*, 2017, **12**, e0183109.
76. Z. León, J. C. García-Cañaveras, M. T. Donato and A. Lahoz, *Electrophoresis*, 2013, **34**, 2762.
77. N. Bordag, U. Rennefahrt, J. Nachtigall, S. G. Maldonado, R. C. Reszka, T. Ramirez-Hernandez, H. Kamp, E. Fux and B. Ravenzwaay, *Metabolomics*, 2016, **6**, 1.
78. C. Simó, C. Ibáñez, A. Gómez-Martínez, J. A. Ferragut and A. Cifuentes, *Electrophoresis*, 2011, **32**, 1765.
79. A. Valdés, V. García-Cañas, C. Simó, C. Ibáñez, V. Micol, J. A. Ferragut and A. Cifuentes, *Anal. Chem.*, 2014, **86**, 9807.
80. B. Cao, J. Aa, G. Wang, X. Wu, L. Liu, M. Li, J. Shi, X. Wang, C. Zhao, T. Zheng, S. Guo and J. Duan, *Anal. Bioanal. Chem.*, 2011, **400**, 2983.
81. T. Soga, Y. Ohashi, Y. Ueno, H. Naraoka, M. Tomita and T. Nishioka, *J. Proteome Res.*, 2003, **2**, 488.
82. K. Urakami, V. Zangiacomi, K. Yamaguchi and M. Kusuhara, *Biomed. Res.*, 2013, **34**, 221.
83. S. R. Kim, T. Kubo, Y. Kuroda, M. Hojyo, T. Matsuo, A. Miyajima, M. Usami, Y. Sekino, T. Matsushita and S. Ishida, *J. Toxicol. Sci.*, 2014, **39**, 717.
84. M. Serizawa, M. Kusuhara, V. Zangiacomi, K. Urakami, M. Watanabe, T. Takahashi, K. Yamaguchi, N. Yamamoto and Y. Koh, *Anticancer Res.*, 2014, **34**, 2779.
85. H. Sakagami, C. Shimada, Y. Kanda, O. Amano, M. Sugimoto, S. Ota, T. Soga, M. Tomita, A. Sato, S. I. Tanuma, K. Takao and Y. Sugita, *Toxicol. Rep.*, 2015, **2**, 1281.
86. Y. Hayashi, K. Otsuka, M. Ebina, K. Igarashi, A. Takehara, M. Matsumoto, A. Kanai, K. Igarashi, T. Soga and Y. Matsui, *Proc. Natl. Acad. Sci. U. S. A.*, 2017, **114**, 8289.
87. K. Yoshida, C. K. Imamura, K. Hara, M. Mochizuki and Y. Tanigawara, *Metabolomics*, 2017, **13**, 98.
88. C. Zhao, H. Du, L. Xu, J. Wang, L. Tang, Y. Cao, C. Li, Q. Wang, Y. Liu, F. Shan, J. Feng, F. Xu and P. Gao, *J. Pharm. Biomed. Anal.*, 2015, **115**, 48.
89. J. Hamuro, M. Ueno, K. Asada, M. Toda, M. Montoya, C. Sotozono and S. Kinoshita, *Invest. Ophthalmol. Visual Sci.*, 2016, **57**, 4452.
90. A. Mastrangelo, M. I. Panadero, L. M. Pérez, B. G. Gálvez, A. García, C. Barbas and F. J. Rupérez, *Biochem. J.*, 2016, **473**, 2187.

91. S. Goldberg, *2D PAGE: Sample Preparation and Fractionation*, ed. A. Posch, Humana Press, USA, 2008, vol. 424, ch. 1, pp. 3–22.
92. S. Naz, A. García and C. Barbas, *Anal. Chem.*, 2013, **85**, 10941.
93. Y. Hatazawa, N. Senoo, M. Tadaishi, Y. Ogawa, O. Ezaki, Y. Kamei and S. Miura, *PLoS One*, 2015, **10**, e0129084.
94. S. Urakawa, K. Takamoto, T. Nakamura, S. Sakai, T. Matsuda, T. Taguchi, K. Mizumura, T. Ono and H. Nishijo, *Physiol. Rep.*, 2015, **3**, e12279.
95. Y. Kim, I. S. Lee, K. H. Kim, J. Park, J. H. Lee, E. Bang, H. J. Jang and Y. C. Na, *Am. J. Chin. Med.*, 2016, **44**, 1639.
96. X. Y. Qin, H. Tatsukawa, K. Hitomi, Y. Shirakami, N. Ishibashi, M. Shimizu, H. Moriwaki and S. Kojima, *Cancer Prev. Res.*, 2016, **9**, 205.
97. D. González-Peña, D. Dudzik, A. García, B. de Ancos, C. Barbas and C. Sánchez-Moreno, *Int. J. Mol. Sci.*, 2017, **18**, E267.
98. W. B. Xu, A. Iwasawa and M. Yayota, *Metabolomics*, 2017, **13**, 90.
99. K. Kami, T. Fujimori, H. Sato, M. Sato, H. Yamamoto, Y. Ohashi, N. Sugiyama, Y. Ishihama, H. Onozuka, A. Ochiai, H. Esumi, T. Soga and M. Tomita, *Metabolomics*, 2013, **9**, 444.
100. T. Ogawa, J. Washio, T. Takahashi, S. Echigo and N. Takahashi, *Int. J. Oral Maxillofac. Pathol.*, 2014, **118**, 218.
101. S. Kobayashi, M. Sato, T. Kasakoshi, T. Tsutsui, M. Sugimoto, M. Osaki, F. Okada, K. Igarashi, J. Hiratake, T. Homma, M. Conrad, J. Fujii, T. Soga, S. Bannai and H. Sato, *J. Biol. Chem.*, 2015, **290**, 8778.
102. H. Takahashi, T. Imamura, A. Miyagi and H. Uchimiya, *Metabolomics*, 2012, **8**, 154.
103. H. Takahashi, H. Abe, K. Fujita and K.-T. Sekine, *Metabolomics*, 2017, **13**, 51.
104. S. Kitano, A. Miyagi, Y. Oono, Y. Hase, I. Narumi, M. Yamaguchi, H. Uchimiya and M. Kawai-Yamada, *Metabolomics*, 2015, **11**, 134.
105. A. Miyagi, K. Takahara, I. Kasajima, H. Takahashi, M. Kawai-Yamada and H. Uchimiya, *Metabolomics*, 2011, **7**, 524.
106. A. Anegawa, M. Ohnishi, D. Takagi, C. Miyake, C. Shichijo, K. Ishizaki, H. Fukaki and T. Mimura, *Plant Biotechnol.*, 2015, **32**, 65.
107. J. Zhao, C. Hu, J. Zeng, Y. Zhao, J. Zhang, Y. Chang, L. Li, C. Zhao, X. Lu and G. Xu, *Metabolomics*, 2014, **10**, 805.
108. K. Iino, M. Sugimoto, T. Soga and M. Tomita, *Metabolomics*, 2012, **8**, 99.
109. Y. Ohashi, A. Hirayama, T. Ishikawa, S. Nakamura, K. Shimizu, Y. Ueno, M. Tomita and T. Soga, *Mol. BioSyst.*, 2008, **4**, 135.
110. K. Yoshikawa, T. Hirasawa, K. Ogawa, Y. Hidaka, T. Nakajima, C. Furusawa and H. Shimizu, *Biotechnol. J.*, 2013, **8**, 571.
111. T. Fukui, K. Chou, K. Harada, I. Orita, Y. Nakayama, T. Bamba, S. Nakamura and E. Fukusaki, *Metabolomics*, 2014, **10**, 190.
112. R. H. Nugroho, K. Yoshikawa and H. Shimizu, *J. Biosci. Bioeng.*, 2015, **120**, 396.
113. K. Hayakawa, F. Matsuda and H. Shimizu, *AMB Express*, 2016, **6**, 38.
114. G. A. Canuto, E. A. Castilho-Martins, M. Tavares, A. López-Gonzálvez, L. Rivas and C. Barbas, *Electrophoresis*, 2012, **33**, 1901.

115. G. A. Canuto, E. A. Castilho-Martins, M. F. Tavares, L. Rivas, C. Barbas and A. López-Gonzálvez, *Anal. Bioanal. Chem.*, 2014, **406**, 3459.
116. E. A. Castilho-Martins, G. A. B. Canuto, S. M. Muxel, M. F. Laranjeira da Silva, L. M. Floeter-Winter, C. del Aguila, A. López-Gonzálvez and C. Barbas, *Electrophoresis*, 2015, **36**, 2314.
117. D. Rojo, G. A. B. Canuto, E. A. Castilho-Martins, M. F. M. Tavares, C. Barbas, A. López-Gonzálvez and L. Rivas, *PLoS One*, 2015, **10**, e0130675.
118. G. Jeelani, D. Sato, A. Husain, A. Escueta-de Cadiz, M. Sugimoto, T. Soga, M. Suematsu and T. Nozaki, *PLoS One*, 2012, **7**, e37740.
119. H. Wu, V. Tremaroli and F. Bäckhed, *Trends Endocrinol. Metab.*, 2015, **26**, 758.
120. R. Blumberg and F. Powrie, *Sci. Transl. Med.*, 2012, **4**, 137rv7.
121. J. Gratton, J. Phetcharaburanin, B. H. Mullish, H. R. T. Williams, M. Thursz, J. K. Nicholson, E. Holmes, J. R. Marchesi and J. V. Li, *Anal. Chem.*, 2016, **88**, 4661.
122. D. Rojo, M. J. Gosalbes, R. Ferrari, A. E. Pérez-Cobas, E. Hernández, R. Oltra, J. Buesa, A. Latorre, C. Barbas, M. Ferrer and A. Moya, *ISME J.*, 2015, **9**, 2206.
123. D. Rojo, A. Hevia, R. Bargiela, P. López, A. Cuervo, S. González, A. Suárez, B. Sánchez, M. Martínez-Martínez, C. Milani, M. Ventura, C. Barbas, A. Moya, A. Suárez, A. Margolles and M. Ferrer, *Sci. Rep.*, 2015, **5**, 8310.
124. S. Serrano-Villar, D. Rojo, M. Martínez-Martínez, S. Deusch, J. F. Vázquez-Castellanos, R. Bargiela, T. Sainz, M. Vera, S. Moreno, V. Estrada, M. J. Gosalbes, A. Latorre, J. Seifert, C. Barbas, A. Moya and M. Ferrer, *Exp. Biol. Med.*, 2016, **8**, 203.

CHAPTER 3

Capillary Electrophoresis–Mass Spectrometry Using Non-covalently Coated Capillaries for Metabolic Profiling of Biological Samples

RAWI RAMAUTAR

Leiden Academic Center for Drug Research, Leiden University, Einsteinweg 55, 2333 CC Leiden, The Netherlands
*E-mail: r.ramautar@lacdr.leidenuniv.nl

3.1 Introduction

In metabolomics, CE-MS is a very useful analytical technique for the highly efficient and selective profiling of polar and charged metabolites in various biological samples.[1–3] In comparison to chromatographic-based separation techniques, CE-MS provides useful complementary and/or additional information on the metabolic composition of biological samples.[4,5] Furthermore, the costs of CE accessories are less than in LC, due to very low organic solvent consumption, the small amount of reagents needed, and the use of capillaries instead of more expensive LC columns. CE is especially suited for samples of which only small volumes are available. However, the general use of

New Developments in Mass Spectrometry No. 6
Capillary Electrophoresis–Mass Spectrometry for Metabolomics
Edited by Rawi Ramautar

Published by the Royal Society of Chemistry, www.rsc.org

CE-MS in metabolomics is still hindered by reproducibility problems.[6–8] For example, the stability of analyte migration times is of utmost importance in (especially those) metabolomics studies where multiple biological/clinical samples have to be analyzed and compared. When standard bare fused-silica capillaries are employed, the analysis of biological samples with minimal sample pretreatment may lead to adsorption of matrix components to the capillary wall causing detrimental changes in the electro-osmotic flow (EOF) and, therefore, the analyte migration times. Moreover, the separation efficiencies may be compromised as a result of adverse analyte–capillary wall interactions.

Over the past few years, various strategies have been developed to correct for migration time shifts and for aligning electropherograms in CE-MS-based metabolomics studies.[9–13] However, these procedures are not very effective for aligning compounds showing strong migration time shifts among different samples, especially for late-migrating metabolites. Determination of relative migration times using one or more internal standards, which is needed to correct for variations in sample injection volume for improved quantitative performance, can also be considered to reduce migration time variability with RSD below 1%.[14] Alternatively, the use of capillary coatings in CE-MS has been shown to provide better long-term migration time stability,[15] especially when analyzing highly saline biological samples that vary considerably in ionic strength. In this context, an attractive procedure to minimize variation in migration times is the use of non-covalently coated capillaries, *i.e.* dynamical coating of bare fused-silica capillaries with charged polymers.[16–19] So far, various CE-MS methods employing non-covalently coated capillaries have been developed for the highly efficient and repeatable analysis of proteins, peptides and metabolites in various matrices.[20–28]

Recently, the utility of CE-MS using non-covalently coated capillaries with layers of charged polymers has been demonstrated for the highly repeatable analysis of metabolites in biological samples.[29] Capillaries were coated with a bilayer of polybrene (PB) and poly(vinyl sulfonate) (PVS), or with a triple layer of PB, dextran sulfate (DS) and PB. The bilayer and triple layer coatings have a negative and positive outside layer, respectively. The main advantage of such capillaries is a very stable and a pH-independent EOF and, as a result, reproducible migration times. Moreover, direct injection of biological samples such as, for example, urine or cerebrospinal fluid (CSF) is often possible with these approaches. These coatings have been evaluated at low and high pH separation settings, thereby providing separation conditions for basic and acidic metabolites.

This chapter provides an overview of recently developed CE-MS systems employing non-covalently coated capillaries for metabolomics. Special attention is devoted to the practical aspects of using bilayer and triple layer capillary coatings in CE-MS for the profiling of anionic and cationic metabolites in biological samples. It is shown that the use of these easy to produce capillary coatings of charged polymers improves the performance of CE-MS for metabolic profiling studies.

3.2 Capillary Coatings in CE-MS

A stable CE-MS method is key to obtain reliable and robust results. For example, the stability of migration times is of pivotal importance in metabolic profiling studies where often a large number of biological/clinical samples have to be analyzed and compared. Moreover, for CE-MS a constant and appreciable EOF in the direction of ESI-MS is often needed to obtain acceptable and reproducible spray conditions. When conventional bare fused-silica capillaries are used, the separation efficiencies may be compromised due to adverse analyte–capillary wall interactions. Furthermore, the analysis of biological samples with minimal sample pretreatment is often not possible as a result of adsorption of matrix components to the capillary wall resulting in intolerable changes in the EOF and analyte migration times. So far, various chemical coatings have been proposed and evaluated in order to minimize analyte adsorption to the inner surface of the capillary and to improve the performance of the CE-MS method. However, not all coating strategies are appropriate or even compatible with ESI-MS and further requirements for the properties of capillary coatings hold when it comes to achieving efficient and stable CE-MS systems. Apart from only reducing analyte adsorption to the inner capillary wall, capillary coatings may also provide the means to create conditions for effective interfacing and MS detection. For an overview of the use of capillary coatings in CE-MS, readers are referred to recent reviews in ref. 15 and 30.

During CE-MS method development, a capillary coating should be evaluated for its ability to minimize compound adsorption and potential analyte loss in order to achieve high separation efficiencies. In this context, the different possible interactions between the analyte and the coating have to be considered, as well as the ability of the coating to shield analyte molecules from the fused-silica capillary wall. These aspects refer to both the composition and surface of the coating and the properties of the compounds. In addition, the coating should provide a strong compatibility with sample matrix components in such a way that, *e.g.*, biological samples can be analyzed with minimal pretreatment. The coating should also be stable over time, allowing multiple analyses and the use of different BGEs with respect to the nature of the buffer, concentration, pH, and additives. The capillary coating should preferably be inexpensive, be easy to prepare regarding both its synthesis and its introduction onto the capillary wall, and allow easy regeneration.

All capillary coatings used in CE-MS should not interfere with ESI-MS detection. Coating material entering the MS instrument can give rise to severe background noise, suppression of analyte signals, and/or contamination of the ion source and MS optics. Obviously, this requirement limits the usefulness of coating agents that are present in the BGE. When covalent or adsorbed coatings are used, bleeding of the coating material during CE-MS analyses needs to be minimized in order to avoid the risk of reduced detection sensitivity.

For stable CE-MS operation, an appreciable CE effluent flow towards the ion source is preferable. In cases where a sheath-liquid interface is employed for coupling CE to MS, very low or zero EOFs may cause moving ionic boundaries and pH shifts inside the capillary as counter-ions from the sheath liquid can enter the capillary.[33] With a sheathless interface, a small EOF in the direction of the capillary outlet is required to achieve proper spray conditions and analyte ionization, except in cases where an additional external pressure is used. Overall, this means that the properties regarding EOF formation have to be taken into account when selecting an appropriate capillary coating for CE-MS.

So far, coating materials for permanent modification of the inner wall of fused-silica capillaries by chemical reactions or for physical adsorption *via* electrostatic interactions have been often used in CE-MS. The coating stability has to be high enough to prevent bleeding into the MS instrument, preferably also in the presence of organic solvents in the BGE. This also includes stability at the pH range of interest.

3.3 Non-covalently Coated Capillaries for CE-MS-based Metabolomics

A common problem in CE is migration time shift between injections, especially in the case of analyzing biological samples. An effective sample pretreatment strategy and a proper rinsing procedure between runs are often required to minimize adsorption of sample matrix components to the inner wall of fused-silica capillaries. Another way to minimize adsorption of matrix components to the capillary wall is by using coated capillaries, as outlined in this section. Migration time shifts between sample injections may also be the result of variations in the ionic strength of a sample, leading to different stacking conditions and, therefore, to different effective field strengths within the separation capillary.[31] Desalting, in addition to removing proteins, may resolve this issue; however, such a strategy is not feasible in CE-MS-based metabolomics studies, as common procedures employed for desalting may also result in the concurrent loss of highly polar and charged metabolites.

Herein, the utility of CE-MS using non-covalently coated capillaries with layers of charged polymers is shown for metabolic profiling of biological samples. Katayama and co-workers were the first to illustrate the potential of such an approach for bio-analytical studies by CE and CE-MS.[16,17] They have shown that the use of a non-covalent capillary coating based on a triple layer of PB, DS and PB was significantly more stable as compared to using a conventional PB-coated capillary. The triple layer capillary coating endured

600 replicate analyses and also showed strong stability against rinsing with 1 M sodium hydroxide and 0.1 M hydrochloric acid. Overall, these studies revealed that non-covalent capillary coatings based on multiple layers have favorable characteristics for metabolic profiling of biological samples by CE-MS.

Soga *et al.* were the first to demonstrate the utility of the PB–DS–PB coated capillary for the global profiling of anionic metabolites by CE-MS.[20] Apart from improving the repeatability of the CE-MS analyses, a main reason to use such a capillary was to reverse the EOF in order to allow the fast migration of anionic metabolites to the MS when using CE in reverse polarity mode. However, this approach was not suited for the analysis of anionic compounds carrying multiple negative charges under the employed separation conditions such as, for example, nucleotides and di- or tri-carboxylic acids. In order to prevent the adsorption of compounds like ATP and ADP onto the fused-silica capillary wall, Soga and co-workers preconditioned the capillary with a BGE containing phosphate in order to mask the silanol groups.[32] During preconditioning, the nebulizer gas was switched off to avoid contamination of the MS ion source and optics with phosphate ions.

A recent cross-platform comparison for the analysis of primary metabolites involving GC-MS, reversed-phase LC-MS and CE-MS indicated that the use of the latter method with a PB–DS–PB coated capillary for anionic metabolic profiling was quite complicated and not reliable.[33] Soga *et al.* developed an improved CE-MS approach for the global analysis of anionic metabolites using a platinum ESI spray needle instead of a stainless steel needle.[34] In this study it was shown that the material of the spray needle had a significant effect on the analysis of anionic compounds in negative ionization mode when using CE in reversed polarity. A reversed polarity is required when using positively charged capillary coatings for analysis of anionic metabolites. The authors found that under these conditions, the stainless steel spray needle, which acted as the anode, showed oxidation and corrosion due to electrolysis. The resulting precipitation of iron oxides plugged the capillary outlet and consequently shortened the lifetime of the capillary. Some anionic metabolites also appeared to form complexes with iron oxides and nickel ions from the steel needle. The metal-metabolite complex formation caused ionization suppression and significantly reduced the detection sensitivity for the anionic metabolites. Overall, the use of a platinum ESI needle prevented both oxidation of metals and needle corrosion and greatly improved the method performance and long-term stability when employing CE with cationic coatings under reversed polarity for anionic metabolic profiling. However, a platinum needle is not required for anionic metabolic profiling under normal polarity CE conditions since electrolytic corrosion at the anode is avoided.

Ramautar *et al.* used the bilayer (PB–PVS) and triple layer (PB–DS–PB) capillary coatings in CE-MS for the analysis of (endogenous) metabolites in biological samples.[23,29] The use of these approaches under both low and high pH separation conditions allows the global profiling of acidic and basic metabolites and, therefore, a wider metabolic coverage. Moreover, both capillary coatings generate a relatively strong and constant EOF, *i.e.*, nearly independent of the pH of the BGE, which can be directed to the MS entrance to achieve stable ESI conditions and as a result effective analyte ionization and MS detection. An overview of CE-MS systems employing bilayer and triple layer capillary coatings for the profiling of anionic and cationic metabolites is shown in Figure 3.1. With these systems, cationic metabolites have been analyzed under low pH separation conditions using 1 M formic acid (pH 1.8) as the BGE, whereas anionic metabolites have been analyzed under high pH separation conditions using 50 mM ammonium acetate (pH 8.5) as the BGE.

As indicated in the previous sections, the use of PB–DS–PB CE-MS under high pH separation conditions in reversed polarity mode for anionic metabolic profiling is complicated and not robust. In this context, Ramautar *et al.* have especially evaluated the utility of this approach for cationic metabolic profiling.[29] In comparison to PB–PVS CE-MS, a wider separation window for cationic metabolites can be obtained with the PB–DS–PB CE-MS method. A representative metabolite mixture and spiked urine samples applying no sample pre-treatment were used for the evaluation of the four CE methods. The migration time repeatability (RSD < 2%) and plate numbers (N, 100 000–400 000) were similar for the analysis of metabolite test mixtures

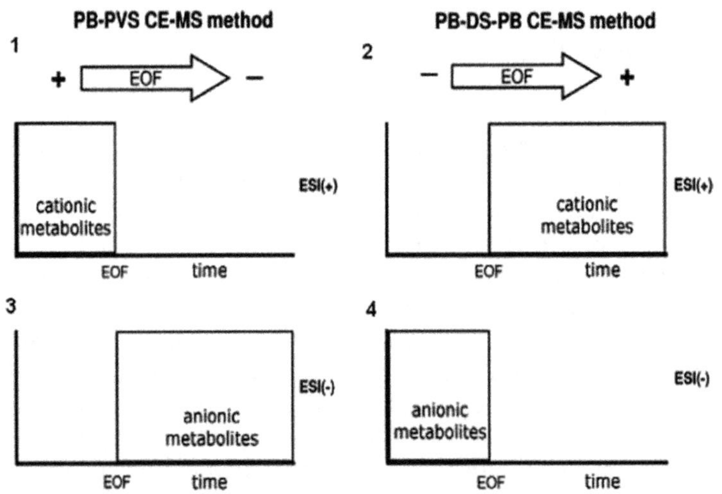

Figure 3.1 A scheme showing the EOF direction, the sign of the electrode potential, the ionization mode in MS and the expected separation window for cationic and anionic metabolites in the PB–PVS and PB–DS–PB CE-MS methods under low and high pH separation conditions. (1) and (2) show the conditions at pH 2.0; (3) and (4) show the conditions at pH 9.0. Reproduced with permission from ref. 29. Copyright © 2010 WILEY-VCH Verlag GmbH & Co. KGaA, Weinheim.

by both CE-MS methods. When applied to rat urine, the analysis of cationic compounds with the PB–DS–PB CE method at low pH (*i.e.* migration after the EOF time) provided a wider separation window and larger number of separated peaks in rat urine as compared to the analysis with the PB–PVS CE method at low pH (*i.e.* migration before the EOF time), as shown in Figure 3.2. Approximately 600 molecular features were observed in rat urine by the PB–DS–PB CE-MS method, whereas about 300 features were detected with the PB–PVS CE-MS method. For anionic metabolic profiling, the PB–PVS CE-MS system may be considered; however, in this case it would be more

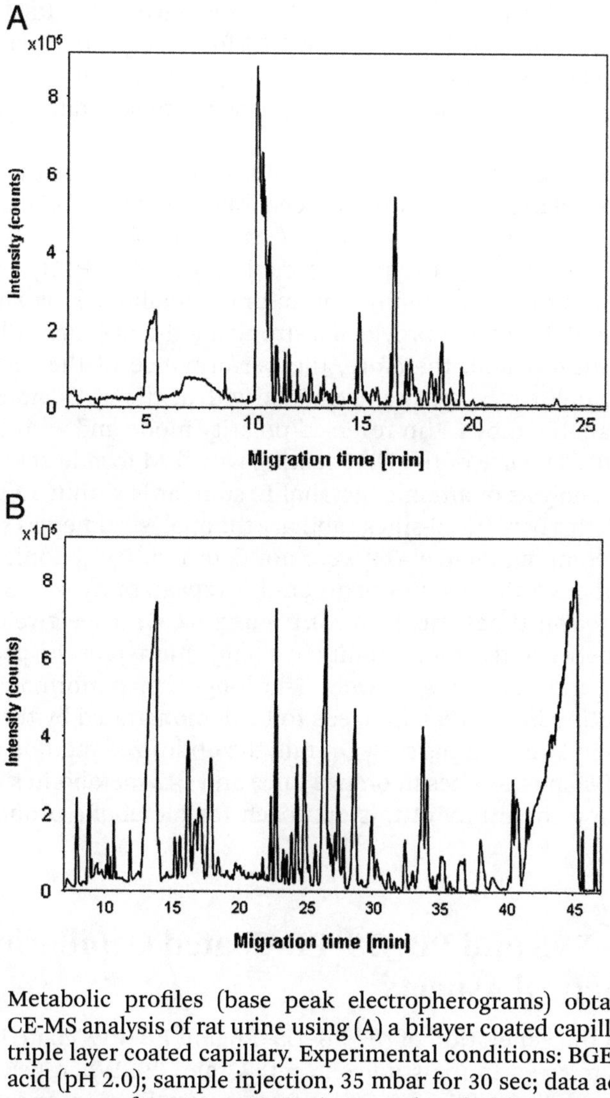

Figure 3.2 Metabolic profiles (base peak electropherograms) obtained during CE-MS analysis of rat urine using (A) a bilayer coated capillary and (B) a triple layer coated capillary. Experimental conditions: BGE, 1 M formic acid (pH 2.0); sample injection, 35 mbar for 30 sec; data acquired for a mass range from 50 to 1000 *m*/*z*. Reproduced with permission from ref. 29. Copyright © 2010 WILEY-VCH Verlag GmbH & Co. KGaA, Weinheim.

practical to simply use a bare fused-silica capillary as under high pH conditions a maximum EOF is generated in such a system allowing most anionic metabolites (*i.e.* migration after the EOF time) to be analyzed within a reasonable time.

A CE-MS method using a PB–PVS-coated capillary has been used for the relatively fast and direct analysis of amino acids in human urine under low pH separation conditions.[24] The PB–PVS coating provided stable CE-MS profiles of amino acids in urine with favorable migration time repeatability (RSD < 1.5%). For quantitative purposes, the stability of the PB–PVS CE-MS method was also determined on the basis of the intraday and interday variability of the peak areas of compounds spiked into human urine. The RSD of the peak areas of amino acids in urine was below 10% for ten repeated analyses, which is acceptable for bio-analytical studies. Though the system allowed relatively fast profiling of amino acids, it was not possible to separate isomeric compounds such as isoleucine and leucine.

Apart from the PB–PVS and PB–DS–PB coated capillaries, other non-covalently coated capillaries have also been evaluated in CE-MS-based metabolomics studies. For example, Acunha *et al.* assessed poly-*N*,*N*,*N*′,*N*′-tetraethyldiethylenetriamine, *N*-(2-hydroxypropyl) methacrylamide (PTH) as a cationic capillary coating for the profiling of anionic metabolites by sheath-liquid CE-MS.[35] The PTH coating provided a strong anodic EOF under low pH separation conditions and, therefore, the performance of the method using anionic metabolite mixtures was evaluated with acidic BGEs. Anionic metabolites were analyzed by CE in reversed polarity mode and with MS in negative-ion mode. The use of the PTH coating with 1 M formic acid as the BGE allowed the analysis of anionic metabolite standards within 12 min (Figure 3.3), while with a bare fused-silica capillary the analysis time was significantly longer and compounds like ATP were not detected. For anionic metabolite standards, the within-day and between-day repeatability (n = 3 injections) for the migration times was below 0.1% and 0.6%, respectively. The within-day and between-day repeatability (n = 3 injections) for the peak area was below 0.5% and 7.3%, respectively. The long-term performance of using the PTH coating in CE-MS still needs to be demonstrated by the analysis of metabolites in large sample sets. Overall, the optimized method allowed the detection of 87 metabolites in orange juice and 142 metabolites in red wine, demonstrating the utility of this approach for metabolic profiling of food products.

3.4 PB–PVS and PB–DS–PB Coated Capillaries: Practical Aspects

Concerning the generation of the PB–DS–PB and PB–PVS coated capillaries, readers are referred to the studies in ref. 23 and 29. When using such systems for metabolic profiling studies, a few (practical) aspects need to be carefully considered. For example, upon generation, rinsing of a bare fused-silica

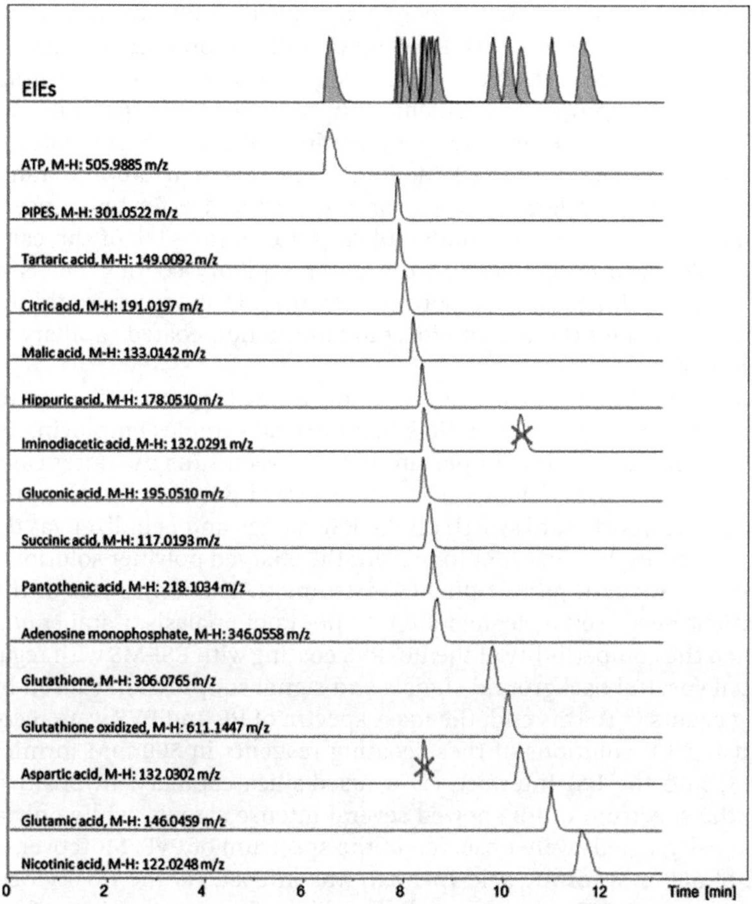

Figure 3.3 Extracted ion electropherograms obtained with sheath-liquid CE-MS using a cationic capillary coating for an anionic metabolite mixture containing ATP (0.3 mM), nicotinic acid (7.8 mM), glutamic acid (0.5 mM), aspartic acid (1.0 mM), oxidized glutathione (0.3 mM), reduced glutathione (0.6 mM), iminodiacetic acid (2.7 mM), AMP (0.2 mM), pantothenic acid (0.2 mM), succinic acid (1.4 mM), gluconic acid (0.1 mM), hippuric acid (0.4 mM), malic acid (0.3 mM), citric acid (0.2 mM), tartaric acid (0.2 mM), and 1,4 piperazinediethanesulfonic acid (2.8 mM). Experimental conditions: CE analysis under reverse polarity (−20 kV); BGE, 1 M formic acid (pH 2.4); sample injection, 35 mbar for 80 sec. Reproduced from *Journal of Chromatography A*, **1428**, T. Acunha, C. Simo, C. Ibanez, A. Gallardo and A. Cifuentes, Anionic metabolite profiling by capillary electrophoresis–mass spectrometry using a non-covalent polymeric coating. Orange juice and wine as case studies, 326–335.[35] Copyright 2015, with permission from Elsevier.

capillary with 0.1 or 1 M sodium hydroxide solution is needed to ensure that the inner wall of the fused-silica capillary is fully negatively charged. Only then will the positively charged polybrene polymers effectively attach electrostatically to the negatively charged inner wall. For practical reasons, when rinsing the capillary with solutions of charged polymers, a lower pressure is applied to ensure proper attachment of the second or third polymer layer to the previous layer *via* electrostatic interactions. If properly generated, then the use of these capillaries in CE-MS for the analysis of metabolite standards needs to provide a migration time variation below 1% for ten consecutive analyses using an injection volume of *circa* 15 nL (or ~1% of the capillary volume). Between each injection the outer capillary coating needs to be renewed *via* flushing with the polymer solution. In cases where these data are not obtained for the metabolite standards, a new coated capillary needs to be prepared.

In order to obtain repeatable migration times for metabolite standards and for the profiling of metabolites in biological samples employing multiple sample injections, it is important that between runs the outer capillary coating is renewed by flushing with the charged polymer solution. In this context, it is important to switch off the ESI voltage and nebulizer gas during the rinsing procedure in order to prevent the charged polymer solution from entering the vacuum part of the MS instrument. The ion source of the MS instrument needs to be cleaned after 24 hours of analysis. Catai *et al.* have evaluated the compatibility of the PB–PVS coating with ESI-MS with regard to potential spectral background signals and signal suppression caused by the coating agents.[19] To this end, the mass spectra of PB and PVS were recorded by infusing 1% solutions of these coating reagents in 500 mM formic acid (pH 2.5) into the ESI interface *via* a fused-silica capillary. In positive ion mode, the spectrum of PB showed several intense signals, while only a few low-intensity signals were observed in the spectrum of PVS. Moreover, when a BGE of 500 mM formic acid (pH 2.5) was infused *via* the PB–PVS coated capillary into the MS using a sheath-liquid interface at a flow rate of 170 nL min^{-1} while the mass spectra were recorded, no signals from PB or PVS were observed in the mass spectra, indicating that the PB–PVS coated capillary could be used in combination with ESI-MS without any complications.

Though repeatable migration times can be obtained for metabolic profiling studies, the analytical performance and durability/lifetime of a single PB–DS–PB or PB–PVS coated capillary still needs to be examined for the analysis of a large set of biological/clinical samples. Only by performing such a study will it be clear how each method needs to be adapted when analyzing large numbers of samples, notably in terms of frequency of rinsing and equilibration.

3.5 Conclusions

In this chapter, the utility of non-covalently coated capillaries using multiple layers of charged polymers has been demonstrated for metabolic profiling of biological samples by CE-MS. An improved repeatability of the migration time can be obtained by these approaches, which is of pivotal importance for comparative metabolic profiling studies. However, for actual metabolomics studies, which often involve hundreds of biomedical/clinical samples, the usefulness of using non-covalently coated capillaries still needs to be demonstrated. When employing the rinsing and clean-up procedures, as outlined in Section 3.4, in a proper manner, it is anticipated that the PB–PVS and PB–DS–PB CE-MS methods will provide the required migration repeatability for metabolic profiling of large sample sets.

Overall, the use of bilayer and triple-layer coatings in CE-MS can be considered a flexible approach, as it may be possible to use one capillary for obtaining metabolic profiles by both the bilayer and triple-layer approach, thereby extending the metabolic coverage. Furthermore, by switching the separation conditions between runs, the same sample can be analyzed in both the positive and negative ion mode. The real potential and reproducibility of such a general approach need to be investigated.

Acknowledgements

Dr Rawi Ramautar would like to acknowledge the financial support of the Veni and Vidi grant scheme of the Netherlands Organisation for Scientific Research (NWO Veni 722.013.008 and Vidi 723.016.003).

References

1. A. Hirayama, M. Wakayama and T. Soga, *TrAC, Trends Anal. Chem.*, 2014, **61**, 215–222.
2. N. L. Kuehnbaum and P. Britz-McKibbin, *Chem. Rev.*, 2013, **113**, 2437–2468.
3. R. Ramautar, *Adv. Clin. Chem.*, 2016, **74**, 1–34.
4. S. Naz, A. Garcia and C. Barbas, *Anal. Chem.*, 2013, **85**, 10941–10948.
5. M. G. Kok, G. W. Somsen and G. J. de Jong, *Talanta*, 2015, **132**, 1–7.
6. A. Garcia, J. Godzien, A. Lopez-Gonzalvez and C. Barbas, *Bioanalysis*, 2017, **9**, 99–130.
7. R. Ramautar, G. W. Somsen and G. J. de Jong, in *Metabolomics in Practice*, Wiley-VCH Verlag GmbH & Co. KGaA, 2013, pp. 177–208.
8. B. Timischl, K. Dettmer, H. Kaspar, M. Thieme and P. J. Oefner, *Electrophoresis*, 2008, **29**, 2203–2214.
9. E. Nevedomskaya, R. Derks, A. M. Deelder, O. A. Mayboroda and M. Palmblad, *Anal. Bioanal. Chem.*, 2009, **395**, 2527–2533.
10. I. Garcia-Perez, P. Whitfield, A. Bartlett, S. Angulo, C. Legido-Quigley, M. Hanna-Brown and C. Barbas, *Electrophoresis*, 2008, **29**, 3201–3206.
11. M. G. Kok, M. M. Ruijken, J. R. Swann, I. D. Wilson, G. W. Somsen and G. J. de Jong, *Anal. Bioanal. Chem.*, 2013, **405**, 2585–2594.
12. C. Barbas, E. P. Moraes and A. Villasenor, *J. Pharm. Biomed. Anal.*, 2011, **55**, 823–831.

13. R. Baran, H. Kochi, N. Saito, M. Suematsu, T. Soga, T. Nishioka, M. Robert and M. Tomita, *BMC Bioinf.*, 2006, **7**, 530.
14. A. N. de Macedo, K. Teo, A. Mente, M. J. McQueen, J. Zeidler, P. Poirier, S. A. Lear, A. Wielgosz and P. Britz-McKibbin, *Anal. Chem.*, 2014, **86**, 10010–10015.
15. C. Huhn, R. Ramautar, M. Wuhrer and G. W. Somsen, *Anal. Bioanal. Chem.*, 2010, **396**, 297–314.
16. H. Katayama, Y. Ishihama and N. Asakawa, *Anal. Chem.*, 1998, **70**, 5272–5277.
17. H. Katayama, Y. Ishihama and N. Asakawa, *Anal. Chem.*, 1998, **70**, 2254–2260.
18. J. R. Catai, H. A. Tervahauta, G. J. de Jong and G. W. Somsen, *J. Chromatogr. A*, 2005, **1083**, 185–192.
19. J. R. Catai, J. S. Torano, G. J. de Jong and G. W. Somsen, *Electrophoresis*, 2006, **27**, 2091–2099.
20. T. Soga, Y. Ueno, H. Naraoka, Y. Ohashi, M. Tomita and T. Nishioka, *Anal. Chem.*, 2002, **74**, 2233–2239.
21. R. Haselberg, G. J. de Jong and G. W. Somsen, *Anal. Chim. Acta*, 2010, **678**, 128–134.
22. R. Haselberg, V. Brinks, A. Hawe, G. J. de Jong and G. W. Somsen, *Anal. Bioanal. Chem.*, 2011, **400**, 295–303.
23. R. Ramautar, O. A. Mayboroda, A. M. Deelder, G. W. Somsen and G. J. de Jong, *J. Chromatogr. B: Anal. Technol. Biomed. Life Sci.*, 2008, **871**, 370–374.
24. R. Ramautar, O. A. Mayboroda, R. J. Derks, C. van Nieuwkoop, J. T. van Dissel, G. W. Somsen, A. M. Deelder and G. J. de Jong, *Electrophoresis*, 2008, **29**, 2714–2722.
25. A. Elhamili, J. Samuelsson, J. Bergquist and M. Wetterhall, *Electrophoresis*, 2011, **32**, 647–658.
26. A. Elhamili, M. Wetterhall, M. Sjodin, R. Sebastiano and J. Bergquist, *Electrophoresis*, 2010, **31**, 1151–1156.
27. A. Elhamili, M. Wetterhall, B. Arvidsson, R. Sebastiano, P. G. Righetti and J. Bergquist, *Electrophoresis*, 2008, **29**, 1619–1625.
28. C. A. Lucy, A. M. MacDonald and M. D. Gulcev, *J. Chromatogr. A*, 2008, **1184**, 81–105.
29. R. Ramautar, J. S. Torano, G. W. Somsen and G. J. de Jong, *Electrophoresis*, 2010, **31**, 2319–2327.
30. L. Hajba and A. Guttman, *TrAC, Trends Anal. Chem.*, 2017, **90**, 38–44.
31. J. C. Reijenga, J. H. P. P. Martens, A. Giuliani and M. Chiari, *J. Chromatogr. B*, 2002, **770**, 45–51.
32. T. Soga, T. Ishikawa, S. Igarashi, K. Sugawara, Y. Kakazu and M. Tomita, *J. Chromatogr. A*, 2007, **1159**, 125–133.
33. J. M. Buscher, D. Czernik, J. C. Ewald, U. Sauer and N. Zamboni, *Anal. Chem.*, 2009, **81**, 2135–2143.

34. T. Soga, K. Igarashi, C. Ito, K. Mizobuchi, H. P. Zimmermann and M. Tomita, *Anal. Chem.*, 2009, **81**, 6165–6174.
35. T. Acunha, C. Simo, C. Ibanez, A. Gallardo and A. Cifuentes, *J. Chromatogr. A*, 2016, **1428**, 326–335.

CHAPTER 4

Capillary Electrophoresis–Mass Spectrometry for Metabolomics Using New Interfacing Designs

RAWI RAMAUTAR

Leiden Academic Center for Drug Research, Leiden University, Einsteinweg 55, 2333 CC Leiden, The Netherlands
*E-mail: r.ramautar@lacdr.leidenuniv.nl

4.1 Introduction

Capillary electrophoresis–electrospray ionization–mass spectrometry (CE-MS) can be considered a powerful tool for the analysis of charged compounds in the field of bioanalysis.[1–3] The intrinsic high separation efficiency of CE in combination with the selectivity and sensitivity of modern MS instruments provides a very attractive tool for the analysis of (a wide range of) charged compounds in complex sample matrices. However, in comparison to other analytical techniques, especially chromatographic-based separation techniques hyphenated to MS, the use of CE-MS in the field of bioanalysis is still limited. Restrictions like poor concentration sensitivity, migration time variability, and lack of standard operating procedures including support from vendors may have hindered the widespread adoption of the CE-MS approach within the separation science community. In this context, it is important to stress that CE-MS can be used as a reliable tool for bio-analytical studies. For example, in a recent inter-laboratory study, in which 13 independent

New Developments in Mass Spectrometry No. 6
Capillary Electrophoresis–Mass Spectrometry for Metabolomics
Edited by Rawi Ramautar

Published by the Royal Society of Chemistry, www.rsc.org

laboratories from academia and industry participated, the robustness of CE-MS using a conventional sheath-liquid interface was assessed for peptide mapping.[4] The study revealed an excellent reproducibility for relative migration times and peak areas, thereby demonstrating that the CE-MS approach is robust enough to allow method transfer across multiple laboratories for peptide mapping. Other recent studies have highlighted the value of CE-MS for the reliable profiling of native peptides in large cohorts of human urine samples in a clinical context.[5–8]

Until recently, the co-axial sheath-liquid interface was primarily used for the coupling of CE (referring here to capillary zone electrophoresis and denoted as CE in this chapter) to MS.[9–13] In this configuration (Figure 4.1), which is basically a triple-tube interface design and was originally developed by Richard Smith and co-workers,[14] the CE capillary is surrounded by a tube of larger diameter *via* which the sheath-liquid is delivered to the end of the CE capillary as a terminal electrolyte reservoir, thereby providing a closed electrical contact. To facilitate ESI spray formation, a nebulizer

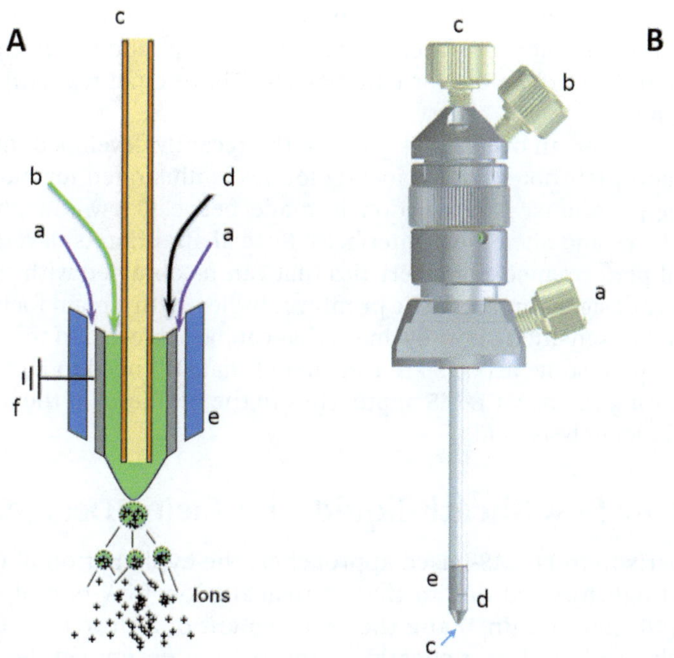

Figure 4.1 (A): Graphic representation of an Agilent co-axial sheath-liquid CE-MS interface; (a) nebulizing gas, (b) sheath-liquid, (c) CE capillary with BGE, (d) stainless steel spray needle (0.4 mm i.d., 0.5 mm o.d.), (e) outer tube, (f) ground connection. (B): Engineering sketch of the co-axial sheath-liquid CE-MS interface. Reprinted from *Chromatographia*, Developments in Interfacing Designs for CE-MS: Towards Enabling Tools for Proteomics and Metabolomics, **78**, 2015, 377, P. W. Lindenburg, R. Haselberg, G. Rozing and R. Ramautar, (© Springer-Verlag Berlin Heidelberg 2014), with permission from Springer.

gas is applied *via* the outermost tube. A strong feature of the sheath-liquid interface is that it allows independent optimization of the background electrolyte (BGE) conditions for separation from the sheath-liquid composition for analyte ionization.[15] For many bio-analytical applications, the use of this interface in CE-MS provides an acceptable analytical performance and robustness.[16–18] However, in this design the sheath-liquid significantly dilutes the CE effluent, thereby compromising the achievable concentration sensitivity. The overall detection sensitivity in sheath-liquid CE-MS can be improved by using front-end preconcentration approaches, such as chromatographic- and/or electrophoretic-based techniques. The reader is referred to Chapters 5 and 6 of this book for a comprehensive overview on these topics.

Recently, various new interfacing designs have been developed for coupling CE to MS, which operate ideally under low flow-rate separation conditions. These advances have resulted in increased interest in the CE-MS approach for proteomics and metabolomics studies, notably for the analysis of low abundance compounds in limited amounts of biological samples. In this context, the sheathless porous tip interface of Mehdi Moini,[19] the flow-through microvial interface of David Chen,[20] and the electrokinetic sheath-liquid interface of Norman Dovichi[21] have emerged as important developments.

In this chapter, an overview is given of the recently developed interfacing designs and their potential for metabolomics is illustrated for biomedical/clinical applications. A distinction is made between low-flow sheath-liquid interfaces and sheathless interfaces. Special attention is devoted to the analytical performance characteristics that can be obtained with these new interfacing designs for metabolic profiling studies, with a main focus related to the level of sensitivity improvement that can be achieved. In the final part of this chapter, some aspects are considered that still need to be tackled in order to bring the new CE-MS approaches to the next level in the metabolomics and bioanalysis field.

4.2 Low-flow Sheath-liquid Interfacing Designs

In comparison to LC-MS-based approaches, the hyphenation of CE to MS is not straightforward due to the intrinsically low flow-rate of CE separations (10–100 nL min^{-1}) and the requirement of a closed electrical circuit. A sheath-liquid or a sheathless interfacing design can be used for the hyphenation of CE to MS. Obviously, in the sheathless configuration no sheath-liquid or liquid modifier is applied and the electrical contact is often provided close to the BGE at the exit of the CE capillary.[10] For an in-depth overview of technical details regarding interfacing designs developed for CE-MS, the reader is referred to more dedicated reviews on this topic.[9,12,22–24] In this section, an overview is given of recently developed low-flow sheath-liquid interfacing designs and their utility for metabolomics studies is outlined.

Over the past few years, the group of Norman Dovichi has developed a set of electrokinetically pumped sheath-liquid nano-ESI interfaces to couple CE to MS (Figure 4.2).[25] In this design, a CE capillary is threaded *via* a cross into a glass emitter. Fluidic contact with a sheath buffer reservoir is provided by a side arm, which is connected to a power supply. The potential applied to the sheath buffer drives the EOF in the emitter, which provides the sheath-liquid at nL min^{-1} flow-rates. The most recent design, designated by the Dovichi group as a third-generation electrokinetic-based interface, uses a larger diameter emitter orifice with a very short distance between the capillary tip and the emitter orifice (Figure 4.2). The analytical performance of CE-MS using this interface has been evaluated by analyzing bovine serum albumin digests using a fused-silica capillary for 5000 min, which corresponded to 127 runs over 4 days. A 10% relative standard deviation (RSD) in peak area, an average of 160 000 theoretical plates, and very low carry-over (<1%) was observed with this approach, demonstrating an acceptable robustness. Moreover, this approach allows ultrasensitive proteomic analyses; for example, *E. coli* digests have been analyzed using samples at

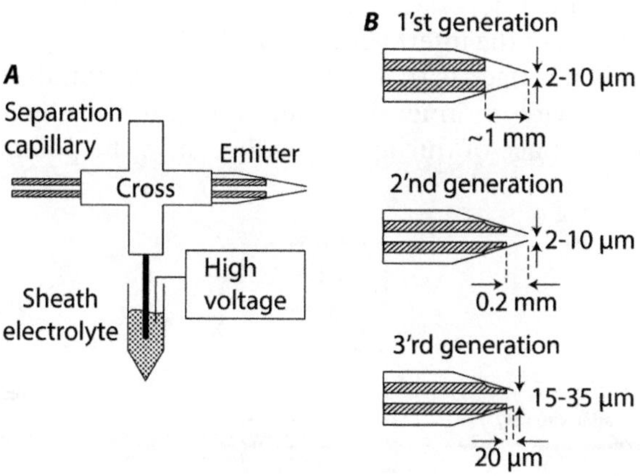

Figure 4.2 Comparison of electrokinetically pumped nanospray CZE interfaces. (A) Overall design of the interface. The separation capillary is threaded through a PEEK cross (1.25 mm through hole) into a glass emitter. A tube (~1 mm i.d.) connects a side arm of the cross to a sheath electrolyte reservoir, which is connected through a platinum electrode to a high-voltage power supply. The mass spectrometer inlet is held at ground potential. (B) Designs of three generations of interface. The original interface uses a flat separation capillary, which is able to approach within ~1 mm of the emitter orifice; typical orifice diameters are 2 to 10 µm. The second generation uses a separation capillary with an etched tip, which approaches within ~200 µm of the emitter orifice; typical orifice diameters are 2 to 10 µm. The third-generation interface also uses an etched separation capillary but with a much larger exit orifice. The etched tip approaches within a few micrometers of the orifice; typical orifice diameters are 15 to 35 µm. Reproduced with permission from ref. 25. Copyright (2015) American Chemical Society.

femtogram-amounts.[26] Recently, CMP Scientific has commercialized this interface design as the EMASS-II ion source and, therefore, the approach is now available to the scientific community. Though very promising results have been obtained, especially in the proteomics field, the utility of this approach for metabolomics still needs to be demonstrated. Moreover, until now, this interface has mostly been used by the Dovichi lab and, therefore, performance evaluation by other research groups is critical in order to stimulate the widespread use of this approach within the separation science community.

The group of David Chen developed a flow-through microvial interface by inserting a CE capillary into a tapered stainless steel hollow ESI emitter for CE-MS.[20] The small volume between the CE capillary end and the inner walls of the needle electrode tip forms a flow-through microvial, which functions as both the outlet vial and the terminal electrode. The flow-through microvial also allows the addition of a chemical modifier solution at low flow-rates in order to provide a stable flow of solution to the needle tip and to increase the compatibility of the CE effluent with ESI while minimizing sample dilution. The metal emitter tip was beveled in order to improve spray stability and also to move the site of ionization away from the CE outlet and the emitter tip orifice. A scheme of the interfacing design is shown in Figure 4.3. A major advantage of this interface is that it allows the use of standard CE capillaries, *i.e.* capillaries with an inner diameter of 50 μm, thereby improving the stability and ease-of-use of this approach. Recently, the potential of CE-MS

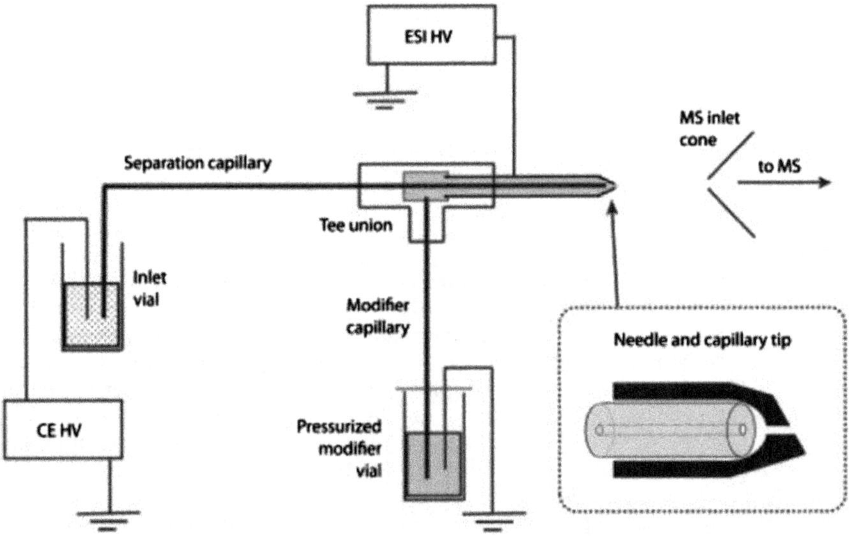

Figure 4.3 Schematic illustration of the flow-through microvial interface approach, including a dissected view of the needle tip with the inserted capillary (inset). Reproduced with permission from ref. 20. Copyright 2010 WILEY-VCH Verlag GmbH & Co. KGaA, Weinheim.

using the flow-through microvial interface was evaluated for metabolic profiling studies.[27] A comparison between the flow-through microvial interface, employing 500 nL min^{-1} as the flow-rate for the sheath-liquid, and a classical co-axial sheath-liquid CE-MS interface (employing 4 μL min^{-1} as the flow-rate) for the analysis of a cationic metabolite mixture is shown in Figure 4.4. The limits of detection (LODs) obtained with the flow-through microvial interface ranged from 0.01 to 3 μM, which was a five-fold improvement as compared to the results obtained by conventional CE-MS. Therefore, this interfacing design shows strong potential for highly sensitive profiling of cationic metabolites; however, the long-term performance of this approach still needs to be examined by the analysis of large numbers of biological samples. Moreover, the flow-through microvial interface does not tolerate high CE currents as the electrolysis gases formed may accumulate in the spray needle, thereby potentially hindering the electrical contact. As such, selection and optimization of the BGE is a critical aspect when using this interface for CE-MS studies.

In order to enable the analysis of neurotransmitters at the single cell level, Nemes and co-workers have developed a CE-MS approach using a

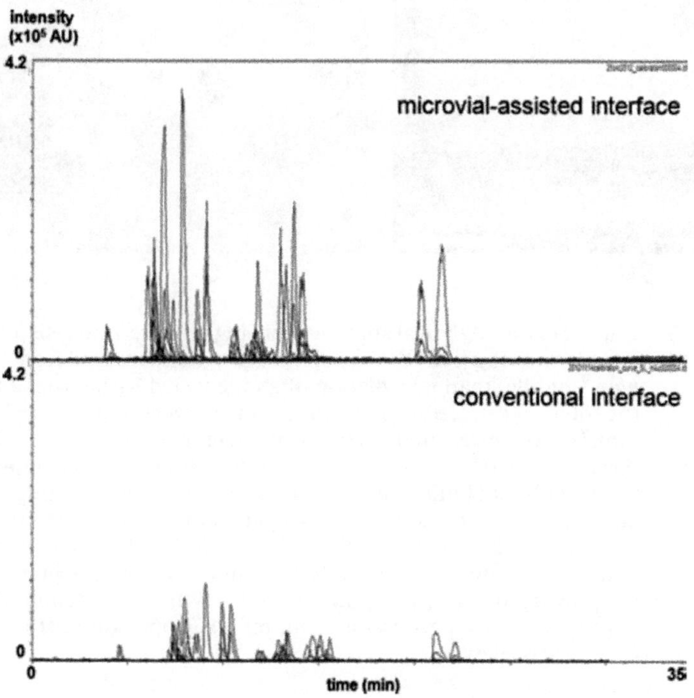

Figure 4.4 Multiple extracted ion electropherograms of 35 cationic metabolites (25 μM) obtained by CE-MS using a flow-through microvial interface (upper electropherogram) and a classical sheath-liquid interface (lower electropherogram). Reproduced with permission from ref. 27. Copyright 2013 WILEY-VCH Verlag GmbH & Co. KGaA, Weinheim.

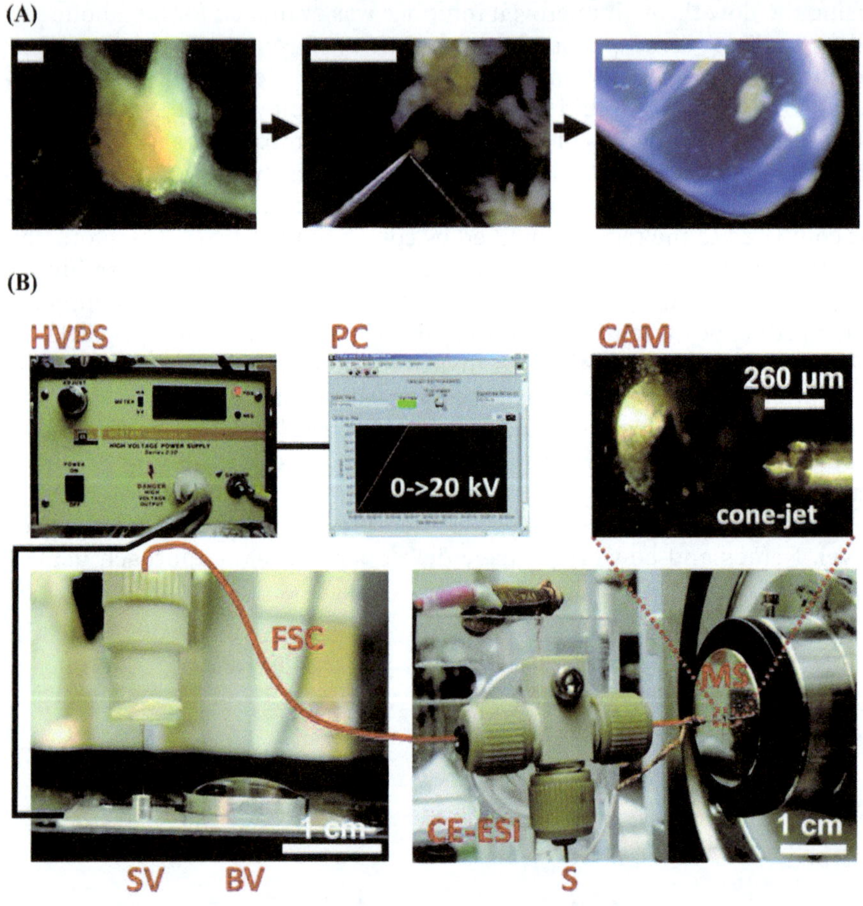

Figure 4.5 Outline of the analytical approach for single-neuron analysis. (A) Optical images showing (left) the abdominal ganglion removed from *A. californica*, (middle) manual isolation of an identified R2 neuron, and (right) the solution-extraction of the intracellular analytes. White scale bars = 1 mm. (B) Single-cell analysis platform including a regulated high voltage power supply (HVPS) remotely controlled by a personal computer (PC), SST sample and buffer vials (SV and BF, respectively) for sample-loading and separation in the fused silica capillary (FSC), a custom-designed coaxial sheath-flow (S) CE–ESI interface (CE–ESI) hyphenated to MS, and a 6.3× stereomicroscope (CAM) to observe and operate the electrospray in the cone-jet spraying mode. White bars denote the scale. Reproduced with permission from ref. 28. Copyright (2010) American Chemical Society.

miniaturized sheath-liquid interface.[28] A scheme of the analytical platform for single neuron analysis is shown in Figure 4.5. For coupling CE to MS, a home-made miniaturized sheath-liquid interface was employed. In this design, the sheath-liquid was provided at a flow-rate of 750 nL min^{-1}, which was significantly lower as compared to the flow-rates delivered in standard sheath-liquid interfaces (often between 2 and 10 μL min^{-1}). A volume of 6 nL from each neuron extract was hydrostatically introduced into the CE capillary using a custom-designed injection platform, consisting of a 500 nL stainless steel sample loading vial. More than 300 distinct ion signals (*m/z* values) could be observed by this CE-MS approach in an extract from a single neuron. Metabolite features were evaluated among six different neuron types (B1, B2, left pleural 1 (LPl1), metacerebral cell (MCC), R2, and R15), which were specifically selected for their various physiological functions and representative examples are depicted in Figure 4.6. The obtained results identified chemical similarities among some neuron types (B1 to B2 and LPl1 to R2) and distinctive features for other neurons (MCC and R15 cells). Overall, the developed home-made CE-MS approach shows strong potential for probing biochemical changes in ultra-small sample amounts. The reader is referred to Chapter 10 of this book for more details about the utility of CE-MS for single cell metabolomics.

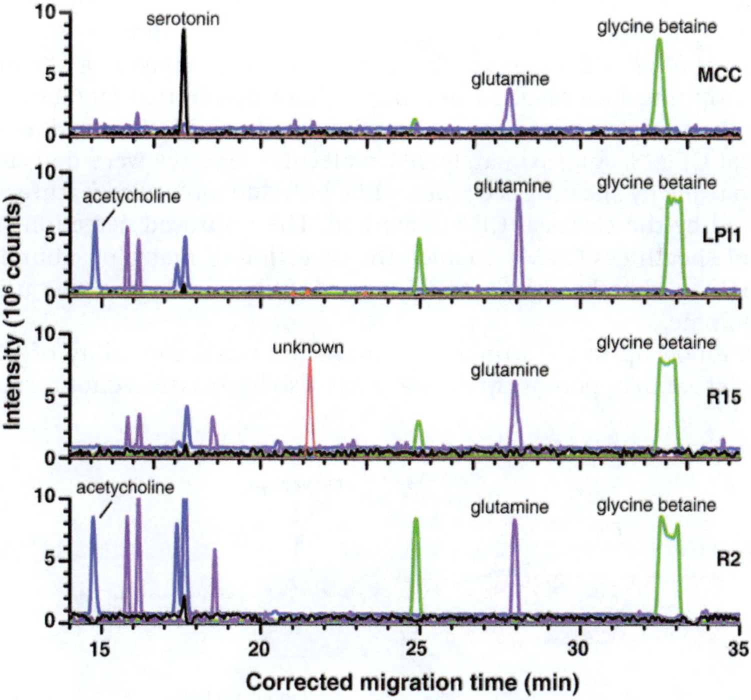

Figure 4.6 Metabolic differences observed among neuronal types by CE-MS. Representative selected-ion electropherograms are shown for a ±5 mDa *m/z* window (signal multiplied by indicated factor): *m/z* 146.118 (1×, acetylcholine, compound 9); *m/z* 177.103 (2×, serotonin, 13); *m/z* 267.133 (2×, unknown); *m/z* 147.077 (5×, glutamine, 26); and *m/z* 118.087 (0.2×, glycine betaine, 32). Reproduced with permission from ref. 28. Copyright (2010) American Chemical Society.

4.3 Sheathless Interfacing Designs

In a sheathless interface design, the CE voltage is applied directly to the buffer at the capillary outlet. This can be achieved by applying a metal coating to the end of a tapered separation capillary or by connecting a metal-coated, full metal or conductive polymeric sprayer tip to the CE outlet. Another way to make a closed electrical circuit is by insertion of a metal micro-electrode through the capillary wall into the buffer end or by direct introduction of a micro-electrode into the end of the capillary.[13] Ramautar and co-workers have developed a CE-MS approach using a sheathless porous tip interface, which was invented by Moini,[19] for the highly sensitive detection of polar and charged metabolites in human urine.[29] The porous tip interface was created by removing the polyimide coating of the outlet section of the capillary and etching the capillary wall with a 49% solution of hydrofluoric acid to a thickness of *circa* 5 μm (Figure 4.7). In this configuration, the electrical connection to the capillary outlet was attained by inserting the etched conductive section into an ESI needle and filling this part with BGE.[30] The sheathless porous tip design is particularly useful for interfacing narrow (<30 μm id) capillaries and for low flow-rate (<100 nL min^{-1}) nano-ESI-MS analyses. The use of CE-MS employing a sheathless porous tip interface resulted in a highly information-rich metabolic profile for human urine as compared to the metabolic profile obtained with classical CE-MS. Approximately 900 molecular features were detected in human urine by sheathless CE-MS, while only 300 molecular features were observed by the classical CE-MS method. The improved detection sensitivity of sheathless CE-MS enabled the detection of many low-abundance metabolites, thereby significantly improving the coverage of the urinary metabolome.

As a follow-up to the urinary metabolomics work, the utility of CE-MS using a sheathless porous tip interface has also been assessed for metabolic

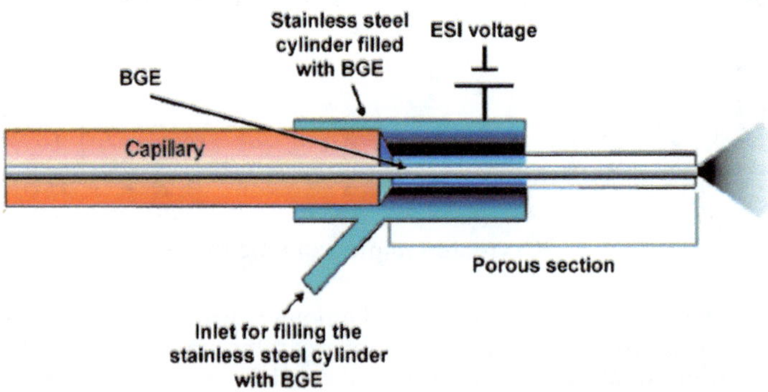

Figure 4.7 Schematic of the high-sensitivity porous sprayer sheathless interface. Reproduced with permission from ref. 30. Copyright (2010) American Chemical Society.

profiling of minute amounts of biological samples. For this purpose, cerebrospinal fluid (CSF) from mice was selected as a typical example of a volume-restricted biological sample, as only a few microliters can be obtained under proper conditions.[31] Figure 4.8 shows a metabolic profile obtained by the sheathless CE-MS approach for mouse CSF after only dilution with water (1 : 1 ratio). By using an injection volume of *circa* 45 nL from a vial containing only 2 µL of diluted CSF, more than 300 compounds could be detected. As only 45 nL of the sample was used for one analytical run, the sheathless CE-MS method allows multiple analyses on a single highly valuable mouse CSF sample, enabling repeatability studies and the analysis of the same sample under different separation conditions in order to further increase the metabolic coverage.

Shyti *et al.* used the same sheathless CE-MS method for the profiling of cationic metabolites in deproteinized plasma samples from mice.[32] In order to increase the injection volume while still maintaining the high separation efficiency of CE, transient-isotachophoresis was used as an in-capillary preconcentration technique. The sheathless CE-MS method was used for metabolic profiling of plasma samples obtained from wild-type mice

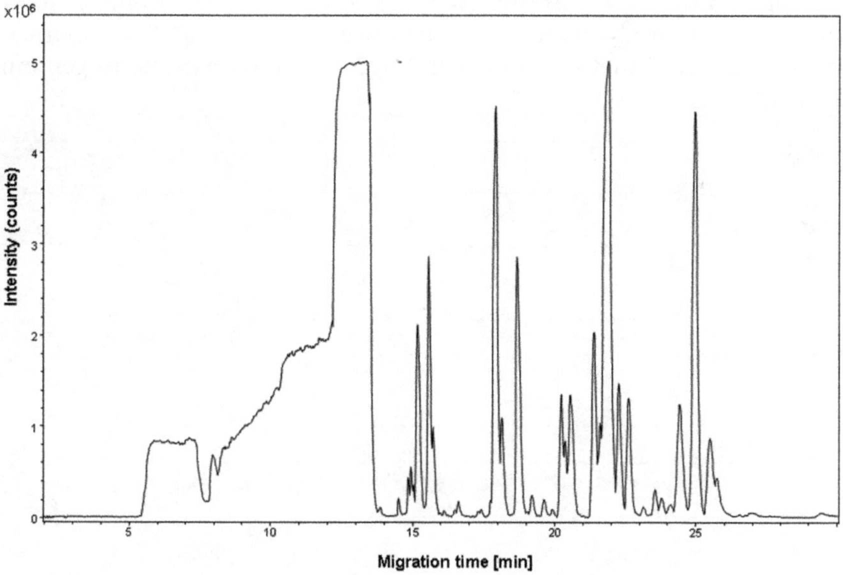

Figure 4.8 Metabolic profile (*m/z* 65–1000) of mouse cerebrospinal fluid obtained with sheathless CE-MS using a porous tip sprayer. Conditions: separation buffer, 10% acetic acid (pH 2.2); electrophoretic separation at +30 kV; sample injection, *circa* 45 nL. Reproduced from *Anal Bioanal Chem*, Metabolic profiling of mouse cerebrospinal fluid by sheathless CE-MS, **404**, 2012, 2895, R. Ramautar, R. Shyti, B. Schoenmaker, L. de Groote, R. J. Derks, M. D. Ferrari, A. M. van den Maagdenberg, A. M. Deelder and O. A. Mayboroda,[31] (© Springer-Verlag Berlin Heidelberg 2012) with permission from Springer.

and transgenic hemiplegic migraine mice after experimentally induced cortical spreading depression. Chemometric analysis of the recorded data revealed a clear distinction between the metabolic profiles of wild-type and transgenic mice after CSD, and in particular the compounds L-lysine and its by-product pipecolic acid were found to be mainly responsible for this classification. According to the authors, these compounds may suggest a compensatory increase in GABAergic neurotransmission upon enhanced excitatory neurotransmission. Overall, this study clearly illustrated the utility of sheathless CE-MS for providing potentially relevant biochemical information in the case of a volume-restricted biomedical problem/case.

Next to volume-restricted samples, such as mouse CSF, there is at present strong interest in analytical techniques capable of providing highly sensitive metabolic profiles for microscale cell culture samples (*i.e.*, 100 to 100 000 cells as the starting material). Therefore, the analytical performance of CE-MS using a sheathless porous tip interface method has also been assessed for the profiling of metabolites in an extract from *circa* 10 000 HepG2 cells.[33] A reasonable number of cationic metabolites could be detected by sheathless CE-MS when using an injection volume of *circa* 60 nL, as shown in Figure 4.9, which in this case corresponded to a content/aliquot of only 5 HepG2 cells. In another study, Yuan *et al.* used the same sheathless CE-MS approach for the analysis of methylcytosine and hydroxymethylcytosine in genomic

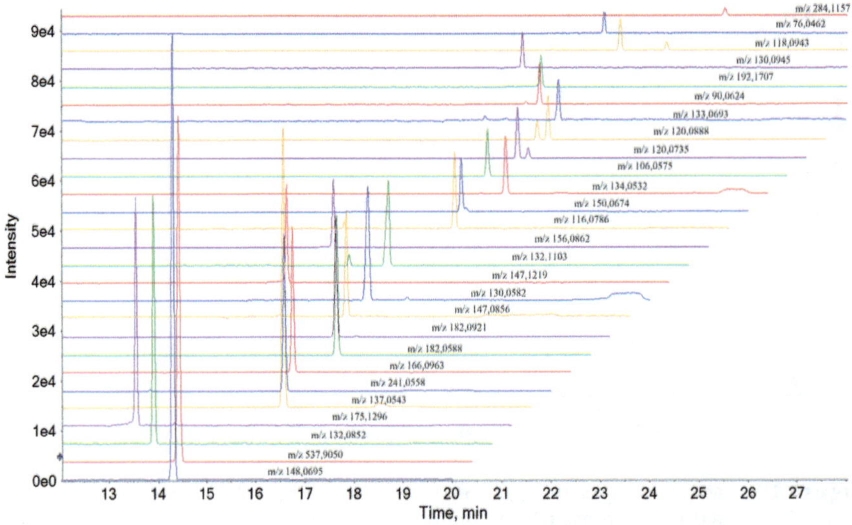

Figure 4.9 Multiple extraction ion electropherograms for a selected number of metabolites obtained with sheathless CE-MS using a porous tip sprayer in an extract of HepG2 cells. Conditions: separation buffer, 10% acetic acid (pH 2.2); electrophoretic separation at +30 kV; sample injection, *circa* 60 nL.

DNA.[34] The analysis of these compounds is considered a challenging task due to their low amounts within genomic DNA. An injection volume of only *circa* 25 nL provided LODs of 50 pM and 100 pM for methylcytosine and hydroxyl-methylcytosine standards, respectively, which was significantly lower as compared to the LODs obtained by LC-MS methods employing chemical derivatization, the approach that is normally applied for this purpose. Sheathless CE-MS was used for the determination of these compounds in a genomic DNA extract from mouse embryonic stem cells. The within- and between-day RSDs for the peak areas were all below 8% and the method allowed the detection of methylcytosine and hydroxymethylcytosine in 125 pg of genomic DNA material dissolved in 16 μL of buffer, corresponding to the amount of *circa* 20 cells. This was significantly less as compared to the amount of genomic DNA material (10 to 200 ng) required for analysis by LC-MS.

In order to further improve the detection sensitivity of the sheathless CE-MS method for the profiling of trace-level metabolites in ultra-small samples, strategies need to be developed for more efficient injection of the analytes of interest into the CE capillary. Nevertheless, the results obtained so far clearly indicate the utility of CE-MS using a sheathless porous tip interface for the profiling of metabolites in limited amounts of biological samples.

So far, most metabolomics studies performed by sheathless CE-MS have focused on the analysis of cationic metabolites, including compounds like amino acids, amines, nucleosides and small peptides. However, many metabolites of central carbon metabolism have an acidic nature, such as, for example, organic acids, sugar phosphates and nucleotides. As the sheathless CE-MS method provides a very good detection sensitivity for the profiling of cationic metabolites, the potential of this approach has recently also been evaluated for the profiling of anionic metabolites in biological samples using exactly the same separation conditions, only with the MS detection and CE separation voltage polarity reversed.[35] A broad array of anionic metabolites could be profiled under these conditions, including sugar phosphates, nucleotides and organic acids (Figure 4.10). An injection volume of *circa* 20 nL resulted in nanomolar detection limits, which corresponded to a significant improvement as compared to the micromolar detection limits typically obtained with conventional sheath-liquid CE-MS approaches. Structural isomers of phosphorylated sugars as well as isobaric metabolites could be selectively analyzed by the sheathless CE-MS method without using any derivatization strategy. The applicability of the method was demonstrated for the profiling of anionic metabolites in extracts from the glioblastoma cell line. Using an injection volume of 20 nL, about 100 compounds could be observed in this cell line extract, which in this case corresponded to the content of *circa* 400 cells. It is important to stress here that the separation conditions, as used for cationic metabolic profiling, may not be considered optimal for the global profiling of anionic metabolites, as under the employed separation

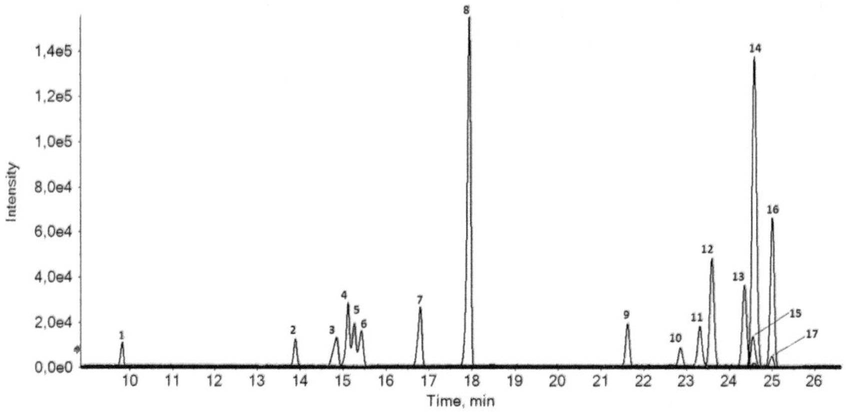

Figure 4.10 Performance of CE-MS using a sheathless porous tip sprayer for anionic metabolic profiling. Multiple extracted ion electropherograms for the metabolite test mixture (25 μM) obtained with sheathless CE-MS in negative ion mode using a porous tip sprayer. Peaks: 1, 2-naphtol-3,6-disulfonic acid; 2, D(+)2-phosphoglyceric acid; 3, D-ribose-5-phosphate; 4, D-glucose-1-phosphate; 5, D-glucose-6-phosphate; 6, D-fructose-6-phosphate; 7, inosine 5′-monophosphate (IMP); 8, guanosine 3′,5′-cyclic monophosphate (cGMP); 9, guanosine 5′-monophosphate; 10, citric acid; 11, trimesic acid; 12, isocitric acid; 13, gluconic acid; 14, adenosine 3′,5′-cyclic monophosphate (cAMP); 15, 2-hydroxybutyric acid; 16, b-diphosphopyridine nucleotide (NAD^+); 17, 3-hydroxybutyric acid. Experimental conditions: BGE, 10% acetic acid (pH 2.2); separation voltage, −30 kV (+0.5 psi applied at the inlet of the CE capillary); sample injection, 2.0 psi for 60 s.

conditions only those anionic metabolites that are (partially) negatively charged can be analyzed. Still, a wide range of acidic and basic metabolites can be analyzed by a single sheathless CE-MS approach by only using two sample injections, thereby showing the utility of this approach for global metabolic profiling studies.

The studies discussed in this section clearly indicate the potential of CE-MS employing a sheathless porous tip interface for providing highly useful results in the field of metabolomics, notably for the analysis of highly polar and charged metabolites in material-limited samples. Hirayama and co-workers have recently shown that a single sheathless porous tip capillary emitter could be used for more than 180 consecutive runs of a ten-fold-diluted human urine sample.[36] In order to determine whether this approach can also be used for actual metabolomics studies, the reproducibility and robustness need to be assessed in more extended studies analyzing large numbers of clinical samples. It is also important to note that the use of standard fused-silica capillaries (using dimensions of 50 or 75 μm) is not possible due to the configuration of the interface. Therefore, replacement capillaries have to be acquired from the vendor, which may be considered as a drawback of the CE-MS approach employing a sheathless porous tip interface, as a major advantage of CE over other separation techniques, *i.e.* the use of relatively cheap capillaries, is discarded in this way. Moreover, the sheathless porous tip interface should preferably be used under relatively low CE currents (below 10 μA) and, consequently, the composition and concentration of the BGE should be carefully considered.

Overall, various low-flow sheath-liquid interfacing designs have been developed over the past years and their utility has been especially explored for proteomics and metabolomics studies. Still, the development or further improvement of interfacing techniques for CE-MS remains an active research area,[37–45] as there is still a strong demand for an easy-to-use, flexible (that is, compatible with all commercial CE and MS instruments), relatively cost-effective, and especially robust interfacing design, which will also allow the use of various buffer compositions and capillary diameters.

4.4 Conclusions

Over the past few years, CE-MS has gained increased attention for metabolomics studies, especially with the development of new low-flow interface designs to improve concentration sensitivity. In this chapter, the utility of CE-MS employing these recently developed interfacing designs has been demonstrated for metabolic profiling of biological samples. Though promising results have been obtained with these approaches, notably in terms of sensitivity enhancement, the figures of merit for large-scale clinical studies still need to be determined. Therefore, at this stage, the new CE-MS approaches are in their infancy for metabolomics studies.

It is expected that CE-MS using a sheathless porous tip interface will play an important role in micrometabolomics, *i.e.* metabolic profiling of limited sample amounts including applications such as single cell or sub-cellular metabolic profiling. Apart from using a low-flow interfacing technique, attention should also be paid to the development of reliable sampling techniques for volume-restricted or mass-limited biological samples, not only for further enhancing the concentration sensitivity but also for improving the robustness of the overall system. In this context, Nemes and co-workers have recently developed a microprobe single-cell CE-MS method, which integrates capillary microsampling, microscale metabolite extraction, and CE-MS analysis, for *in situ* mass spectrometric profiling of metabolites in single cells (see Chapter 10). Further improvements in this particular field may eventually result in a viable CE-MS approach for highly sensitive metabolic profiling of especially volume-restricted samples.

Acknowledgements

Dr Rawi Ramautar would like to acknowledge the financial support of the Veni and Vidi grant scheme of the Netherlands Organisation for Scientific Research (NWO Veni 722.013.008 and Vidi 723.016.003).

References

1. J. Schappler, V. González-Ruiz and S. Rudaz, in *Capillary Electrophoresis-Mass Spectrometry (CE-MS)*, Wiley-VCH Verlag GmbH & Co. KGaA, 2016, pp. 129–157.
2. S. S. Zhao, X. Zhong, C. Tie and D. D. Chen, *Proteomics*, 2012, **12**, 2991–3012.
3. N. L. Kuehnbaum and P. Britz-McKibbin, *Chem. Rev.*, 2013, **113**, 2437–2468.
4. C. Wenz, C. Barbas, A. Lopez-Gonzalvez, A. Garcia, F. Benavente, V. Sanz-Nebot, T. Blanc, G. Freckleton, P. Britz-McKibbin, M. Shanmuganathan, F. de l'Escaille, J. Far, R. Haselberg, S. Huang, C. Huhn, M. Pattky, D. Michels, S. Mou, F. Yang, C. Neusuess, N. Tromsdorf, E. E. Baidoo, J. D. Keasling and S. S. Park, *J. Sep. Sci.*, 2015, **38**, 3262–3270.
5. C. Pontillo and H. Mischak, *Clin. Kidney J.*, 2017, **10**, 192–201.
6. K. Markoska, M. Pejchinovski, C. Pontillo, P. Zurbig, L. Jacobs, A. Smith, J. Masin-Spasovska, O. Stojceva-Taneva, M. Polenakovic, F. Magni, H. Mischak and G. Spasovski, *Nephrol., Dial., Transplant.*, 2017, DOI: 10.1093/ndt/gfx263.
7. J. Siwy, P. Zurbig, A. Argiles, J. Beige, M. Haubitz, J. Jankowski, B. A. Julian, P. G. Linde, D. Marx, H. Mischak, W. Mullen, J. Novak, A. Ortiz, F. Persson, C. Pontillo, P. Rossing, H. Rupprecht, J. P. Schanstra, A. Vlahou and R. Vanholder, *Nephrol., Dial., Transplant.*, 2017, **32**, 2079–2089.
8. C. Pontillo, S. Filip, D. M. Borras, W. Mullen, A. Vlahou and H. Mischak, *Proteomics: Clin. Appl.*, 2015, **9**, 322–334.
9. G. Bonvin, J. Schappler and S. Rudaz, *J. Chromatogr. A*, 2012, **1267**, 17–31.
10. J. M. Risley, C. A. G. De Jong and D. D. Y. Chen, in *Capillary Electrophoresis-Mass Spectrometry (CE-MS)*, Wiley-VCH Verlag GmbH & Co. KGaA, 2016, pp. 7–39.
11. R. Ramautar and G. J. de Jong, *Bioanalysis*, 2014, **6**, 1011–1026.
12. P. Hommerson, A. M. Khan, G. J. de Jong and G. W. Somsen, *Mass Spectrom. Rev.*, 2011, **30**, 1096–1120.
13. P. W. Lindenburg, R. Haselberg, G. Rozing and R. Ramautar, *Chromatographia*, 2015, **78**, 367–377.
14. R. D. Smith and H. R. Udseth, *Nature*, 1988, **331**, 639–640.
15. C. W. Klampfl and M. Himmelsbach, in *Capillary Electrophoresis-Mass Spectrometry (CE-MS)*, Wiley-VCH Verlag GmbH & Co. KGaA, 2016, pp. 41–65.
16. T. Soga, Y. Ohashi, Y. Ueno, H. Naraoka, M. Tomita and T. Nishioka, *J. Proteome Res.*, 2003, **2**, 488–494.
17. A. Stalmach, H. Husi, K. Mosbahi, A. Albalat, W. Mullen and H. Mischak, *Methods Mol. Biol.*, 2015, **1243**, 187–205.
18. H. Mischak, A. Vlahou and J. P. Ioannidis, *Clin. Biochem.*, 2013, **46**, 432–443.

19. M. Moini, *Anal. Chem.*, 2007, **79**, 4241–4246.
20. E. J. Maxwell, X. Zhong, H. Zhang, N. van Zeijl and D. D. Chen, *Electrophoresis*, 2010, **31**, 1130–1137.
21. R. Wojcik, O. O. Dada, M. Sadilek and N. J. Dovichi, *Rapid Commun. Mass Spectrom.*, 2010, **24**, 2554–2560.
22. R. Ramautar, A. A. Heemskerk, P. J. Hensbergen, A. M. Deelder, J. M. Busnel and O. A. Mayboroda, *J. Proteomics*, 2012, **75**, 3814–3828.
23. R. Ramautar, *Adv. Clin. Chem.*, 2016, **74**, 1–34.
24. E. J. Maxwell and D. D. Chen, *Anal. Chim. Acta*, 2008, **627**, 25–33.
25. L. Sun, G. Zhu, Z. Zhang, S. Mou and N. J. Dovichi, *J. Proteome Res.*, 2015, **14**, 2312–2321.
26. L. Sun, G. Zhu, Y. Zhao, X. Yan, S. Mou and N. J. Dovichi, *Angew. Chem., Int. Ed. Engl.*, 2013, **52**, 13661–13664.
27. P. W. Lindenburg, R. Ramautar, R. G. Jayo, D. D. Y. Chen and T. Hankemeier, *Electrophoresis*, 2014, **35**, 1308–1314.
28. P. Nemes, A. M. Knolhoff, S. S. Rubakhin and J. V. Sweedler, *Anal. Chem.*, 2011, **83**, 6810–6817.
29. R. Ramautar, J. M. Busnel, A. M. Deelder and O. A. Mayboroda, *Anal. Chem.*, 2012, **84**, 885–892.
30. J. M. Busnel, B. Schoenmaker, R. Ramautar, A. Carrasco-Pancorbo, C. Ratnayake, J. S. Feitelson, J. D. Chapman, A. M. Deelder and O. A. Mayboroda, *Anal. Chem.*, 2010, **82**, 9476–9483.
31. R. Ramautar, R. Shyti, B. Schoenmaker, L. de Groote, R. J. Derks, M. D. Ferrari, A. M. van den Maagdenberg, A. M. Deelder and O. A. Mayboroda, *Anal. Bioanal. Chem.*, 2012, **404**, 2895–2900.
32. R. Shyti, I. Kohler, B. Schoenmaker, R. J. Derks, M. D. Ferrari, E. A. Tolner, O. A. Mayboroda and A. M. van den Maagdenberg, *Mol. BioSyst.*, 2015, **11**, 1462–1471.
33. R. Ramautar, *LC GC Eur.*, 2017, **30**, 658–661.
34. F. Yuan, X. H. Zhang, J. Nie, H. X. Chen, Y. L. Zhou and X. X. Zhang, *Chem. Commun.*, 2016, **52**, 2698–2700.
35. M. C. Gulersonmez, S. Lock, T. Hankemeier and R. Ramautar, *Electrophoresis*, 2016, **37**, 1007–1014.
36. A. Hirayama, M. Tomita and T. Soga, *Analyst*, 2012, **137**, 5026–5033.
37. V. Gonzalez-Ruiz, S. Codesido, J. Far, S. Rudaz and J. Schappler, *Electrophoresis*, 2016, **37**, 936–946.
38. V. Gonzalez-Ruiz, S. Codesido, S. Rudaz and J. Schappler, *Electrophoresis*, 2017, DOI: 10.1002/elps.201700427.
39. J. Krenkova, K. Kleparnik, J. Grym, J. Luksch and F. Foret, *Electrophoresis*, 2016, **37**, 414–417.
40. A. Tycova, M. Vido, P. Kovarikova and F. Foret, *J. Chromatogr. A*, 2016, **1466**, 173–179.
41. A. Tycova, J. Prikryl and F. Foret, *Electrophoresis*, 2016, **37**, 924–930.
42. T. T. Nguyen, N. J. Petersen and K. D. Rand, *Anal. Chim. Acta*, 2016, **936**, 157–167.

43. Y. Yin, G. Li, Y. Guan and G. Huang, *Rapid Commun. Mass Spectrom.*, 2016, **30**(1), 68–72.
44. X. Guo, T. L. Fillmore, Y. Gao and K. Tang, *Anal. Chem.*, 2016, **88**, 4418–4425.
45. R. T. Johnson, N. H. To, J. F. Stobaugh and C. E. Lunte, *Chromatographia*, 2017, **80**, 1061–1067.

CHAPTER 5

Stacking or On-line Sample Concentration in CE-MS for Metabolomics

ALAIN WUETHRICH[a] AND JOSELITO P. QUIRINO*[b]

[a]Centre for Personalised Nanomedicine, Australian Institute for Bioengineering and Nanotechnology (AIBN), University of Queensland, Building 75, Brisbane, QLD 4072, Australia; [b]Australian Centre for Research on Separation Science (ACROSS), School of Physical Sciences-Chemistry, University of Tasmania, Private Bag 75, Hobart, TAS 7001, Australia
*E-mail: joselito.quirino@utas.edu.au

5.1 Introduction

Metabolomics targets the study of small molecules and their involvement in biological systems. Most of the metabolites are charged and hydrophilic, which defines the analytical conditions for their interrogation. A powerful analytical workflow is required to deal with the complexity of biological samples. For profiling strategies, adequate separation is needed to yield a profile of significant depth. For identification, the metabolite is preferably detected free of interferences and with a sufficient signal intensity. Capillary electrophoresis (CE) separates charged analytes (and even neutral analytes) based on the differences in migration velocity under the influence of an electric field. The most commonly used mode of CE in metabolomics is capillary zone electrophoresis (CZE), where a simple buffer is used as

New Developments in Mass Spectrometry No. 6
Capillary Electrophoresis–Mass Spectrometry for Metabolomics
Edited by Rawi Ramautar

Published by the Royal Society of Chemistry, www.rsc.org

a background solution (BGS). The metabolites separate due to differences in electrophoretic mobility, and the separation can be tuned by manipulating the electroosmotic flow (EOF). Another mode is electrokinetic chromatography (EKC) where additives known as the pseudostationary phase are added to the BGS. The pseudostationary phase interacts with the metabolites, and the analytes separate due to differences in effective electrophoretic mobility. Neutral analytes and analytes with very similar electrophoretic mobilities that are not separated by CZE can be separated by EKC. The most common pseudostationary phase is micelles from sodium dodecyl sulphate (SDS).

The high electric field strengths in CE provide a high resolution for metabolite analysis and the compact CE setup is suitable for small samples with minimum sample requirements. Mass spectrometry (MS) provides essential mass-to-charge information and has become an indispensable and universal detector for many laboratories. Thus, the hyphenation of CE with MS is particularly useful for metabolomics because of the high separation power of CE and unique sample information from MS. In practice, the EOF must be controlled such that electrical connectivity is maintained throughout the CE run.

Stacking or on-line sample concentration in CE is a family of techniques that concentrate a dilute sample into a very narrow zone. This possibility to focus the injected analytes in a sample by manipulation of the sample and/or separation conditions was recognised early-on and was also central to the exploitation of the full potential of CE.[1,2] Although CE provides a high separation efficiency, an inherent limitation is the small diameter of the separation capillary, which allows only the injection of nanolitre volumes. The minute sample injection leads to poor detection sensitivity, which is often referred to in combination with UV detection and cited as a main drawback of CE analysis. Stacking can provide improvement in analyte signals by many orders of magnitude and can be integrated with relative ease into an existing separation system. The need for sensitive analysis and the range of established stacking techniques is also reflected in their wide popularity and applications.[3]

MS is ~1–2 orders of magnitude more sensitive than absorbance, and thus the need for stacking might not be obvious in CE with MS detection. Stacking, however, can offer more advantages than the improvement in detection sensitivity. Stacking is a form of sample preparation that can also be used to clean up the sample and to yield a clearer electropherogram. The concentration range of metabolites covers nine orders of magnitude or pM to mM levels.[4] Although MS is a sensitive detector, trace levels of some metabolites might go undetected. Furthermore, the wide concentration range of many metabolites can be a challenge for MS detection as the dynamic range is around 4 to 5 orders of magnitude only. For low abundance metabolites, the mass sensitivity in MS can be adjusted by reducing the mass scan range of the mass analyser, so that more selected target ions are passed to the detector.

For instance, a small mass range of 30 *m/z* was used to achieve the required sensitivities for the analysis of a *Bacillus subtilis* extract by conventional CE using a quadrupole MS.[5] To cover the whole *m/z* range the same sample was repeatedly injected for a total time of 16 h. Stacking might provide a viable strategy to reach relevant sensitivities without time-consuming and repetitive injection.

In this chapter, an overview of the stacking techniques and a brief introduction to the mechanisms are provided in Section 5.2. In Section 5.3, the integration of stacking for a CE-MS workflow is described. In Section 5.4, representative applications of stacking in CE-MS are discussed with some focus on method development. Stacking in CE-MS is then summarised in Section 5.5.

5.2 Stacking Techniques – an Overview

Sample stacking in CE is the term for techniques that improve the sensitivity by allowing higher sample volume loads to be introduced. A long sample injection would typically result in broad peaks and compromised separation. To counteract the adverse effect of high sample loads, stacking focuses the long sample zone inside the capillary into narrow zones. The most basic explanation to describe the focusing mechanism of all stacking techniques is the velocity difference of the analytes in the sample and stacking zone. Stacking is performed by changing the sample and/or separation conditions to induce a change in analyte velocity. The implementation of stacking into a CE run can be performed with ease and has been demonstrated in many ways. Six main stacking mechanisms have been established and various hyphenated or sequential stacking techniques have been proposed. In some instances, two or more stacking mechanisms occur during the analyte focusing step.

Table 5.1 shows the stacking techniques, type of analyte concentrated, mechanism, and typical sensitivity improvement or signal enhancement factor (SEF). The SEF is commonly expressed as the increase in peak signal and calculated as the ratio of the peak heights of stacking to typical injection. Typical injection is a short pressure-injected sample plug (*e.g.*, 3–5 s at 50 mbar), which translates to an injection length of <1% (or ~3 mm) of the total capillary length. The SEF may also be calculated using the limit of detection (LOD) ratio of stacking to typical injection. The change in sample velocity of the analyte in the sample and separation buffer or BGS is induced by electrophoretic or (electrokinetic) chromatographic effects. Stacking by electrophoretic effects occurs in field-amplification or enhancement, transient isotachophoresis (tITP), and dynamic pH junctions. Chromatographic effects due to the presence of a pseudostationary phase (*i.e.*, micelles) and an electric field are relevant in sweeping, analyte focusing by micelle collapse (AFMC), and micelle to solvent stacking (MSS).

Table 5.1 Overview of the main sample stacking techniques, including the mechanism, and typical sensitivity enhancement factor (SEF).

Technique	Analytes	Mechanism	Typical SEFs
1. Field amplification/ enhancement	Charged	Analyte electrophoretic velocity difference at a boundary between a high conductivity BGS and low conductivity zone sample	10^1–10^3
2. Transient isotachophoresis	Charged	Concentration adjustment between a leading and terminating electrolyte	10^1–10^3
3. Sweeping	Charged/ Neutral	Picking and accumulation of analytes by micelles penetrating the sample zone	10^1–10^3
4. Dynamic pH junction	Charged	Electrophoretic mobility difference between zones with different pH	10^1–10^2
5. Analyte focusing by micelle collapse	Charged/ Neutral	Analyte transport and release at a dilution zone with the aid of micelles	10^1–10^2
6. Micelle-to-solvent stacking	Charged	Effective electrophoretic mobility reversal in the presence of micelles and organic solvent	10^1–10^2

5.2.1 Field Amplification or Enhancement

In the presence of an electric field and under discontinuous conditions, field amplification occurs at the zone inside the capillary with a lower conductivity. The electrical resistance is higher in the low conductivity zone. As a consequence, the local electric field in the low conductivity zone is higher than that in the high conductivity zone. The velocity of ions is directly proportional to the electric field strength, and thus the ion velocity in the low conductivity zone is faster than that in the high conductivity zone, which results in the concentration of ions at the boundary between the two zones. More importantly, according to the Kohlrausch Regulation Function, the total sum of ions moving in and out of a zone remains the same.[6] The utility of the difference in the conductivity of two adjacent zones for stacking was initially introduced by Mikkers group[7,8] and then elaborated on in the early works of Chien and Burgi.[9,10] The sample for stacking with field amplification/enhancement is prepared in a low conductivity diluent. The analytes are focused based on changes in the electrophoretic velocity of the analytes at the boundary between the low and high conductivity zone. The analyte velocity goes from fast in the sample zone to slow in the background solution (BGS), causing the analytes to concentrate at the boundary.

The application of stacking by field amplification in metabolomics is quite obvious since most metabolites are small ionic molecules. However, since biological samples contain salts, the conductivity of metabolomic samples has to be adjusted or lowered in order to utilise this form of stacking. The conductivity of the BGS may alternatively be increased without causing issues with the ESI-MS or decreasing the resolution power of CE due to Joule heating.

This form of stacking is typically performed with hydrodynamic field-amplified or enhanced sample stacking (FASS/FESS) or electrokinetic field-amplified or enhanced sample injection (FASI/FESI). Simultaneous electrokinetic and hydrodynamic injection of a sample has also been performed, with the idea being to further increase the sample load. In all stacking injections, the sample and separation buffer are of low and high conductivity, respectively. For FESS, both cationic and anionic samples are injected where the injection length needs to be controlled. This is because the length of the injected plug reduces the length for sample separation (effective separation length), and more importantly there is a broadening effect caused by the mismatch of local EOFs. The EOF velocity in the low conductivity sample zone is faster than that in the BGS, but the apparent or overall flow is the same inside the capillary. The sample zone pushes the BGS zone, which causes a disturbance and dispersive effect at the concentration boundary. Because of this, a typical long sample injection in FESS is only approximately 10× that of the typical injection, which therefore provides a SEF of ~10. This simple FESS can be easily adopted in CE-MS for metabolomics, because the only difference with typical injection CE-MS is that the sample is loaded 10× more; however, the sample should have a conductivity that is at least 10× lower than that of the BGS.

A strategy to overcome the dispersive effect of the EOF mismatch in FESS is to remove the sample matrix of the long injection, for instance by counter-pressure or counter-EOF. This provides higher SEF values depending on the volume/length of the sample injected. The removal of the sample matrix also returns part of the lost separation length due to the long injection. These modes of FESS with counter-pressure and counter-EOF are often referred to as large volume sample stacking (LVSS) and reversed electrode polarity switching mode (REPSM).[11] LVSS is the case where EOF is used to pump the sample matrix and where the polarity of the separation is the same as in the stacking; it is also called large volume sample stacking with EOF pump (LVSEP). In REPSM, the stacking and separation voltages are of different polarity. The stacking is halted and a separation voltage is applied when the current during stacking reaches >95% of the separation current. These two modes, however, permit the introduction of either cations or anions only, since the target analyte must move in the opposite direction to the counter-pressure and counter-EOF. The implementation of LVSS or REPSM to CE-MS is quite problematic, because there is no BGS vial to replenish the BGS inside the capillary that is moving to the inlet vial during stacking. A solution for this was proposed by Kim and co-workers, where an Eppendorf tube containing the BGS was placed at the outlet end of the capillary in the CE-ESI-MS, and after matrix removal, the outlet was re-configured to the ESI-MS.[12] The sheath liquid may also serve to replenish the BGS during stacking.[13] The demand for high throughput in metabolomics might hinder the use of these effective sample concentration techniques in CE-MS for metabolomics.

In FESI, the electrokinetic injection is selective and depends on the analyte's (*e.g.*, metabolites') mobility.[14] The amount of injected sample is not limited by the volume of the sample introduced into the capillary, and therefore the amount can be very high and SEFs of up to three orders of magnitude can be achieved. The conductivity difference, however, must be very high to achieve high SEFs, since the theoretical SEF is dependent on the conductivity ratio (conductivity of the BGS/conductivity of the sample). It will be interesting to study the effect of simple dilution of the metabolomic sample on the LODs that can be obtained with FESI and then compare them to the LODs with typical injection of the original metabolomic sample. Simple dilution of the sample and FESI might be a useful strategy to improve the sensitivity of low abundance metabolites.

The stacking boundary in FESI is at the capillary inlet end and this boundary needs to be maintained for successful injection. The pressure injection of a short water plug prior to FESI can improve the stability of injection. The field enhancement is higher if only a small fraction of the capillary is filled with the low conductivity sample or water plug. Capillary coatings that control the EOF or counter-pressure can also be used to assist in sample injection. FESI is sometimes referred to as head column FESS, but FESI is more abundant in the literature since FESS was already originally defined to be performed with hydrodynamic injection.

5.2.2 Transient Isotachophoresis (tITP)

In isotachophoresis (ITP), the sample is concentrated in bands between a leading (LE) and terminating (TE) electrolyte. An important requirement is that the electrophoretic mobility order is LE > sample > TE. The concentration of the sample is adjusted to the concentration of the LE. t-ITP is when ITP is used as the on-line sample concentration technique prior to CZE separation. The diluted analytes are stacked or focused behind the LE during the ITP step. tITP is suitable for charged analytes and SEF values of two orders of magnitude are typical for this technique. The theoretical foundation of tITP was pioneered by Verheggen and Beckers and later demonstrated for trace analysis of proteins by Foret and co-workers.[15–17]

Figure 5.1 shows one strategy to implement tITP in CZE. The LE is added to the sample and separation buffer and the sample is injected into the capillary (a). Prior to voltage application, the sample at the capillary inlet is replaced with separation buffer that contains the TE (b). It is important that the charge signs of the LE, TE, and sample are the same – either a positive or negative sign for all stacking participants in tITP. Examples of LE ions with a high mobility are H^+ and Na^+ for cations and Cl^- and PO_4^{3-} for anions. Examples of ions with a low mobility are cationic β-alanine and glutamic acid and anionic glycinate and histidine. Voltage application follows, which causes the alignment of the LE, sample, and TE zones (c). The concentration

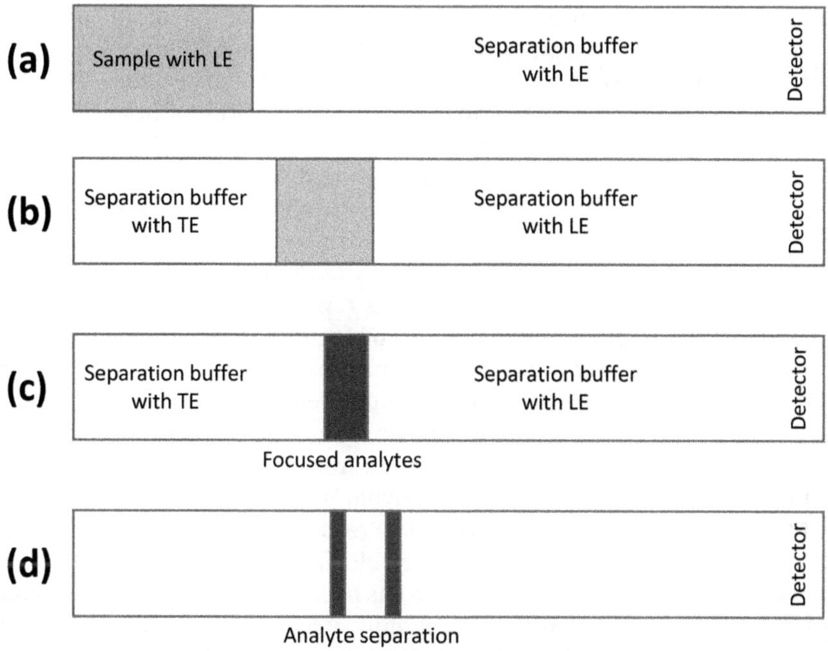

Figure 5.1 Stacking based on transient isotachophoresis (tITP). (a) The sample is injected into a capillary with separation buffer. The sample and separation buffer contain the leading electrolyte (LE). (b) The sample solution at the capillary inlet is replaced with separation buffer containing the terminating electrolyte (TE). The sample, LE and TE have the same sign. (c) During voltage application, the analytes in the sample will stack between the LE and TE with decreasing electrophoretic mobility. (d) The stacked zones are then separated by electrophoresis.

of the sample is dictated by the LE, and thus dilute samples can be efficiently concentrated at high concentrations of LE. The focusing process stops once all of the zones are formed with the same concentration as the LE. After this step, the ITP diffuses and separation of the analytes in the sample zone is performed *via* electrophoresis (d).

5.2.3 Dynamic pH Junctions

The mobility of an analyte is dependent on its mass-to-charge ratio. Manipulation of this ratio by changing the pH environment is used to concentrate the analytes in dynamic pH junctions. This mode is suitable for ionisable analytes such as weakly acidic and basic molecules and amphoteric compounds. The change in ionisation affects the migration velocity and direction, which is used for stacking. This technique is suitable for high salt-containing samples such as those typically found in metabolomics. Careful manipulation of the charge state is key in obtaining high analyte enrichment and SEFs of one to three orders of magnitude have been reported.[18] The use of this pH effect (before coining of the term dynamic pH junction) was first reported in 1990 by Aebersold and Morrison for the five-fold enrichment of peptides. The term dynamic pH junction and its fundamental investigation by Britz-McKibbin and Chen occurred 10 years later.[19,20]

In a dynamic pH junction, the change in analyte mobility for focusing is induced by titration with H^+ or OH^-. For acidic analytes, a long sample plug is introduced into the capillary that contains an acidic separation buffer. The sample solution has a basic pH value, where the analytes are negatively charged. After injection, the sample vial is replaced with the acidic separation buffer. Upon application of a voltage of negative sign at the detector side of the capillary, the negatively charged analytes move towards the capillary inlet. H^+ ions from the acidic separation buffer enter the sample zone, which causes a pH decrease and reverses the analyte charge and migration direction. The boundary between the acidic separation buffer and the sample is the pH junction and this boundary is dynamic since it moves through the sample zone and titrates the analytes. The resulting focusing at the dynamic pH junction is the consequence of the velocity change of the analyte and completed once the whole sample zone has been titrated. The same principle applies for basic analytes, where the sample is prepared in acidic conditions and using a basic separation buffer. OH^- ions sweep through the sample zone instead of the H^+ ions used for acidic analytes. In many instances, ITP effects occur where the H^+ or OH^- acts as the terminating ion, as these ions are neutralised at the pH boundary. Focusing by dynamic pH junction only occurs when the charge state of the weakly acidic/basic or amphoteric metabolite is changed at the pH boundary.

5.2.4 Sweeping

In sweeping, the migration of an additive that picks up the analytes through an additive-free sample zone is responsible for the focusing. The additive is also referred to as the pseudostationary phase – which is the same as in EKC because it is reminiscent of the partitioning of analytes with the stationary phase in liquid chromatography. The pseudostationary phase can be charged or neutral and includes surfactants, bile salts (*e.g.*, sodium cholate), complexing agents (*e.g.*, EDTA), and starch derivatives (*e.g.*, cyclodextrins). Surfactants are widely used and examples of this group are the anionic sodium dodecyl sulfate (SDS), cationic cetyltrimethylammonium bromide (CTAB), neutral Brij-35, and zwitterionic sulfobetaine. Sweeping was introduced in 1998 for the trace analysis of pollutants.[21] In theory, analyte focusing by sweeping is unlimited since this stacking process is governed by the affinity of the analyte to the pseudostationary phase. The focusing is described as the compression of the analyte band after injection and is inversely proportional to the analyte's retention factor with the pseudostationary phase. Thus, the compressed analyte zone will be very narrow for analytes with a high affinity towards the pseudostationary phase.

For sweeping to occur, the fundamental requirement is that the sample is prepared in a solution free of pseudostationary phase. A long sample plug is then injected into a capillary filled with additive containing separation buffer. When surfactants are used as the pseudostationary phase, the concentration is above the critical micelle concentration (cmc). After injection, separation buffer is placed at the capillary inlet and outlet prior to voltage application. The additive sweeps through the injected sample zone and interacts with the analyte. The pseudostationary phase–analyte interaction is a "collection process" where the collected analytes migrate bound to the pseudostationary phase according to their effective electrophoretic mobility. It is important that the sample upon interaction with the pseudostationary phase migrates towards the detector. For cationic analytes, neutral and negatively-charged additives are used for more effective sweeping. For anionic analytes, neutral and positively-charged additives are used. For neutral analytes, both negatively- and positively-charged additives are suitable. The focusing is completed when the sample zone has been swept by the pseudostationary phase. After focusing, the separation of the analytes is normally achieved by EKC.

Sweeping in EKC will be useful for metabolites that can interact strongly with the pseudostationary phase. The derivatisation of metabolites to form hydrophobic derivatives that can interact strongly with the additives is an approach to increase the SEFs.[22] The selectivity of the sweeping process for highly retained metabolites to tailored or even commonly used pseudostationary phases could be harnessed for targeted metabolite fingerprinting or determination. In CE-MS, partial filling techniques in EKC or the use of MS-compatible pseudostationary phases will be required.[23]

5.2.5 Analyte Focusing by Micelle Collapse (AFMC)

In AFMC, charged micelles serve as the analyte carrier to transport the analyte to the stacking boundary. At the boundary, dilution of the micelles to below the cmc causes micellar collapse and release of the analytes. The continuous collapse and migration of micelles results in analyte concentration at the stacking boundary. This stacking technique is suitable for neutral and charged analytes. Examples of charged micellar carriers include surfactants and ionic liquids.[24] The first demonstration of AFMC was in 2008 for sensitive detection of steroids and provided analyte enrichment of more than two orders of magnitude.[25] For focusing by AFMC, the sample is prepared in a micellar solution with a micelle concentration slightly above the cmc. The role of the separation buffer, apart from supporting the separation, is to facilitate the collapse of the micelles by having a lower conductivity than the micellar sample solution. AFMC is completed when all the micelles have collapsed and released the analyte. Subsequent separation is based on CZE or MEKC. For application of AFMC in CE to metabolomics, the analytes should have sufficient affinity with the micelles, as in sweeping.

5.2.6 Micelle to Solvent Stacking (MSS)

In MSS, the analytes are concentrated at the micelle to solvent stacking boundary (MSSB) that is between a micellar and organic solvent rich zone. The analyte must have a charge that is opposite to that of the micelles (*e.g.*, cationic analytes are stacked by anionic micelles). Central to the MSS concentration mechanism is that the analytes' migration direction reverses at the MSSB. Also, the collapse of the micelles is not required since the effective electrophoretic mobility of a charged analyte could reverse once the interaction of the micelles with the analytes is insignificant. Thus, the migration is only dictated by the electrophoretic mobility in the presence of micelles and organic solvent. After the release of the micelle-bound analyte, the micelles migrate in the opposite direction to the analyte, which makes MSS suitable for coupling with ESI-MS. The concept of MSS in CZE-ESI-MS for cationic adrenergic β-blockers was conceptualised in 2009.[26] The SEF by MSS is typically equivalent to more than one order of magnitude. For application of MSS in CE to metabolomics, the analytes should also have sufficient affinity with the micelles, as in sweeping and AFMC. There is the possibility for the use of MSS for simultaneous sample clean-up and preconcentration of metabolites, since MSS is selective for anionic or cationic analytes. Under co-EOF CZE conditions, where the EOF has the same direction as the targeted charged analytes, the unwanted components of the sample can be removed from the sample zone if they have strong affinity with the micelles during MSS while the targeted analytes are focused. Then, only the targeted analytes are detected by ESI-MS since the unwanted components will migrate in the direction of the micelles and away from the detector.

Figure 5.2 shows a MSS schematic for stacking of cationic analytes. The analyte is prepared in a sample which contains negatively charged micelles (*e.g.*, SDS) at a concentration slightly above the cmc. The BGS contains a sufficient amount of organic solvent (*e.g.*, 30% acetonitrile). The sample is injected *via* hydrodynamic pressure or voltage and after injection the separation buffer vial is placed at the capillary inlet (Figure 5.2a). Upon application of a voltage, the MSSB is formed between the separation buffer and injected sample zone (Figure 5.2b). This boundary migrates with the EOF towards the detector. The micelle-bound analytes migrate towards the MSSB dictated by the effective electrophoretic mobility of the micelles ($\mu_{eff}(A)$). At the MSSB, the organic solvent content causes reversal of the analytes' migration direction towards the detector side ($\mu(A)$) (Figure 5.2c). MSS can be implemented by injecting an organic solvent rich plug after the micellar sample solution, instead of having a high amount of organic solvent in the BGS. The analyte focusing is completed once all the micelles from the sample have migrated across the MSBB (Figure 5.2d). Separation of the focused analytes by CZE principles follows (Figure 5.2e).

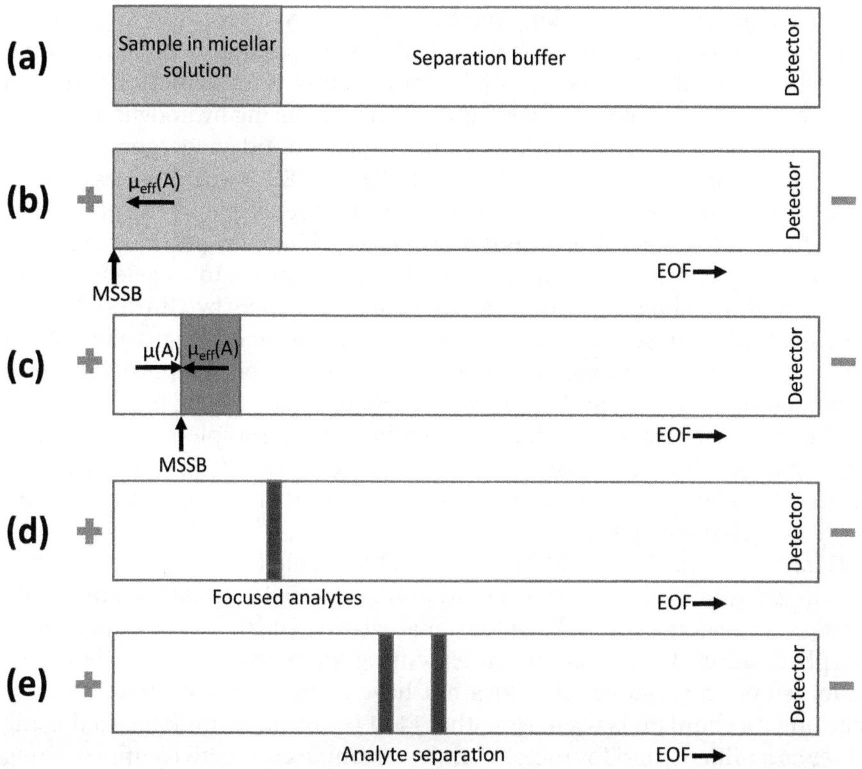

Figure 5.2 Schematic for MSS of cationic analytes. (a) The sample is prepared in a micellar solution and injected into a capillary containing the separation buffer. The sample vial is then replaced with separation buffer. (b) Upon application of a voltage, the micelle-bound analytes migrate towards the positive sign (μ_{eff}(A)). The introduction of separation buffer due to the cathodic EOF results in the micelle to solvent stacking boundary (MSSB), which migrates towards the detector. (c) At the MSSB, the migration direction of the analytes in the sample is reversed from μ_{eff}(A) to μ(A) due to micellar dilution and analyte release. (d) MSS is completed when all the micelle-bound analytes have been released and formed the zone with focused analytes. (e) The focused analytes are then separated based on electrophoresis.

5.2.7 Hyphenated Stacking

Hyphenated stacking is defined as the combination of two or more stacking techniques, which can be used to further improve the detection sensitivity and in some cases also to clean-up the sample within the same capillary. Hyphenated stacking includes synergistic and sequential stacking. Synergistic stacking is when the stacking techniques work simultaneously while sequential (or stepwise) stacking occurs one stacking approach at a time. Hyphenated stacking enables the capillary to be overloaded with the sample, since multiple stacking mechanisms are involved. FESI in combination with another stacking technique is the most popular hyphenated strategy. This is because the amount of sample that is electrokinetically loaded onto the capillary is substantially higher than when using hydrodynamic injection. Up to now, the hyphenation of stacking has found wide interest. These hyphenations include FESI-MSS, FESI-AFMC, FESI-sweeping, sweeping-AFMC, sweeping-MSS, dynamic pH junctions-sweeping, and FESI-tITP. The hyphenation of FESI-tITP is referred to as electrokinetic supercharging (EKS) as coined by Hirokawa[27] and FESI-sweeping is referred to as selective (cation or anion) exhaustive injection-sweeping as coined by Quirino.[28] More recently three step stacking has also been demonstrated.[29,30] Analyte SEFs of 1000–10 000 are common and even million-folds have been reported.[28] When approaching very high SEF values with FESI, however, the injection can suffer from poor repeatability due to depletion of the sample in proximity to the capillary inlet. Today the hyphenation of stacking has yet to be demonstrated for simultaneous on-line sample concentration of positively- and negatively-charged species.

Hyphenated stacking by FESI-MSS and EKS should be attractive for metabolomic analysis by CE-MS. In order to implement field enhancement/amplification, the sample must have low conductivity, which can be achieved by simply diluting the sample or implementing an appropriate sample preparation step. The conductivity does not have to be very low since a second stacking mechanism is used. Note that FESI is normally implemented using water as a diluent, and for metabolomics a sample conductivity approaching that of water is impossible. In FESI-MSS, dilution with an organic solvent will both decrease the conductivity of the sample and render the sample ready for MSS.

5.3 Considerations for Stacking in CE-MS-based Metabolomics Studies

The requirements for mass spectrometry detection primarily dictate the procedure for the implementation of stacking. ESI is the most common interface in CE-MS and is also appropriate for metabolites; thus, this section discusses the stacking requirements for ESI-MS. Basically, the same guidelines can be followed for the implementation of stacking in ESI-MS as for CE-MS, except that the separation window should additionally be considered. The separation window is the time for separation between the completion of the stacking process and prior to detection of the EOF. Stacking of long injections will reduce the separation window. Ion suppression and ESI source contamination are two factors that negatively affect the signal and thus these should be minimised. Source contamination can be avoided by using volatile reagents. Ion suppression is in general less significant using low background electrolyte concentrations and when performing electrophoretic separations under very low flow rate conditions.[31] The use of a sheath liquid can help to stabilise the ionisation process; however, it might also lead to a decreased sensitivity. Non-volatile electrolytes and buffer additives can be used when a stacking strategy is applied that prohibits or separates the entrance of the non-volatiles with the analyte into the ESI source. Based on these requirements, stacking is implemented in two ways; (1) directly, where the stacking technique is directly compatible with ESI-MS, or (2) sequentially, where the stacking is not ESI-MS friendly but its components are controlled during electrophoresis such that they do not enter the detector with the focused analytes.

5.3.1 Stacking with MS-compatible Reagents

In this stacking mode, the separation buffer and sample enter the ESI, and thus the separation and stacking systems are exclusively made from volatile compounds. The stacking techniques that can be used are FESI, FESS, dynamic pH junctions, and tITP. AFMC and sweeping are also possible when volatile or ESI-MS-friendly pseudostationary phases are used. Examples of volatile buffer systems for anionic analytes are ammonium formate and ammonium bicarbonate adjusted to basic pH with ammonia or triethylammonium acetate. Buffer examples for cationic analytes are formic acid and acetic acids and their acidic volatile salts (*i.e.*, ammonium acetate, ammonium formate). In FESS, the conductivity of the sample should be 10× lower than the BGS.[32] Since the volume of sample injected should be at the most 10× more than typical injection, the gain in SEF will almost be negligible for metabolomics samples. A simple sample preparation step that does not require extensive salt removal will be required for metabolomics analysis that is aided by FESS.

For FESI, the sample is prepared in water and, if required, the pH value is adjusted to ionise the analyte or the sample is prepared in highly diluted separation buffer (*e.g.*, 1/100 dilution). As discussed earlier, low conductivity metabolomics samples are not typically encountered. A second stacking mechanism of t-ITP (EKS) can be implemented using volatile LE and TE to increase the SEF with FESI. The control of the EOF is important; when the analyte and EOF have the same migration direction (co-EOF CZE), then the electrokinetic analyte injection is compromised by introduction of sample matrix that is carried by the EOF (EOF mismatch and lower field enhancement at the injection point). When the analyte and EOF have opposite migration directions, the analyte injection is reduced by the EOF magnitude. Also, since the EOF is pumping the liquid out of the capillary, electrical disconnection might occur. The use of pressure to counter-act the EOF and EOF suppression by capillary coatings or acidic buffers supports FESI. Aside from the Eppendorf tube approach by Chung, Moini used an adjustable porous tip that can be pulled inside the sheath liquid for FESI, preserving the electrical conductivity during injection with reversed EOF.[33]

For dynamic pH junctions, the sample and separation buffer have different pH values, but not necessarily the same conductivity. The pH of the sample and separation buffer is adjusted using volatile acids or bases. The pH value is set so that the analyte in the sample and separation buffer provides the highest difference in analyte electrophoretic mobility. The isoelectric point (pI) and acidity constant (*i.e.*, pK_a) of the analyte are considered to find suitable pH values for these solutions. In tITP, a convenient strategy is to condition the capillary with a volatile separation buffer containing the TE and adding the LE to the sample or by injection of a plug of LE prior to sample injection. Examples of cationic LEs are H^+ and NH_4^+ with reported electrophoretic mobilities of 362.5 and 76.2 ln [10^{-9} m^2 V^{-1} s^{-1}].[34] Volatile anionic LEs include OH^- and CO_3^{2-} with mobilities of −202.5 and −71.8 ln [10^{-9} m^2 V^{-1} s^{-1}], respectively. Cationic and anionic TEs include imidazole and 2-(*N*-morpholino)ethanesulfonic acid with mobility values of 52.0 and −28.0 ln [10^{-9} m^2 V^{-1} s^{-1}], respectively. Alternatively, the neutralisation of H^+ and OH^- with ions from the separation electrolyte can be used in tITP. For both dynamic pH junctions and tITP, the samples are typically hydrodynamically injected into the capillary.

At first, surfactant-assisted stacking techniques seem inappropriate for use with MS since most surfactants strongly suppress the analyte signal and contaminate the ESI source. The advances in material science have led to the introduction of MS-compatible pseudostationary phases such as nanoparticles and surfactants of low boiling point. Examples of the latter include the relatively volatile surfactant perfluorooctanoic acid (boiling point = 189 °C), lauric acid, and polymeric surfactants.[35,36] Nanoparticles made from latex or polymethacrylate have also been demonstrated.[37,38] The introduction of the particles from the ESI to the MS was avoided by using an orthogonal sprayer design, or the particles did enter the MS but did not suppress the analyte signal. The pseudostationary phases were used to couple MEKC with ESI-MS; however, the implementation of sweeping and AFMC in full capillary filling mode has yet to be demonstrated.

5.3.2 Stacking with Non-volatile Reagents

In this mode, the stacking process is completed and the analytes are detected separated from the stacking process. This separation of the analyte and stacking process allows the use of non-volatiles. For instance, the stacking techniques that rely on non-volatiles, such as common surfactants, are MSS and AFMC. For MSS, micelles are used, which have an opposite charge to the analyte. Upon release of the analyte at the MSSB, the micelles' electrophoretic mobility directs them away from the detector. In AFMC, the collapse of the micelles to the surfactant monomer at the MSSB causes the release and accumulation of the analytes. After the analyte concentration step is completed, the surfactant monomer does not enter the MS with the analyte. For both MSS and AFMC, the surfactant counter ion should be volatile to reduce cluster formation and ion suppression. For instance, ammonium lauryl sulfate can be used instead of SDS.

5.4 Applications of Stacking in CE-MS-based Metabolomics Studies

The small sample volumes and the complex nature of most metabolomics samples require dedicated handling, high separation power and sensitive detection. In the pursuit to interrogate single biological entity levels, the need for highly sensitive and powerful analytical strategies will continue to increase. The use of stacking as part of the analytical strategy can improve separation and give orders of magnitude improvements in sensitivity. Table 5.2 shows examples of stacking strategies in CE-MS implemented in metabolomics or biologically-relevant investigations. This section first discusses the single stacking strategies followed by the hyphenated approaches. For sweeping, AFMC, and MSS in CE-MS, there were no specific metabolomics reports; however, examples were selected that dealt with similar sample complexities or instrumental setups.

5.4.1 Applications of Field Amplification/Enhancement

In stacking by field amplification, the sample needs to be of sufficiently low ionic strength (at least 1/10 of the separation buffer), which can be achieved by adequate sample preparation or dilution of the biological sample. Sample preparations that can be used to lower the conductivity include solid-phase extraction (SPE), solid–liquid extraction (SLE), liquid–liquid extraction (LLE), and desalting (*e.g.*, preparative size exclusion chromatography). A simple strategy is sample dilution to lower the ionic strength. This is suitable when the separation buffer has a high ionic strength, so that the sample dilution is sufficient to achieve stacking by field enhancement. For instance, FESS was applied for single neuron metabolomic profiling of *Aplysia californica*.[39] Figure 5.3 shows the metabolite profile of two subcellular regions of a single neuron. The top and bottom extracted ion electropherograms were obtained from the soma and neurite part of the neuron. After dissection of the neurons, the isolated cells were rinsed with purified water to provide the low conductivity matrix for long hydrodynamic injection of 6 nL. FESS and the use of a nebulizer-free sheath flow nano-interface with *time-of-flight* MS were crucial to achieve relevant detection limits of <50 nM for acetylcholine, histamine, dopamine, and serotonin. Human plasma was investigated for arginine and methylated arginine metabolites after simple protein precipitation.[40] FESI directly from the supernatant was possible due to the use of a high conductivity separation buffer (*i.e.*, 1.5 M formic acid). A water plug was followed by electrokinetic injection of the sample for 200 s at 12 kV, and finally a separation buffer plug to push the stacked analytes further into the capillary. An internal standard was added (*i.e.*, deuterated analyte) to account for differences in sample composition and to improve the method performance. Method development for anionic metabolites typically involves a basic separation buffer, which engenders a strong cathodic

Table 5.2 Application examples of stacking in CE-MS.

Analytes	Analyte charge for separation	Sample	Sample preparation	Stacking	Injection (% length of capillary)	LOD	References
32 organic acids, sugar phosphates, and nucleotides	Anionic	Human U-87 MG glioblastoma cell	Lysis, LLE	FESS	60 s at 138 mbar (3.0%)	10 to 200 nM	43
32 carboxylic acids, saccharides, nucleotides, and nicotinamide and flavin adenine coenzymes	Anionic	*Bacillus subtilis* cells	Lysis, LLE	FESS	50 mbar for 30 s (2.6%)	0.3 to 6.7 μM	41
3,4-Methylenedioxyamphetamine, 3,4-methylenedioxymethamphetamine, methadone, 2-ethylidene-1,5-dimethyl-3,3-diphenylpyrrolidine morphine, codeine, 6-monoacetylmorphine, cocaine	Cationic	Hair	LLE	FESI	30 s at 7 kV	0.002 to 0.01 ng mL^{-1}	44
Nicotine and nicotine metabolites	Cationic	Urine	SPE	FESI + FESS	12 s at 25 mbar and 30 kV	0.55 and 11.25 ng mL^{-1}	46
Cocaine, benzoylecgonine, cocaethylene, ecgonine methyl ester, norcocaine	Cationic	Urine	SPE	FESI	60 s at 10 kV	1.5 to 10 ng mL^{-1}	45
L-Arginine, mono-methyl-arginine, dimethylarginine isomers	Cationic	Plasma	Protein precipitation	FESI	200 s at 12 kV	0.3 to 3 μM	40
Peptides from tryptic digest	Cationic	*E. Coli* and bovine serum albumin	Tryptic digest	tITP	15 psi 90 s (30.2%)	~20 nM	48

Carnosine, L-carnitine, creatinine, citrulline, 4-aminohippuric acid, L-tyrosine, L-phenylalanine, L-lysine, L-arginine, L-isoleucine, L-leucine, L-histidine, cystathionine, homoserine, L-aspartic acid, L-hydroxyproline, hippuric acid, dopamine, phenylalanylglycine, L-cysteine, guanosine	Cationic	Urine	Centrifugation	tITP	25 psi 30 s (14.0%)	0.6 to 24 nM	49
Varenicline, 2-hydroxy-varenicline	Cationic	Urine	None	tITP	10 μL	52 pg mL^{-1}	50
4 synthetic peptides (L-Ala–L-Ala, L-Leu–D-Leu, Gly–D-Phe, Gly–Gly–L-Leu)	Cationic	Urine	Dilution and filtration	Dynamic pH junction	300 s at 50 mbar (16.2%)	0.2 to 2.0 nM	51
Leucine enkephalin, methionine enkephalin, dynorphin A, beta-endorphin, angiotensin II	Cationic	HL-1 cardiac muscle cells	Lysis, SPE	Dynamic pH junction	90 s at 280 mbar (27.8%)	1 to 200 nM	53
Recombinant Arg–Gly–Asp–hirudin	Cationic	Urine	Dilution and filtration	Dynamic pH junction	300 s at 50 mbar (26.8%)	11 nM	52
Untargeted metabolites	Cationic	*E. Coli*	Dilution	Dynamic pH junction	75 s at 50 mbar (7.8%)	~1 to 25 nM	54
Androstenedione, testosterone, epitestosterone	Neutral	Standards	None	Sweeping	2.5 psi for 10 s (2.4%)	0.75 to 5 μg mL^{-1}	57
Diethylphthalate, dipropylphthalate, dibutylphthalate	Neutral	Standards	None	AFMC	23 s at 1 bar (20.9%)	~3 μg mL^{-1}	58
Pindolol and metoprolol	Cationic	Standards	None	MSS	60 s at 50 mbar (2.6%)	0.03 and 0.01 μg mL^{-1}	26

(*continued*)

Table 5.2 *(continued)*

Analytes	Analyte charge for separation	Sample	Sample preparation	Stacking	Injection (% length of capillary)	LOD	References
L-Alanine, L-arginine, L-glutamic acid, L-isoleucine, L-leucine, L-lysine, L-methionine, L-phenylalanine, L-tyrosine, L-valine	Cationic	Urine	None	pH-mediated stacking	90 s at 90 mbar (4%)	20 to 215 nM	65
22 amino acids and 2 organic acids	Cationic	Urine	Protein precipitation	pH-mediated stacking	50 mbar for 40 s (5.4%)	0.004 to 1.27 μM	66
Atorvastatin, fluvastatin, gemfibrozil, pravastatin, rosuvastatin	Anionic	Wastewater	LLE	FESI-tITP	−10 kV for 170 s	0.18 ng mL^{-1}	61
Oxacillin, cloxacillin, penicillin G, sulfamethizole, sulfaquin-oxaline, sulfamethoxazole, sulfadimethoxine	Anionic	Seawater	SPE	FESI-MSS	−10 kV for 180 s	0.002 to 0.011 ng mL^{-1}	70

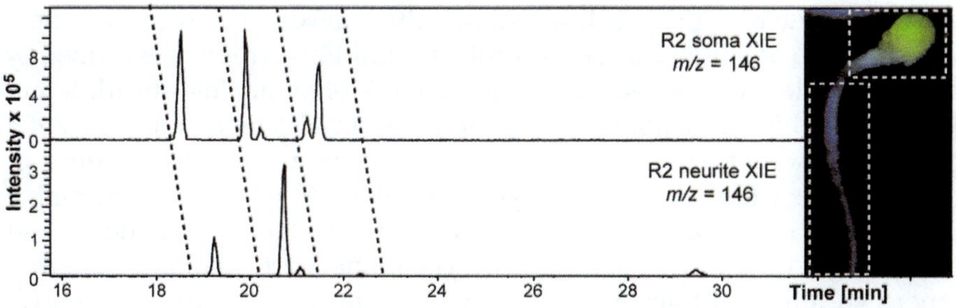

Figure 5.3 Extracted ion electropherograms from two subcellular regions of a single neuron from *Aplysia californica*. Reprinted with permission from ref. 39. Copyright (2009) American Chemical Society.

EOF opposite to the direction of metabolite migration. In reversed polarity separations with CE-ESI-MS, the cathodic EOF can pump out the separation buffer from the capillary and thereby cause instability of the separation and ESI process. Capillary coatings that provide an anodic EOF or zero EOF are often used for stacking and separation of negatively charged metabolites, or separations are performed in co-EOF or no-EOF, respectively. For instance, a successive multiple ionic layer (SMIL) capillary coating with an end coating of cationic polymer provided an anodic EOF that was required to investigate the negatively charged metabolites from glycolysis intermediates of *Bacillus subtilis*.[41] The SMIL was based on sandwiching layers of cationic (*i.e.*, dextran sulfate) and anionic (*i.e.*, polybrene) polymers on the bare fused silica capillary wall.[42] The extracted cell lyophilisate was reconstituted in purified water and injected for 30 s at 50 mbar for FESS. The metabolite detection sensitivity ranged from 0.3 to 6.7 μM.

Another strategy for analysis of anionic metabolites is to apply a small hydrodynamic pressure at the inlet of a bare fused-silica capillary in order to counter-act the cathodic EOF. For instance, a pressure of 35 mbar at the inlet enabled the analysis of negatively charged model metabolites (*i.e.*, organic acids, sugar phosphates, and nucleotides) from glioblastoma cell extracts.[43] The separation buffer was 10% acetic acid at pH 2.2. The acidic pH of the separation buffer also suppressed the EOF. FESS was performed from reconstituted aqueous cell extract for 60 s at 138 mbar (2 psi). This provided a 10-fold enrichment and enabled detection sensitivities of 10 to 200 nM. The reported advantage of this strategy was that the same method can be used to analyse cationic metabolites by simply switching the polarity of the separation voltage. The combination of off-line sample concentration and stacking by FESI has been reported for trace metabolite analysis. Off-line SLE and LLE was applied to extract illicit drugs and their metabolites from hair samples.[44] The extracted amphetamines, opiates, and cocaine were reconstituted in purified water to fulfil the criteria for field enhancement. First, a short water plug was injected (1 s at 35 mbar) followed by injection for 30 s at 7 kV. The ammonium salts of formate, tris(hydroxymethyl)aminomethane, and non-volatile borate and phosphate were probed. Interestingly, the non-volatile inorganic buffers were found to be suitable under controlled conditions. This was explained by the favourable low flow rates in CE and the use of an organic sheath liquid (*i.e.*, acidified isopropanol/water mixture, 50:50 v/v). The hair samples of a drug addict were then analysed and μg mL^{-1} to ng mL^{-1}-concentrations of cocaine, benzoylecgonine, cocaethylene, ecgonidine methylester, ecgonine methylester, morphine, 6-monoacetylmorphine, codeine, methadone, and 2-ethylidene-1,5-dimethyl-3,3-diphenylpyrrolidine were found. In another report, cocaine and four of its metabolites in urine were also targeted *via* SPE prior to electrophoretic separation with ion trap MS detection.[45] The weak basic analytes were separated with a separation buffer of 15 mM ammonium formate at pH 9.5. At this pH value, four of the analytes were negatively charged and one was partially negatively charged (*i.e.*, benzoylecgonine). To facilitate the detection of benzoylecgonine, a slight hydrodynamic pressure at the capillary inlet was applied. Also, the capillary was coated with polyvinyl alcohol to suppress the strong EOF and reduce analyte–capillary wall adsorption. FESI was performed by applying a voltage of 10 kV for 60 s. The analysis of this sheath flow-assisted approach was completed in less than 15 min. A similar strategy of SPE followed by FESI was applied for nicotine and eight nicotine metabolites.[46] SPE provided ~200-fold analyte preconcentration and FESI provided another ~100-fold increase in sensitivity.

5.4.2 Applications of tITP

In tITP for MS, a common pair of leading and terminating ions for cationic metabolites is NH_4^+ and H^+ (from acetic acid), respectively. The H^+ is neutralised by the buffering counter ion in the LE.[47] For anionic metabolites, the LE and TE pair frequently used is OH^- and formate, correspondingly. tITP can deal with complex biological samples; thus, laborious offline sample treatment can be avoided or reduced to simple filtration. This is particularly useful for cases where a minute volume of a complex sample needs to be analysed and off-line sample preparation will be extremely difficult. tITP together with a sheathless ESI are well suited for very small sample volumes such as those typically obtained from *in vivo* studies. For instance, a sheathless porous capillary interface and tITP were used to analyse tryptic digests of *E. coli* and bovine serum albumin.[48] The separation buffer was 10% acetic acid at pH 2.2. The tryptic digest was prepared in LE (*i.e.*, 50 mM ammonium acetate at pH 4). One third of the capillary was filled with the sample, which provided SEF values of 10- to 86-fold. This tITP strategy was also applied for human urine profiling.[49] A minimal centrifugation step and dilution of the sample in separation buffer was sufficient prior to injection and stacking by tITP. This combination of tITP and sheathless ESI was also beneficial to improve the sensitivity and in providing a more comprehensive urine profile.

ITP was used as a sample preparation step in two-dimensional CE, with CZE as the second CE step. Stacking and sample clean-up of varenicline and its metabolite 2-hydroxy-varenicline in the first capillary of untreated urine by ITP was reported.[50] The capillary in the first dimension had a high internal diameter of 800 μm i.d., which allowed a sample injection of 10 μL. The LE and TE was 10 mM ammonium acetate in 20 mM acetic acid at pH 4.5 and 10 mM acetic acid at pH 3.1, respectively. The drug and its metabolite were stacked between the LE and TE prior to column switching to introduce the stacked analyte into the second capillary for separation of metabolites by CZE. For CZE, the volatile TE (*i.e.*, 10 mM HAc at pH 3.1) was used prior to tandem MS detection.

5.4.3 Applications of Dynamic pH Junctions

A dynamic pH junction was used to analyse metabolites where the charge state could be manipulated due to a change in pH environment. Examples include peptides, amino acids, or other metabolites with an isoelectric point (pI) or suitable pK_a value. The pH value of the sample is adjusted so that the metabolite has no charge or opposite charge compared to the metabolite in the separation buffer. This is to maximise the sample enrichment by the migration velocity change at the stacking boundary. The analytes are stacked at the dynamic pH junction that sweeps through the sample. The separation is preferably performed under conditions where metabolite wall adsorption is suppressed, *e.g.*, with the use of a neutral capillary coating or acidic BGS. Sensitive analysis of four small peptides with a similar pI of ~5 in urine was obtained without laborious sample preparation.[51] The stacking strategy involved an acidic separation buffer of 0.5 M formic acid at pH 2.2. The sample was injected for 300 s at 50 mbar or 16.2% of the total capillary length. After sample injection, a positive separation voltage was applied at the detection side of the capillary with BGS at the inlet.

Figure 5.4 shows the effect of sample pH on peptide separation. At a pH of 4.0 (a), the peptides were mostly cationic, which resulted in insufficient migration velocity change at the pH boundary created by the sample and BGS. The peaks were broad due to sample overloading. As the pH was gradually increased to 7.5 (c), the peptides became negatively charged, which caused a change in migration direction at the dynamic pH junction boundary and thus efficient sample stacking. The negatively charged peptides in the sample migrated towards the anode or capillary inlet. H^+ ions from the separation buffer titrated through the sample and caused the peptides to become positively charged. This reversal of the migration direction resulted in the focusing of the peptides and the focusing process was completed once the whole sample zone had been titrated with H^+. At a pH of 9.0 (d), the peak focusing and separation became compromised due to the strong alkalinity of the sample. The SEF for the peptides was ~100×. Similar enrichment was also achieved for larger peptides such as recombinant Arg–Gly–Asp–hirudin (M_w of ~7 kDa).[52]

Dynamic pH junctions can tolerate high ionic strength samples and the sample matrix can contain non-volatile electrolytes. The cell lysate obtained from HL1 cells was reconstituted in a non-volatile sample diluent (*i.e.*, 100 mM

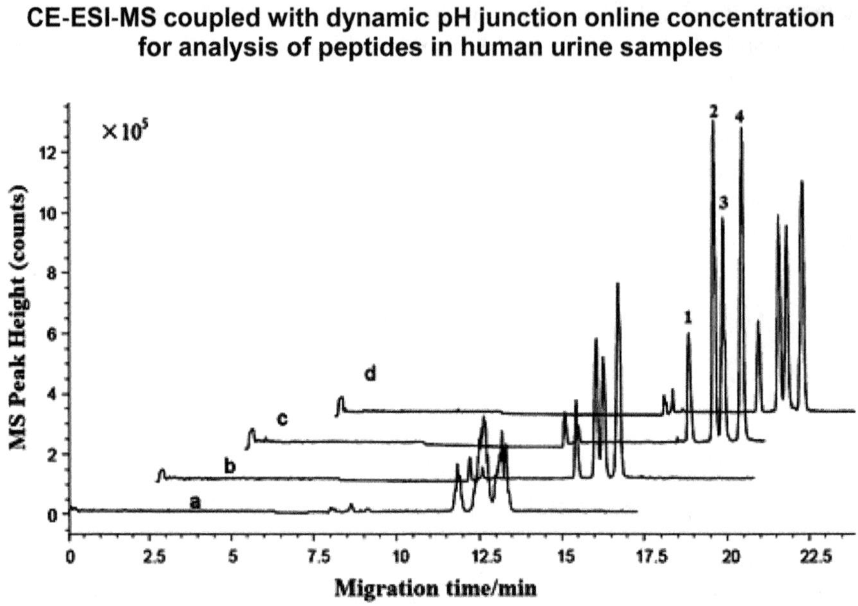

Figure 5.4 Effect of sample pH on stacking by dynamic pH junction. The separation buffer was 0.5 M formic acid at pH 2.2. The sample contained four synthetic peptides with a pI of ~5. The sample was prepared in 50 mM ammonium acetate with a pH of 4.0 (a), 6.0 (b), 7.5 (c), and 9.0 (d). Reprinted with permission from ref. 51. Copyright © 2010 WILEY-VCH Verlag GmbH & Co. KGaA, Weinheim.

borate buffer (pH 10.0) with 50% v/v acetonitrile).[53] The lysate was analysed for intracellular angiotensin II and five model peptides. The separation buffer was 1 M formic acid at pH 2.0. After stacking, the positively charged peptides migrated to the detector, while the non-volatile borate did not interfere with the detection. High SEF values of 3–4 orders of magnitude were obtained. Unknown nutrients and cationic metabolites in broth of *Escherichia coli* were directly analysed by diluting the broth in 200 mM ammonium acetate at pH 7.0.[54] The separation electrolyte was 1 M formic acid at pH 1.8. The combination of stacking and tandem-MS enabled sensitive *de novo* characterization and quantification of these metabolites at the nM-level.

5.4.4 Applications of Sweeping, AFMC and MSS

In sweeping, the concentration efficiency is in favour of analytes with high affinity to the pseudostationary phase, such as hydrophobic analytes. The sample is prepared in a matrix devoid of pseudostationary phase. When ESI is used, sweeping can be integrated into a partial filling technique to avoid source contamination. Hereby, only a fraction of the capillary contains the pseudostationary phase for sample concentration, while the other fractions contain a volatile separation buffer. In the first report on sweeping CE-MS by

Isoo and colleagues, the capillary was completely filled with micellar solution.[55] This was possible due to the use of an atmospheric pressure chemical ionisation interface that can tolerate SDS and non-volatile buffers. However, lower SEF values of ~100 were obtained for the environmental pollutants compared to the SEF of three orders of magnitude typically obtained by sweeping with UV. The reason for the lower SEF was due to the dilution of the ionisation source with the sheath liquid, incomplete analyte ionisation, and ion suppression. Partial filling was used for the analysis of dialkylphthalates and steroids using an ESI source.[56,57] For the three dialkylphthalates, the capillary was first flushed with 30 mM ammonium acetate at pH 7 followed by injection of a micellar solution (*i.e.*, 25 mM SDS in 10 mM ammonium acetate at pH 7 with 20% ACN) for partial filling with a plug of <1% of the total capillary length. This micellar plug was enough to achieve an SEF of ~100. A similar method was applied for steroids and compared to UV detection, where UV detection provided LODs 10-fold lower than those with MS.

In AFMC, this technique was used by a partial filling strategy with a micellar solution.[58] For instance, neutral alkylphenylketones and dialkylphthalates were on-line concentrated up to more than 100-fold. The capillary was first conditioned with 30 mM ammonium acetate at pH 5 containing 30% methanol. A 5.6 cm injection of micellar solution (*i.e.*, 20 mM SDS and 15 mM ammonium acetate at pH 5 in 30% methanol) was followed by injection of 5.6 cm of sample diluent (*i.e.*, 5 mM SDS in 100 mM ammonium acetate at pH 5), and 19.6 cm of sample prepared in the sample diluent. Upon application of a positive voltage at the detector side, the focusing boundary was formed at the sample zone boundary, which was closer to the capillary inlet side. The EOF velocity was faster than that for the analytes and SDS micelles, which resulted in transport of the micelles and analytes towards the MS. The analytes are separated and detected (devoid of surfactant) in the ESI-MS before the micelle/monomer zones.

In MSS, the opposite charge of the micelles to the analyte is beneficial for integration with MS detection. In co-EOF conditions, for instance in uncoated fused silica capillaries and for cationic analytes, the focused analytes migrate to the detector due to their apparent electrophoretic mobility. The cationic drugs pindolol and metoprolol were dissolved in a micellar diluent (*i.e.*, 10 mM SDS with 13 mM ammonium acetate at pH 5).[26] First a short plug of micellar diluent (*i.e.*, 15 s at 50 mbar) was injected prior to the sample (*i.e.*, 60 s at 50 mbar) in order to complete the analyte transport by the micelles to the MSSB. This long injection provided SEF values of 16 and 15 for pindolol and metoprolol, respectively. MSS was also demonstrated for sensitive detection at ng mL^{-1} concentrations for three cationic alkaloids in plasma using a volatile separation buffer.[59] The micellar diluent was 15 mM NH_4HCO_3 and 6 mM SDS at pH 7.5 and the separation buffer consisted of 50 mM NH_4HCO_3 at pH 7.5 with 40% acetonitrile. SEFs values in the range of 47 to 53 were obtained for berberine, palmatine and jatrorrhizine. For anionic metabolites, the potential of MSS is yet to be demonstrated in combination with MS, although MSS can be integrated in a similar fashion as for UV detection. MSS-CE-UV was demonstrated for anionic hypolipidaemic drugs using a capillary coating for EOF reversal and cationic CTAB as a micellar carrier.[60] The cationic coating for EOF reversal was hexadimethrine bromide.[61] The micellar diluent was 9 mM CTAB and 10 mM NH_4HCO_3 at pH 9.4 and the separation buffer was 50 mM NH_4HCO_3 at pH 9.4 with 50% methanol. This strategy provided ~10-fold enrichment after sample injection for 30 s at 50 mbar.

5.4.5 Applications of Hyphenated Stacking

The hyphenated techniques used in metabolomics studies are stacking by field-amplification/enhancement in conjunction with dynamic pH junctions, t-ITP, and MSS. In the stacking technique called pH-mediated stacking, the focusing of the analytes occurs due to a change in electrophoretic mobility at the boundary of the sample adjacent to an acidic or basic zone.[62,63] The cause of the mobility change is sample titration with H^+ and OH^-, and thus the proposed mechanism for pH-mediated stacking is similar to that at dynamic pH junctions. However, the neutralisation process was believed to decrease the sample zone conductivity, thereby resulting to field enhancement. With this in mind, pH-mediated stacking may be considered as a hyphenated technique with a dynamic pH junction and field enhanced sample stacking involved. On one hand, the presence of high levels of salts (having high electrophoretic mobilities) typically implemented in pH-mediated stacking could also lead to ITP effects. The contribution of each stacking mechanism to the overall enrichment is still unknown, but perhaps it is mainly by ITP. For example, the decrease in sample zone conductivity would not be more than 10× during neutralisation, and thus would not be the main stacking mechanism.

pH-mediated stacking was applied to cationic metabolites from bacterium *Desulfovibrio vulgaris Hildenborough*.[64] The cell extract was reconstituted in dilute methanolic separation buffer. A basic plug of 12.5% ammonium hydroxide was injected for 5 s at 25 mbar (0.1 cm injection length) followed by the sample for 160 s at 50 mbar (7.0 cm injection length), and finally 4 M formic acid in methanol and water (20 : 80, v/v) for 12 s at 50 mbar pressure for 12 s (0.5 cm injection length). The separation buffer was 1.6 M formic acid in methanol and water (20 : 80, v/v). The pH-mediated stacking provided a 20-fold signal enhancement and the detection limits for selected amino acids were 0.1 to 2 μM. Figure 5.5 shows the focusing by pH-mediated stacking (peak 3) of the model metabolite methionine sulfone. The injection volume was 134 nL. This long injection resulted in a broad peak without stacking (peak 1) or in a small peak (peak 2) when typical injection (2.5 nL) was performed.

pH-mediated stacking was also performed with the injection of ammonium hydroxide prior to the sample, but without the acidic plug after the sample. This simpler strategy was applied for urine samples to determine amines and peptides[65] and the metabolomic profile of breast cancer patients.[66]

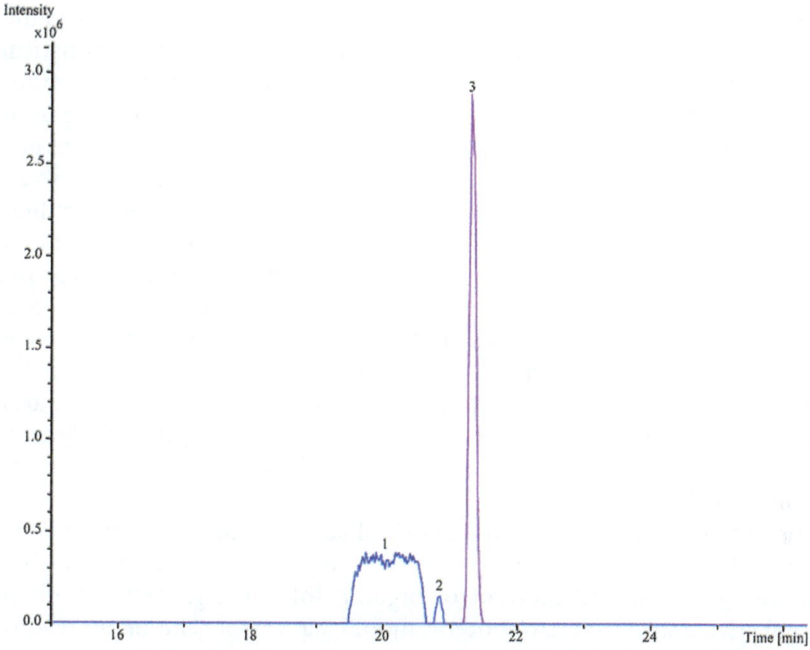

Figure 5.5 Extracted ion mass electropherograms of methionine sulfone from different sample injections. The sample injection volumes were 134 nL in (1 and 3) and 2.5 nL in (2). Peaks (1 and 2) and (3) were obtained in the absence of stacking and by pH-mediated stacking, respectively. Reprinted with permission from ref. 64. Copyright (2008) American Chemical Society.

For the latter, the profile of a patient undergoing chemotherapy could be differentiated from that of healthy subjects. Stacking provided a sensitivity improvement of one order of magnitude. For the peptide and amine samples, a basic plug of 12.5% ammonium hydroxide solution was injected prior to the sample. A long hydrodynamic injection (*i.e.*, 90 s at 90 mbar) of the urine sample diluted in separation buffer (*i.e.*, 1 M formic acid at pH 1.8) followed. To obtain a stable ESI and an adequate EOF velocity under these acidic conditions, the capillary was coated with negatively charged polymer poly(vinyl sulfonate). The EOF reduced the method run time, which was beneficial to achieve a higher clinical sample throughput.

In EKS, FESI is followed by tITP. The basic steps in EKS include conditioning the capillary with BGS containing the leading ion, electrokinetic injection of the low conductivity sample, and injection of TE (by hydrodynamic or electrokinetic injection). The choice of volatile buffer systems, however, which provide a suitable LE and TE is somewhat limited.[67] A MS compatible system for the analysis of five anionic hypolipidaemic drugs in wastewater was demonstrated.[61] To obtain an anodic EOF, the capillary was flushed in-between the runs with a cationic polymer solution of 0.1% hexadimethrine bromide. The wastewater samples were extracted and reconstituted in purified water. The capillary was conditioned with a separation buffer of 60 mM ammonium bicarbonate containing 60% methanol, where the ammonium acted as the leading ion. Then, the sample was injected for 170 s at −17 kV prior to injection of the TE (*i.e.*, 1 mM 3-(cyclohexylamino)-1-propanesulphonic acid). A pressure of 5 mbar at the capillary inlet was applied to augment the flow to the detector. The five drugs were baseline separated in 20 min and SEFs of three orders of magnitude were obtained. SEFs of 1000–10000-fold were also obtained by EKS for the analysis of a tryptic digest of β-lactoglobulin.[68] This study used UV detection; however, the buffer and stacking system were compatible with MS detection. In an EOF-suppressed mode, the capillary was first conditioned with the TE (10% acetic acid) followed by injection of the leading ion (*i.e.*, ammonium ion), and finally sample injection. High SEFs values were obtained because of the long injection for 20 min at 30 kV.

For FESI-MSS, it is important to establish the conditions for FESI prior to MSS. The sample must be prepared in or reconstituted with a diluent containing a sufficient amount of organic solvent (*e.g.*, >50% methanol) for effective analyte electrophoretic mobility reversal. The organic solvent content also contributes to the decrease in the conductivity of the sample, which is desired in stacking with field enhancement. Once the FESI conditions have been selected, MSS is then implemented by injecting a micellar plug before FESI. Alternatively, an organic solvent rich plug may be injected before the sample that may not contain organic solvent. A surfactant with a volatile co-migrating ion (*e.g.*, ammonium lauryl sulfate) can be used instead of SDS where the sodium co-ion might affect MS detection.[69] FESI-MSS provides SEFs of more than two orders of magnitude. Trace levels of antibiotics in seawater were determined by two sample preconcentration steps.[70] The penicillin and sulfonamide antibiotics were ~50 enriched *via* SPE. The SPE extract was then reconstituted in 75% methanol. The anionic surfactant CTAB (*i.e.*, 5 mM CTAB in 50 mM ammonium acetate at pH 9) was injected for 30 s at 50 mbar prior to sample injection for 180 s at −10 kV. An anodic coating of the capillary was necessary to create co-EOF conditions, which supported the electrospray during injection. FESI-MSS provided SEFs of 1629–3328. Tamoxifen and four tamoxifen metabolites from plasma samples were analysed in non-aqueous CE-UV.[71] A 214–625-fold enhancement and LODs of 0.15–0.7 ng mL^{-1} for tamoxifen and its metabolites were obtained. This method could be implemented for MS by applying a partial filling technique with a crown ether modified separation buffer (*i.e.*, 7.5 mM deoxycholic acid/15 mM acetic acid/1 mM 18-crown-6 in methanol). Five antipsychotic drugs in urine samples were analysed *via* FESI-MSS-CZE-UV. More than three orders of magnitude SEFs were obtained. The pretreated urine samples were reconstituted in 5 mM phosphoric acid with 40% acetonitrile. This non-volatile phosphate buffer could be replaced with a volatile ammonium acetate or formate buffer.

5.5 Conclusion

Stacking in CE and CE-ESI-MS is a promising strategy for metabolomics. The analysis of trace metabolites and even targeted metabolites can be significantly improved. Multiple orders of magnitude signal enhancements have been achieved and this was important to enable trace metabolite detection, such as is typically needed for minute and precious biological samples (*e.g.*, single cell analysis). The stacking procedures were mainly limited by the use of ESI-MS compatible reagents, or the use of partial filling approaches that prevented the entry of the non-volatile component into the MS with the targeted analytes. Sample preparation prior to CE is perhaps the most problematic step, since the sample containing mainly highly polar metabolites needs to be modified in order to efficiently perform the stacking. As metabolites are often charged or ionisable species, the implementation of stacking by ITP, field amplification, and pH effects has been easily achieved. Stacking by chromatographic effects (*i.e.*, sweeping, AFMC, and MSS) is yet to be fully explored for metabolomics, but these stacking methods have the promise to also provide sample clean-up during the stacking process. Indeed, some stacking methods have been used to clean-up the sample and thereby minimise off-line sample treatment, *e.g.*, by two-dimensional systems. The capability of stacking for both cleaning-up and concentrating the sample supports efforts to obtain a metabolic profile with more depth and clarity, which is vital in the endeavour to understand the complexity of the metabolome. A rigorous study on the use of different stacking mechanisms for commonly encountered metabolic samples, which will provide a clear method development pathway for a particular sample matrix, will be useful for the future of CE-ESI-MS in metabolomics.

Acknowledgements

AW is grateful to the Swiss National Science Foundation SNSF for the postdoc fellowship (P2SKP2_168309).

References

1. P. Gebauer and P. Boček, *Electrophoresis*, 2009, **30**, S27–S33.
2. S. L. Simpson, J. P. Quirino and S. Terabe, *J. Chromatogr. A*, 2008, **1184**, 504–541.
3. M. C. Breadmore, A. Wuethrich, F. Li, S. C. Phung, U. Kalsoom, J. M. Cabot, A. I. Shallan, A. S. Abdul Keyon, H. H. See, M. Dawod and J. P. Quirino, *Electrophoresis*, 2017, **38**, 33–59.
4. W. B. Dunn and D. I. Ellis, *TrAC, Trends Anal. Chem.*, 2005, **24**, 285–294.
5. T. Soga, Y. Ohashi, Y. Ueno, H. Naraoka, M. Tomita and T. Nishioka, *J. Proteome Res.*, 2003, **2**, 488–494.
6. F. Kohlrausch, *Ann. Phys.*, 1897, **298**, 209–239.
7. F. E. P. Mikkers, F. M. Everaerts and T. P. E. M. Verheggen, *J. Chromatogr. A*, 1979, **169**, 11–20.

8. F. E. P. Mikkers, F. M. Everaerts and T. P. E. M. Verheggen, *J. Chromatogr. A*, 1979, **169**, 1–10.
9. R. Chien, *Anal. Biochem.*, 1991, **63**, 2866–2869.
10. R.-L. Chien and D. S. Burgi, *J. Chromatogr.*, 1991, **559**, 153–161.
11. P. Tůma, M. Šustková-Fišerová, F. Opekar, V. Pavlíček and K. Málková, *J. Chromatogr. A*, 2013, **1303**, 94–99.
12. J. Kim, M. S. Chun, K. Choi and D. S. Chung, *Electrophoresis*, 2009, **30**, 1046–1051.
13. C. W. Wang and G. R. Her, *Electrophoresis*, 2014, 1251–1258.
14. X. Huang, M. J. Gordon and R. N. Zare, *Anal. Chem.*, 1988, **60**, 375–377.
15. T. P. E. M. Verheggen, A. C. Schoots and F. M. Everaerts, *J. Chromatogr. A*, 1990, **503**, 245–255.
16. J. L. Beckers and F. M. Everaerts, *J. Chromatogr. A*, 1990, **508**, 19–26.
17. F. Foret, E. Szökö and B. L. Karger, *Electrophoresis*, 1993, **14**, 417–428.
18. J. B. Kim, P. Britz-McKibbin, T. Hirokawa and S. Terabe, *Anal. Chem.*, 2003, **75**, 3986–3993.
19. R. Aebersold and H. D. Morrison, *J. Chromatogr. A*, 1990, **516**, 79–88.
20. P. Britz-McKibbin and D. D. Y. Chen, *Anal. Chem.*, 2000, **72**, 1242–1252.
21. J. P. Quirino and S. Terabe, *Science*, 1998, **282**, 465–469.
22. A. M. van Wijk, H. A. G. Niederländer, M. D. van Ogten and G. J. de Jong, *Anal. Chim. Acta*, 2015, **874**, 75–83.
23. Y. Liu, B. Wu, P. Wang and S. A. Shamsi, *Electrophoresis*, 2016, **37**, 913–923.
24. D. Abd El-Hady and H. M. Albishri, *Food Chem.*, 2015, **188**, 551–558.
25. J. P. Quirino and P. R. Haddad, *Anal. Chem.*, 2008, **80**, 6824–6829.
26. J. P. Quirino, *J. Chromatogr. A*, 2009, **1216**, 294–299.
27. T. Hirokawa, H. Okamoto and B. Gas, *Electrophoresis*, 2003, **24**, 498–504.
28. J. Quirino and S. Terabe, *Anal. Chem.*, 2000, **72**, 1023–1030.
29. W. Grochocki, M. J. Markuszewski and J. P. Quirino, *J. Chromatogr. A*, 2015, **1424**, 111–117.
30. W. Grochocki, M. J. Markuszewski and J. P. Quirino, *Anal. Methods*, 2016, **8**, 1216–1221.
31. G. Bonvin, J. Schappler and S. Rudaz, *J. Chromatogr. A*, 2012, **1267**, 17–31.
32. D. S. Burgi and R. Chien, *Anal. Chem.*, 1991, **2047**, 2042–2047.
33. M. Moini and B. Martinez, *Rapid Commun. Mass Spectrom.*, 2014, **28**, 305–310.
34. P. Pantůčková, P. Gebauer, P. Boček and L. Křivánková, *Electrophoresis*, 2009, **30**, 203–214.
35. J. Hou, J. Zheng, S. A. A. Rizvi and S. A. Shamsi, *Electrophoresis*, 2007, **28**, 1352–1363.
36. W. K. Goetzinger and H. Cai, *J. Chromatogr. A*, 2005, **1079**, 372–381.
37. C. Palmer and J. Quirino, *Anal. Chem.*, 2010, **82**, 4046–4054.
38. C. Nilsson, P. Viberg, P. Spégel, M. Jömtén-Karlsson, P. Petersson and S. Nilsson, *Anal. Chem.*, 2006, **78**, 6088–6095.
39. T. Lapainis, S. S. Rubakhin and J. V. Sweedler, *Anal. Chem.*, 2009, **81**, 5858–5864.

40. C. Desiderio, D. V. Rossetti, I. Messana, B. Giardina and M. Castagnola, *Electrophoresis*, 2010, **31**, 1894–1902.
41. T. Soga, Y. Ueno, H. Naraoka, Y. Ohashi, M. Tomita and T. Nishioka, *Anal. Chem.*, 2002, **74**, 2233–2239.
42. H. Katayama, Y. Ishihama and N. Asakawa, *Anal. Chem.*, 1998, **70**, 2254–2260.
43. M. C. Gulersonmez, S. Lock, T. Hankemeier and R. Ramautar, *Electrophoresis*, 2016, **37**, 1007–1014.
44. R. Gottardo, I. Mikšík, Z. Aturki, D. Sorio, C. Seri, S. Fanali and F. Tagliaro, *Electrophoresis*, 2012, **33**, 599–606.
45. V. Hezinová, Z. Aturki, K. Klepárník, G. D'Orazio, F. Foret and S. Fanali, *Electrophoresis*, 2012, **33**, 653–660.
46. E. E. K. Baidoo, M. R. Clench, R. F. Smith and L. W. Tetler, *J. Chromatogr. B: Anal. Technol. Biomed. Life Sci.*, 2003, **796**, 303–313.
47. J. P. Quirino and M. C. Breadmore, *J. Sep. Sci.*, 2012, **35**, 60–65.
48. J. M. Busnel, B. Schoenmaker, R. Ramautar, A. Carrasco-Pancorbo, C. Ratnayake, J. S. Feitelson, J. D. Chapman, A. M. Deelder and O. A. Mayboroda, *Anal. Chem.*, 2010, **82**, 9476–9483.
49. R. Ramautar, J. M. Busnel, A. M. Deelder and O. A. Mayboroda, *Anal. Chem.*, 2012, **84**, 885–892.
50. J. Piešťanský, K. Maráková, L. Veizerová, J. Galba and P. Mikuš, *Anal. Chim. Acta*, 2014, **826**, 84–93.
51. H. Ye, S. Xia, W. Lin, L. Yu, X. Xu, C. Zheng, X. Liu and G. Chen, *Electrophoresis*, 2010, **31**, 3400–3406.
52. S. Su and Y. Yu, *J. Chromatogr. A*, 2009, **1216**, 1490–1495.
53. M. N. Hasan, S. H. Park, E. Oh, E. J. Song, E. Ban and Y. S. Yoo, *J. Sep. Sci.*, 2010, **33**, 3701–3709.
54. R. Lee, A. S. Ptolemy, L. Niewczas and P. Britz-McKibbin, *Anal. Chem.*, 2007, **79**, 403–415.
55. K. Isoo, K. Otsuka and S. Terabe, *Electrophoresis*, 2001, **22**, 3426–3432.
56. J. P. Quirino, *Electrophoresis*, 2011, **32**, 665–668.
57. L. K. Amundsen, J. T. Kokkonen and H. Sirén, *J. Sep. Sci.*, 2008, **31**, 803–813.
58. J. P. Quirino and P. R. Haddad, *Electrophoresis*, 2009, **30**, 1670–1674.
59. S. Zhang, R. Ma, X. Yang, C. Wang and Z. Wang, *J. Chromatogr. B: Anal. Technol. Biomed. Life Sci.*, 2012, **906**, 41–47.
60. A. M. Guidote and J. P. Quirino, *J. Chromatogr. A*, 2010, **1217**, 6290–6295.
61. M. Dawod, M. C. Breadmore, R. M. Guijt and P. R. Haddad, *Electrophoresis*, 2010, **31**, 1184–1193.
62. Y. Xiong, S. R. Park and H. Swerdlow, *Anal. Chem.*, 1998, **70**, 3605–3611.
63. Y. Zhao and C. E. Lunte, *Anal. Chem.*, 1999, **71**, 3985–3991.
64. E. E. K. Baidoo, P. I. Benke, C. Neusüss, M. Pelzing, G. Kruppa, J. A. Leary and J. D. Keasling, *Anal. Chem.*, 2008, **80**, 3112–3122.
65. R. Ramautar, O. A. Mayboroda, R. J. E. Derks, C. van Nieuwkoop, J. T. van Dissel, G. W. Somsen, A. M. Deelder and G. J. de Jong, *Electrophoresis*, 2008, **29**, 2714–2722.

66. L. Yu, C. Jiang, S. Huang, X. Gong, S. Wang and P. Shen, *Clin. Biochem.*, 2013, **46**, 1065–1073.
67. M. Dawod and D. S. Chung, *J. Sep. Sci.*, 2011, **34**, 2790–2799.
68. J. M. Busnel, N. Lion and H. H. Girault, *Electrophoresis*, 2008, **29**, 1565–1572.
69. A. Wuethrich, P. R. Haddad and J. P. Quirino, *Chirality*, 2014, **26**, 734–738.
70. A. Wuethrich, P. R. Haddad and J. P. Quirino, *Electrophoresis*, 2016, **37**, 1139–1142.
71. L. Y. Thang, H. H. See and J. P. Quirino, *Electrophoresis*, 2016, **37**, 1166–1169.

CHAPTER 6

On-line Solid-phase Extraction Capillary Electrophoresis–Mass Spectrometry and Data Analysis Approaches for Metabolomics

L. PONT, V. SANZ-NEBOT, J. BARBOSA AND F. BENAVENTE*

Department of Chemical Engineering and Analytical Chemistry, Institute for Research on Nutrition and Food Safety (INSA-UB), University of Barcelona, Martí i Franquès 1-11, 3rd floor, 08028, Barcelona, Spain
*E-mail: fbenavente@ub.edu

6.1 Introduction

Metabolomics has seen a huge expansion over the last decades. Its major purpose is the comprehensive analysis of low molecular mass compounds from biological samples.[1,2] In metabolomics studies, two different approaches can be distinguished, *i.e.* the targeted and untargeted approach. The targeted approach is focused on the analysis of preselected metabolites in a biological sample.[3,4] The untargeted approach involves the profiling of biological samples without having *a priori* knowledge on the nature and the identity of the measured metabolites. This strategy may be referred to as metabolic

New Developments in Mass Spectrometry No. 6
Capillary Electrophoresis–Mass Spectrometry for Metabolomics
Edited by Rawi Ramautar

Published by the Royal Society of Chemistry, www.rsc.org

profiling or fingerprinting and it is a powerful platform to elucidate novel biomarkers, which can be later approached through targeted analysis, and gain a broad insight into disease pathogenesis.[5,6]

Different techniques are currently used for targeted and untargeted metabolomics, including nuclear magnetic resonance (NMR),[7] gas chromatography–mass spectrometry (GC-MS),[8] liquid chromatography-mass spectrometry (LC-MS)[9] and capillary electrophoresis–mass spectrometry (CE-MS).[10,11] In CE-MS, the CE mode with the best performance for metabolite separation and detection in positive or negative electrospray ionisation (ESI) MS modes is capillary zone electrophoresis (CZE). This CE mode is particularly suited in CE-MS for the analysis of a wide range of charged polar low molecular mass compounds, such as many metabolites. In addition to high resolution capabilities, CE has many other desirable characteristics, including a separation mechanism based on electrophoretic mobility (which is related to the ion charge-to-radius ratio), low consumption of sample and reagents, reduced analysis times, instrumental simplicity and full automation.[10,11] However, the low concentration sensitivity for most analytes is very often a limitation that hinders its more widespread application.[12–14] The use of more selective and sensitive detectors (*e.g.* laser induced fluorescence (LIF) and MS detection) is often not enough to decrease the limits of detection (LODs). Therefore, CE-MS has often been combined with different electrophoretic and chromatographic techniques for the on-line preconcentration of the analytes after the injection of a large volume of sample.[12–16] Electrophoretic preconcentration based on stacking or transient isotachophoresis (tITP) methods has been widely used with, in some cases, preconcentration factors higher than 10 000-fold.[12,17] However, these methods are, in general, analyte- and matrix-dependent, and show limited reproducibility. Chromatographic preconcentration techniques using on-line solid-phase extraction capillary electrophoresis (SPE-CE) show a more general applicability and can also be very efficient yielding high concentration factors, on-demand selectivity and better reproducibility.[13–16]

Typically, in SPE-CE a microcartridge inserted near the inlet of the separation capillary contains a sorbent to selectively retain the analyte, hence enabling the introduction of a large volume of sample (~50–100 μL). After washing to remove non-retained molecules, the retained analyte is eluted in a small volume of an appropriate solution (~25–50 nL), resulting in sample clean-up and concentration enhancement before electrophoretic separation and detection. As extraction occurs immediately before the electrophoretic separation without human handling, many pioneering authors in the field have referred to the coupling as "on-line" since the late 80s.[15,18,19] However, other authors have preferred the term "in-line" because the microcartridge is fully integrated into the separation capillary.[13,14] In the most common microcartridge configuration, the sample is introduced in the direction of separation, and this can be detrimental for the separation capillary with certain complex samples. Cruciform and staggered microcartridge configurations have been described to overcome this drawback, where the sample is introduced through a transport tube, orthogonal to the separation direction. These new configurations potentially allow no carry-over of substances into the separation capillary and hence extended lifetime of the capillaries.[20,21] However, they are more difficult to construct than the typical microcartridges presented in this chapter because they require control by valves of the fluid flow in a microfluidics set-up.

Nowadays, conventional chromatographic sorbents (reversed phase, ionic exchange, hydrophilic interaction liquid chromatography (HILIC), size exclusion, *etc.*) and other affinity sorbents (antibodies, aptamers, lectins, immobilised metal affinity, *etc.*) can be applied to SPE-CE.[13,14] However, conventional chromatographic sorbents are the most common ones, as they are commercially available, do not require in-house preparation and fulfil most of the requirements for optimum performance. These sorbents provide a large active surface area, without interfering with the on-line electrophoretic separation. Furthermore, they are commercially available, have been optimised for the analysis of a wide variety of compounds and can be purchased at a reasonable price. In particular, reversed phase silica-based and polymeric sorbents (*e.g.* C_{18} or HLB) are widely recognized for their good extraction capabilities, being the most common chromatographic sorbents used for SPE-CE applications involving low molecular mass compounds (*e.g.* peptides[15,16,22–24] and drugs of abuse and their metabolites[25–27]) in different biological samples.

We demonstrated the potential of C_{18}-SPE-CE-MS, after a sample pretreatment based on precipitation with acetonitrile and centrifugal filtration, to lower the LODs for the analysis of opioid peptides in plasma samples up to 1000-fold compared to the values previously obtained by CE-MS.[22,23] This C_{18}-SPE-CE-MS method was applied later for untargeted metabolomics in plasma samples from mice genetically modified to develop Huntington's disease (HD),[28] an inherited neurodegenerative disorder.[29,30] In this metabolomics study, the limited selectivity of the C_{18} sorbent was an advantage to obtain comprehensive and sensitive characteristic metabolite profiles to differentiate between control and HD plasma samples, and to study the onset and progression of the disease. There are other less hydrophobic chromatographic sorbents than C_{18}, such as C_2, HILIC or ion exchange, which could be excellent for the comprehensive and sensitive analysis of charged polar low molecular mass compounds by SPE-CE-MS. However, to the best of our knowledge, they have been scarcely applied in untargeted metabolomics, neither on-line nor off-line before CE-MS, or in other SPE-CE-MS applications.[31,32] This is probably because the off-line sample preparation procedures traditionally applied in untargeted metabolomics before CE-MS, which include liquid–liquid extraction, centrifugal filtration through molecular mass cut-off filters and even drop dialysis, perform reasonably well.[6,10,11] However, combination of these traditional sample pretreatments with SPE-CE-MS may be regarded as a novel extremely selective and sensitive tool to get a deeper and complementary insight into the metabolome of biological fluids, cells or tissues.

This chapter presents a workflow for untargeted SPE-CE-MS-based metabolomics, which includes sample pretreatment, SPE-CE-MS analysis, feature finding, statistical analysis and metabolite identification. A comparative metabolomics study in mouse plasma samples to investigate the disease progression of HD is used as a case study.

6.2 Experimental Procedures

All solvents, reagents and standard peptides were of analytical reagent or higher grade. The background electrolyte (BGE) was 50 mM of HAc and 50 mM of HFor (pH 3.50). The sheath liquid was 60 : 40 (v/v) 2-propanol : water with 0.05% (v/v) HFor.[28,33]

6.2.1 Mouse Blood Plasma Extraction

Plasma samples from male wild-type mice (wt) and R6/1 transgenic mice (B6CBA background) of different ages (8, 12 and 30 weeks, early, middle and late disease stage, respectively) were kindly supplied by the Department of Cellular Biology, Immunology and Neurosciences (Faculty of Medicine, University of Barcelona). Blood from the mice was collected by cardiac puncture in standard clinical vials and placed on ice. Plasma was separated from the blood cells, pooled, deposited into polyethylene tubes and frozen at −20 °C. All animal procedures were approved by the CEEA committee of the University of Barcelona and were in accordance with the European Communities Council Directive (2010/63/EU).

6.2.2 Sample Preparation

The optimised sample pretreatment proposed for the analysis of low molecular mass compounds in plasma samples by C_{18}-SPE-CE-MS consisted of protein precipitation with acetonitrile followed by centrifugal filtration with 10 000 M_r cut-off cellulose acetate filters (Amicon® Ultra-0.5, Millipore) passivated with 5% (v/v) PEG in water.[33] For precipitation with acetonitrile, 1200 µL of cold solvent was added to 200 µL of plasma sample. The mixture was then vortexed for 30 s and kept at −20 °C for 1 h. The supernatant was collected after centrifugation of the mixture at 700*g* for 8 min. The solid residue was rinsed with 100 µL of acetonitrile. After centrifugation of the mixture at 1200*g* for 8 min, the supernatant was pooled with the first one. Finally, the micropipette tip used to deliver the supernatants was rinsed with 100 µL of acetonitrile and the washing volume was also collected. The pooled solution was evaporated to dryness at room temperature under a stream of air and the solid residue was reconstituted with 100 µL of water. The reconstituted sample was subjected to centrifugal filtration. All the centrifugation steps were performed at a centrifugal force of 11 000*g*. First, the filter was washed with water for 15 min and the filtrate was discarded. Then, the reconstituted sample was centrifuged for 10 min. The filtration residue was washed two times for 10 and 8 min with 50 μL of water. In all cases, the filtrates were pooled together and sufficient water was added until the final volume was adjusted to 200 μL.

For passivation of the 10 000 M_r cut-off cellulose acetate filters, 500 μL of PEG (5% (v/v)) was pipetted into the reservoir of the centrifugal device, capped, and allowed to stand for 24 h at room temperature. The reservoir was uncapped and rinsed thoroughly with tap water. A volume of 500 μL of distilled water was added to remove any residue and the centrifuge was then spun to dead stop (30 min at 11 000*g*). The remaining water was eliminated by inverting the reservoir in a vial and spinning once more at a reduced centrifugal force (2 min at 300*g*).

6.2.3 Apparatus and Procedures: SPE-CE-MS

6.2.3.1 Construction of an SPE-CE Microcartridge

The construction of the SPE microcartridge was carried out, as described elsewhere,[34] using a silica-based C_{18} sorbent obtained from commercial Sep-pak® cartridges (Waters, Milford, MA, USA). For SPE-CE-MS, a 72 cm total length (L_T) × 75 mm internal diameter (id) × 365 mm outer diameter (od) fused-silica capillary (Polymicro Technologies, Phoenix, AZ, USA) was activated off-line by flushing for 20 min with aqueous 1.0 M NaOH and 15 min with water. It was then cut into two pieces of 7.5 and 64.5 cm in order to insert the microcartridge. The fabrication process was monitored under a stereomicroscope. A microfrit (100 μm L_T), obtained from a small fragment of the original filters found in the commercial SPE cartridges, was introduced into the body of the microcartridge (Figure 6.1A), *i.e.* a small piece of fused-silica capillary (7 mm L_T × 250 mm id × 365 mm od, Polymicro Technologies). The body of the microcartridge was coupled to the 7.5 cm piece of capillary using a 0.5 cm Tygon® plastic tube (Figure 6.1B). Therefore, it was easily vacuum-filled with the sorbent from the SPE cartridge. When the tube was almost completely full, a second microfrit (100 μm) was inserted into the end side (Figure 6.1C). The column was extended to full length with the 64.5 cm piece of capillary, which was fitted using another Tygon® plastic tube (Figure 6.1D). The joining ends connected by the plastic sleeves had to be cleanly cut in order to minimise dead volumes. Before the analyses, the system was checked for abnormal flow restriction using a plastic syringe with an appropriate connector.

6.2.3.2 SPE-CE-MS Methodology

The C_{18}-SPE-CE-MS capillaries were first conditioned by consecutive flushes of water (1 min), methanol (1 min), water (1 min) and BGE (3 min) at 930 mbar. Standard peptide mixture (Dyn A, End 1 and Met) or mouse plasma samples were then introduced at 930 mbar for 10 min (approximately 60 μL[35]).

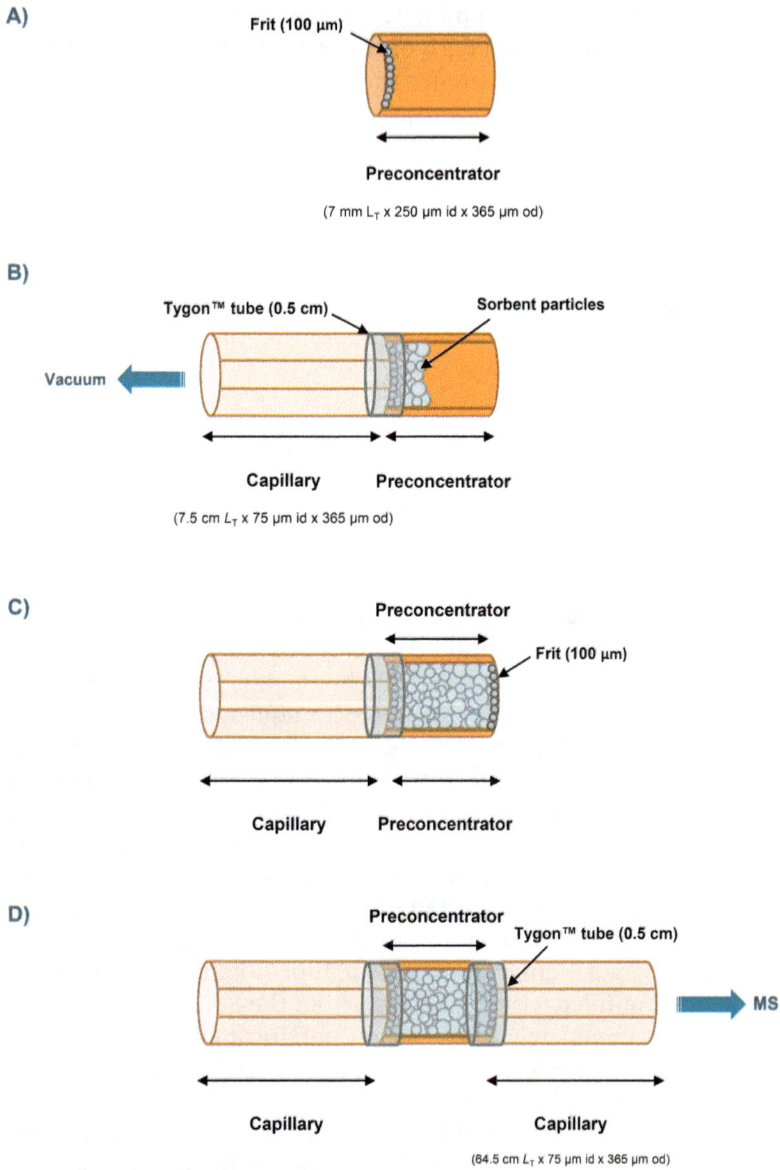

Figure 6.1 Construction procedure of a double-frit SPE-CE microcartridge or analyte preconcentrator.

A final rinse with the BGE (2 min at 930 mbar) eliminated non-retained molecules and equilibrated the capillary before the elution. Retained analytes were eluted by injecting a solution of 60 : 40 (v/v) methanol : water with 50 mM HAc and 50 mM HFor at 50 mbar for 10 s (approximately 50 nL[35]). Separation was carried out at 25 °C by applying a voltage of 17 kV (normal polarity, cathode in the outlet). Between runs, the capillary was rinsed for 2 min with water and 2 min with acetonitrile, in order to avoid carry-over between consecutive analyses. All CE-MS experiments were performed in positive ion mode in an HP3D CE system coupled with an orthogonal G1603A sheath flow interface to a 6220 oa-TOF LC/MS spectrometer (Agilent Technologies, Waldbronn, Germany).[28,33] The sheath liquid was delivered at a flow rate of 3.3 μL min^{-1} by a KD Scientific 100 series infusion pump (Holliston, MA, USA).

6.2.4 Multivariate Data Analysis: MCR-ALS and PLS-DA

Most of the calculations and data analysis were performed using MATLAB R2013a (The Mathworks Inc. Natick, MA, USA). PLS Toolbox 7.3.1 (Eigenvector Research Inc., Wenatchee, WA, USA) was used for the PLS-DA and VIP calculations and the MCR-ALS toolbox was used for the resolution of the electrophoretic and mass spectral metabolite profiles from the full MS scan augmented data matrices.

MCR-ALS is a chemometric method especially useful to analyse multicomponent systems with strongly overlapping contributions, such as those present in CE separations, where the electrophoretic behaviour of metabolites is rather similar.[36,37] In the case of SPE-CE-MS, full scan MS data matrix **D** contains the experimental mass spectra at all migration times in its rows and the electropherograms at all *m*/*z* values in its columns. MCR-ALS analysis of the data matrix **D**, following a bilinear model, gives two factor matrices, **C** and $\mathbf{S}^T$, as shown in eqn (6.1):

$$\mathbf{D} = \mathbf{C}\mathbf{S}^T + \mathbf{E} \tag{6.1}$$

where matrix **C** contains the electrophoretic profiles of the resolved contributions (components), matrix $\mathbf{S}^T$ contains the corresponding mass spectra of the resolved contributions, and matrix **E** contains the residuals unexplained by the model. Different samples can be simultaneously analysed and compared by MCR-ALS using a column-wise augmented data matrix configuration, as shown in eqn (6.2), to obtain similar information from every sample:

$$\mathbf{D}_{\mathbf{aug}} = \begin{bmatrix} \mathbf{D}_1 \\ \vdots \\ \mathbf{D}_{15} \end{bmatrix} = \begin{bmatrix} \mathbf{C}_1 \\ \vdots \\ \mathbf{C}_{15} \end{bmatrix} \mathbf{S}^T + \begin{bmatrix} \mathbf{E}_1 \\ \vdots \\ \mathbf{E}_{15} \end{bmatrix} = \mathbf{C}_{\mathbf{aug}}\mathbf{S}^T + \mathbf{E}_{\mathbf{aug}} \tag{6.2}$$

In this case, the electrophoretic peaks resolved in matrix $\mathbf{C}_{\mathbf{aug}}$ are allowed to vary in position (shifts) and shape among samples because the only requirement for a proper resolution is that the resolved spectra are the same for the common constituents in the different considered samples.[36]

The number of components in MCR-ALS analysis was obtained by single value decomposition, SVD,[38] and the initial estimates were obtained using a purest variable detection method based on the SIMPLISMA algorithm.[39] ALS optimisation was performed under non-negativity constraints for electrophoretic ($\mathbf{C}_{\mathbf{aug}}$) and spectral ($\mathbf{S}^T$) profiles, and spectral normalization (equal height[40]).

Peak areas obtained for the resolved features were autoscaled (mean centred and scaled to unit standard deviation) and PLS-DA[41] was performed later to maximise class separation and rapidly classify the different samples, as well as to identify which components were the most significant to discriminate between the different classes taking into account the variable importance in the projection (VIP)[42] scores of the components in the PLS-DA model. A leave-one-out cross validation of the PLS-DA model was performed during model optimisation.[42]

6.3 Results and Discussion

The workflow discussed in this chapter for untargeted SPE-CE-MS-based metabolomics is summarised in Figure 6.2 and includes sample pre-treatment, SPE-CE-MS analysis, feature finding, statistical analysis and metabolite identification. In particular, this chapter will show, as an example, the potential of C_{18}-SPE-CE-MS to analyse low molecular mass compounds in plasma samples from wt and HD mice in combination with advanced chemometric data analysis tools, such as MCR-ALS and PLS-DA, for the identification of potential metabolite biomarkers involved in HD progression.[28]

6.3.1 Plasma Pretreatment

In SPE-CE-MS-based metabolomics of plasma samples, the sample preparation procedure comprises typically the separation of low molecular mass compounds from large molecules (basically proteins, but also polypeptides and lipids) by deproteinization techniques (see Figure 6.2A).[22,23,28,33] Specifically, solvent precipitation and centrifugal filtration have been widely employed because of their good recoveries, simplicity, high throughput and low cost. Moreover, the sequential combination of both methods extends the centrifugal filter lifespan and produces cleaner extracts, which are particularly required for C_{18}-SPE-CE-MS.[22,23,33]

We have proposed a double-step sample pretreatment for the analysis of low molecular mass compounds in human plasma, which was optimised and validated for the analysis of opioid peptides (*i.e.* Dyn A, End 1 and Met) by CE-MS and C_{18}-SPE-CE-MS. The method consisted of precipitation with acetonitrile followed by centrifugal filtration.[22,23] The first step of protein precipitation was enough before CE-MS (the recoveries were 49, 91 and 102% for Dyn A, End 1 and Met, respectively), but not before C_{18}-SPE-CE-MS, which required the subsequent centrifugal filtration for an improved sample matrix clean-up.[22] The use of 10 000 M_r cut-off cellulose acetate Microcon Amicon® filters (Millipore, Bedford, MA, USA) provided the best peptide recoveries using the complete double-step pretreatment (26, 71 and 72% for Dyn A, End 1, and Met, respectively).[23] These centrifugal filters with the membrane in the horizontal position for direct flow filtration can be substituted with

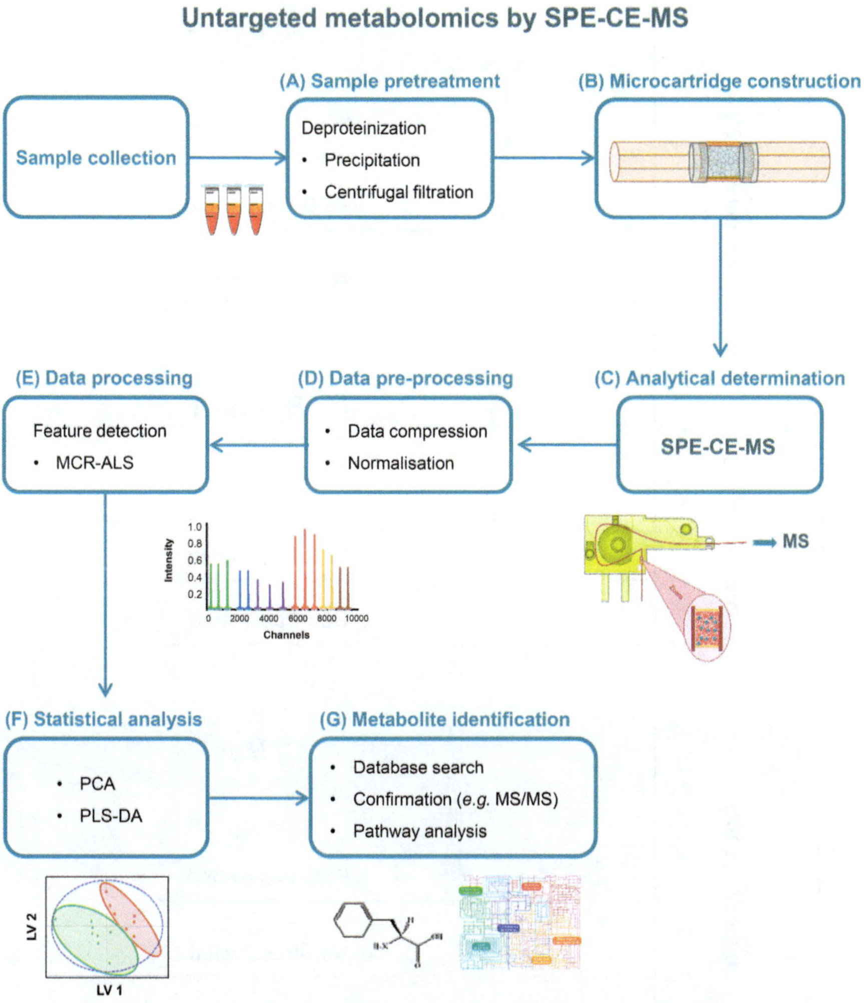

Figure 6.2 Experimental workflow for untargeted metabolomics employing SPE-CE-MS and advanced chemometric data analysis tools.

similar recoveries (23, 69 and 86% for Dyn A, End 1 and Met, respectively) by others with the membrane in the vertical position for tangential flow filtration (Amicon® Ultra-0.5 filters, Millipore).[33]

However, these filters need to be passivated with PEG before use to avoid peptide adsorption of some of the opioid peptides on the inner walls of the plastic sample reservoir (Figure 6.3). Using this double-step sample pretreatment, the LODs for the studied peptides by C_{18}-SPE-CE-MS in plasma samples could be decreased down to 10 ng mL^{-1} for Dyn A and 1 ng mL^{-1} for End 1 and Met, equivalent to 100- and 1000-fold the LODs obtained using CE-MS, respectively.[22,23,33] The preconcentration factors for opioid peptides can be 10 000-fold in plasma if C_{18}-SPE-CE-MS is combined with

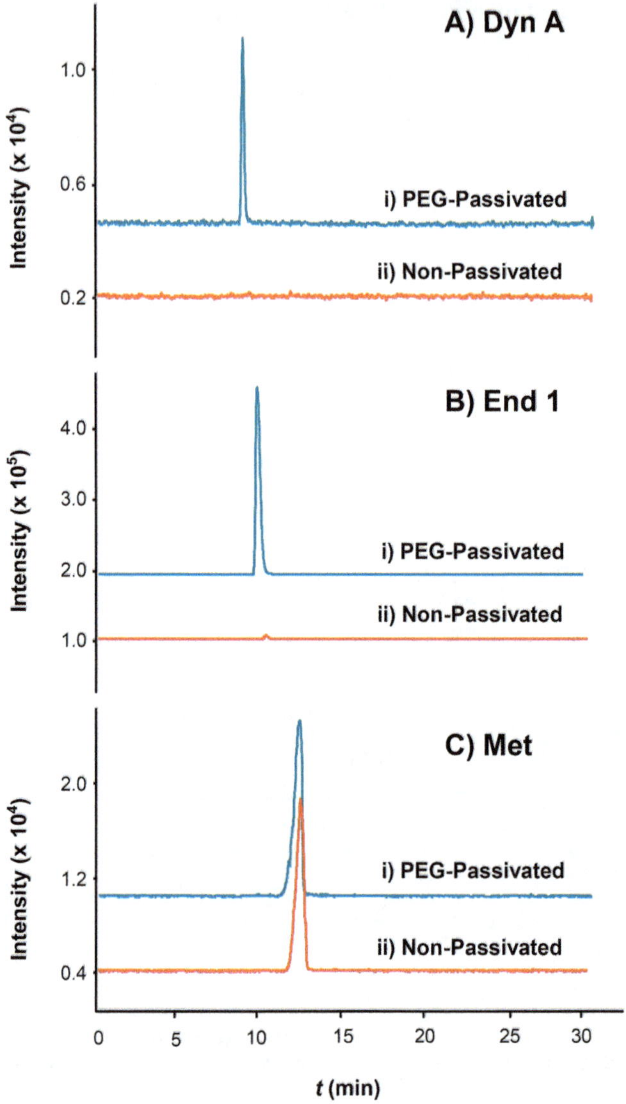

Figure 6.3 Extracted ion electropherograms (EIEs) obtained by C_{18}-SPE-CE-MS for plasma samples spiked with 100 ng mL^{-1} of (A) Dyn A, (B) End 1 and (C) Met before the double-step pretreatment and using (i) PEG-passivated and (ii) non-passivated 10 000 M_r cut-off cellulose acetate filters (Amicon® Ultra-0.5, Millipore).

on-line isotachophoresis (C_{18}-SPE-tITP-CE-MS).[16] This double-step sample pretreatment method can be applied for purification of other low molecular mass compounds from plasma samples, such as metabolites,[28] because solvent precipitation and centrifugal filtration are mostly governed by molecular size.

6.3.2 Analysis of Plasma Samples by C_{18}-SPE-CE-MS

Sorbent selection for untargeted metabolomics analyses by SPE-CE-MS can be quite challenging because sorbents require general conditions for comprehensive and sensitive profiling of low molecular mass compounds from biological samples. The best alternative to obtain the broadest coverage of the metabolome would be the simultaneous use of several sorbents with different selectivities, such as C_{18}, HILIC and ion exchange sorbents. Thus, ion exchange and HILIC sorbents are particularly useful for the retention of highly polar and charged metabolites, while C_{18} is more suitable to retain non-polar or middle-polar metabolites.

In previous studies, we investigated the performance of several commercial sorbents (C_{18}, C_8, C_2, Oasis® HLB, Strata™-X and Oasis® MCX) for the analysis of opioid peptides in human plasma by SPE-CE-MS.[23] The best results were obtained with a C_{18} sorbent, which is our "gold standard" for the analysis of peptides and our first test for a wide range of compounds. In the example presented in this chapter,[28] we show that the conditions developed for preconcentration of opioid peptides by C_{18}-SPE-CE-MS can be applied for metabolite analysis.[23,33] From our experience with peptides and other low molecular mass compounds, a C_{18} sorbent allows good recoveries for compounds with a wide range of polarities. In this case, it was useful to obtain a rich fingerprint of low molecular mass compounds in mouse plasma after the off-line double step sample pretreatment presented before. Figure 6.4 shows the total ion electropherograms (TIEs) obtained for the plasma samples from wt (control) and HD mice of different ages (8, 12 and 30 weeks). As can be observed, all the electropherograms presented a similar complex profile, and advanced chemometric methods were necessary for a high throughput and reliable comparison between the different sets of plasma samples.

6.3.3 Multivariate Data Analysis Tools

6.3.3.1 Feature Detection

Chemometric methods play a crucial role in data processing, exploration, and classification of the massive and complex datasets generated in untargeted metabolomics.[43,44] If the goal of the study is feature detection (Figure 6.2E), different strategies can be employed when working with LC-MS, CE-MS and SPE-CE-MS. Resolution methods such as multivariate curve resolution alternating least squares (MCR-ALS) can be regarded as an excellent alternative.[36,37] Among other advantages, MCR-ALS can resolve overlapped electrophoretic peaks and provide the electrophoretic and mass spectra profiles, as well as the peak areas, of the constituents in the analysed samples. This approach allows problems typically present in CE data, such as retention time shifts, background noise contributions, and differences in signal-to-noise (S/N) ratios among different injections, to be overcome.[36,37,45] Before MCR-ALS, the raw data needs some pre-processing (Figure 6.2D). A column-wise augmented

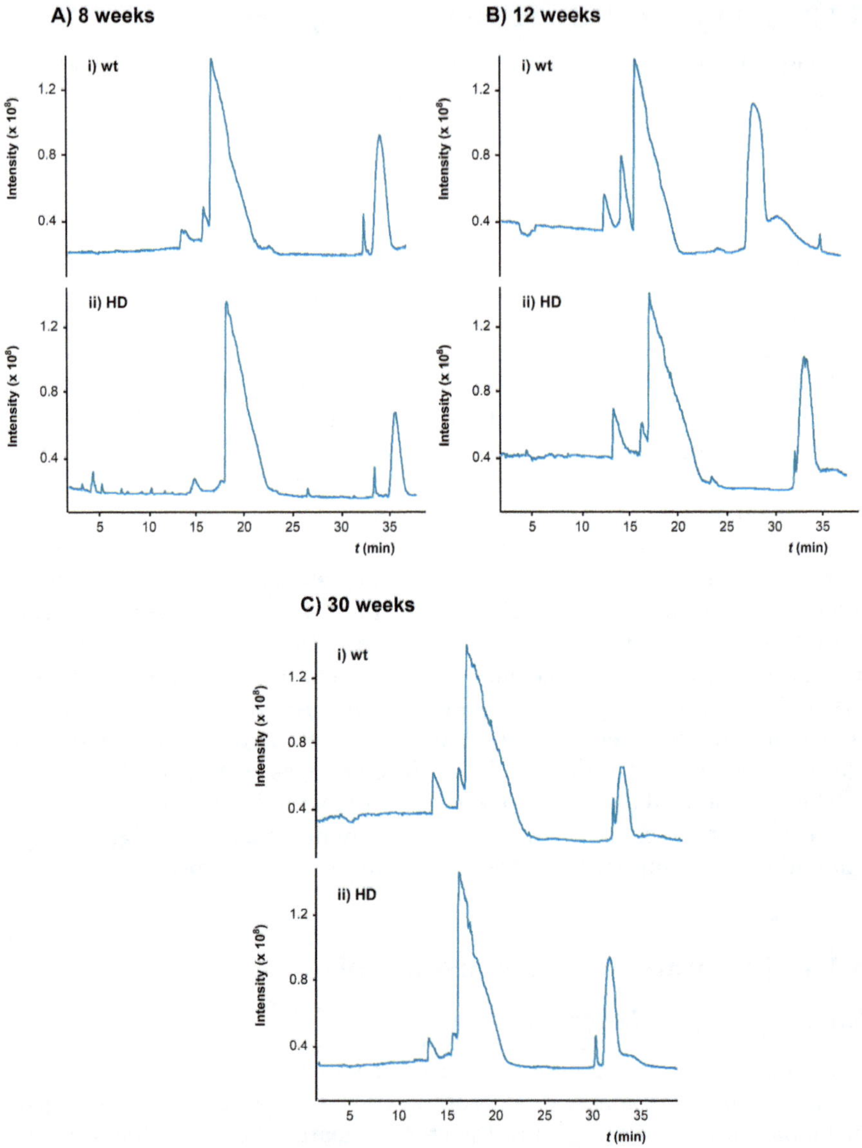

Figure 6.4 Total ion electropherograms (TIEs) obtained by C_{18}-SPE-CE-MS for (i) wild-type (wt) and (ii) Huntington's disease (HD) mouse plasma samples at (A) 8, (B) 12 and (C) 30 weeks after birth.

data matrix simultaneously containing information on all the analysed samples must be generated and the data compressed to a certain extent to facilitate computing. A straightforward option is to reduce the data matrix size in the *m/z* dimension by interpolation, for example from 0.0001 to 0.1 *m/z* resolution.[28] Alternatively, an algorithm based on selection of regions of interest (ROIs) containing features with relevant S/N ratios can also be applied to avoid the potential loss of information by interpolation.[44] After compression, the resulting data matrix can be normalised using the peak area signal of an internal standard that is added to the purified samples immediately before the analysis (*e.g.* L-methionine sulfone).[46] The pre-processed data, depending on the dimensions and the memory and processing capabilities of the computer, can be immediately subjected to MCR-ALS or divided into different time- and/or *m*/*z*-window sub-matrices before resolution.[28]

In this HD study,[28] the column-wise augmented data matrix contained the raw data obtained by C_{18}-SPE-CE-MS for the wt and HD plasma samples at 8, 12 and 30 weeks after birth. This data matrix was compressed in the resolution dimension by interpolation to 0.1 *m*/*z* resolution and MCR-ALS was applied to sub-matrices of 30 *m*/*z* windows within the 10–25 and 30–40 min time ranges where peaks were detected (Figure 6.4). After data pre-processing, a total number of 74 components were resolved by MCR-ALS (*i.e.* features). At this point, the data matrix with the peak areas of the resolved components in the different samples can be directly used to identify through statistical analysis which components are the most significant to discriminate between the different groups of samples. However, before doing this, it is recommended to revise and confirm the reliability of the peak areas and migration times of the features resolved by MCR-ALS. To this end, the peak area matrix for the resolved components can be rebuilt from the full scan raw data using the instrument software, taking as a reference the *m*/*z* value and the migration time of the MCR-ALS resolved features.

6.3.3.2 *Statistical Analysis*

This section describes our recommendation for statistical analysis after SPE-CE-MS, but several other unsupervised and supervised methods can be applied.[43,44] First, we recommend the use of principal component analysis (PCA) for the unsupervised identification of trends and clustering of the data, as well as for the detection of outliers (Figure 6.2F).[47] The main application of PCA is to reduce the dimensionality of the data, which makes most types of plots easier to interpret. Once the data is explored and the classes defined, we can apply partial least squares discriminant analysis (PLS-DA) to build a refined classification model with improved class separation. PLS-DA has the advantage over other supervised classification methods that it can easily and reliably reveal the importance of the different components for discrimination between the different groups of samples (Figure 6.2F).[41] In the HD study,[28] PLS-DA allowed the identification of the most important metabolites that could be useful to discriminate between wt and HD samples, as well as to follow-up HD progression and aging in wt mice. To do this, three different PLS-DA models were built. As can be observed in Figure 6.5, all the PLS-DA models allowed perfect class discrimination (using two latent variables (LVs)). The components with VIP score values higher than 1 were selected as the most relevant biomarker candidates for each PLS-DA model (33 out of 74 features).

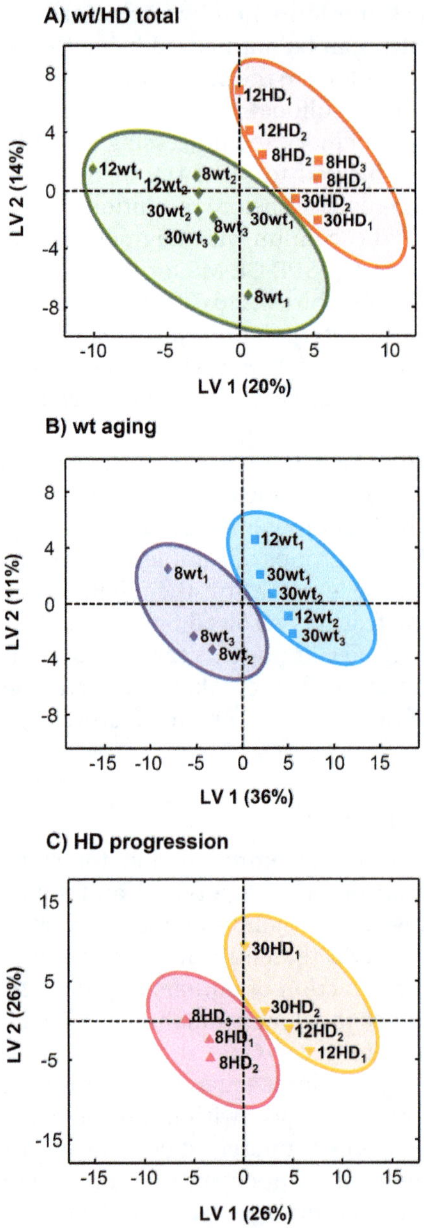

Figure 6.5 PLS-DA score plots for the mouse samples in order to differentiate between (A) all wt and HD plasma samples, (B) wt samples at 8, and from 12 to 30 weeks, and (C) HD samples at 8, and from 12 to 30 weeks. In all cases, a two-class model was used (8 weeks and 12–30 weeks) because of the limited number of samples (a total of at least 3 samples in each class was recommended).

6.3.4 Metabolite Identification

In untargeted metabolomics by SPE-CE-MS, identification of the most relevant compounds is based on comparing their molecular mass with on-line metabolite database resources, such as the METLIN Metabolite Database[48] and the Human Metabolome Database (Figure 6.2G).[49] For this purpose, a mass spectrometer with a high mass accuracy and resolution, such as a TOF mass spectrometer, is very practical as the accurate mass values obtained for unknown compounds considerably reduce the list of possible metabolite candidates delivered by a database. The tentative identity of these metabolites can be confirmed later on using standards, tandem MS/MS measurements and semi-empirical models to predict the electrophoretic migration, which relates the electrophoretic mobility (m_e), or any other relative parameter related to migration, to the charge-to-mass ratio of ionisable compounds, such as many metabolites (*e.g.* $m_e \propto q/M^{1/2}$ in the classical polymer semi-empirical model[50]). The identified metabolites can be subjected afterwards to biological pathway analysis to find relationships with specific disorders or physiological characteristics using web databases such as the Kyoto Encyclopedia of Genes and Genomes (KEGG Database),[51] the MetaCyc and the BioCyc (Figure 6.2G).[52,53]

In the HD study, 29 out of the total 33 relevant features were tentatively identified with a mass error ≤20 ppm (see Table 6.1). Different metabolic pathways were found to be related to 13 of the 29 identified metabolites, but the intracellular signalling pathway was observed to be the most affected, especially after 12 weeks of birth (Table 6.1).[28] This was in agreement with the typically observed dysfunction of specific neurons, altered expression of several types of receptors, and changed expression of neurotransmitters in HD. Furthermore, for the first time, an analytical method was sensitive and comprehensive enough to detect, in periphery, some of the metabolites that were previously described in the striatum of HD mice or other rat models. This paves the way for improvement of prediction of disease onset, and evaluation of HD early progression or response to treatment.

6.4 Conclusions

The low concentration sensitivity of CE-MS very often hinders detection of low abundance metabolites in biological fluids, skewing the metabolome coverage. In the last few years, significant progress has been made in the development of sensitive and reproducible SPE-CE-MS methods, which may allow comprehensive analysis of metabolites found at low concentrations using sorbents such as C_{18}, HILIC and ion exchange. Until now, the analytical performance of most developed SPE-CE-MS methods has been primarily described for preconcentration of target metabolites, whereas the use of this methodology for untargeted metabolomics has been rarely demonstrated. This chapter presents a general workflow for untargeted SPE-CE-MS-based metabolomics and demonstrates the potential of SPE-CE-MS combined with

Table 6.1 Tentative identification, metabolic pathway and PLS-DA model of the relevant metabolites detected by C_{18}-SPE-CE-MS, MCR-ALS and PLS-DA (mass error ≤20 ppm). As it was a tentative identification solely based on the molecular mass, several isobaric metabolites were identified in some cases.

ID	Metabolite	Metabolic pathway	PLS-DA wt/HD total	PLS-DA wt aging	PLS-DA HD progression
1	3-(Methylthio)-1-propene	—	✓	x	✓
2	*m*-Cresol	Degradation of aromatic compounds	✓	x	x
	p-Cresol	Degradation of aromatic compounds/ Protein digestion and absorption			
3	3-Methyl-1,2-cyclopentanedione	—	x	x	✓
	2,5-Dimethyl-3(2H)-furanone				
	(2E,4E)-2,4-Hexadienoic acid				
	(E)-4-Oxo-2-hexen-1-al				
	1,2-Cyclohexanedione				
	trans-1,2-dihydrobenzene-1,2-diol				
4	2-Hepten-4-one	—	x	✓	✓
	(Z)-4-Heptenal				
	3-Hepten-2-one				
	2-Methylcyclohexanone				
	trans-2-*trans*-4-heptadien-1-ol				
	5-Methyl-5-hexen-2-one				
	4-Methylcyclohexanone				
	5-Methyl-3-hexen-2-one				
5	Dimethylbenzimidazole	Riboflavin metabolism/Porphyrin metabolism	x	✓	x
6	Hydantoin-5-propionic acid	—	x	✓	x
	Menadione				
	Methyl (Z)-2-decene-4,6,8-triynoate				
7	(−)-Epinephrine	cAMP signalling pathway[a]/adrenergic signalling in cardiomyocytes[a]/ neuroactive ligand-receptor interaction[a]/tyrosine metabolism	x	✓	✓
	Normetanephrine	Tyrosine metabolism			

8	Vanylglycol	Tyrosine metabolism	x	✓	✓
	Phosphorylcholine	Glycerophospholipid metabolism			
9	3-Indolebutyric acid	Tryptophan metabolism	x	✓	✓
	Glycyl-glutamine Glycyl-gamma-glutamate Asparaginyl-alanine Glutaminyl-glycine Alanyl-asparagine Gamma-glutamyl-glycine	Incomplete breakdown product of protein digestion or protein catabolism with cell signalling effects[a]			
10	Propyl propane thiosulfonate 5-Methoxycanthin-6-one	—	✓	x	✓
11	L-Hexanoylcarnitine	Energy production	✓	✓	✓
12	Histidinyl-histidine	Incomplete breakdown product of protein digestion or protein catabolism with cell signalling effects[a]	✓	x	x
13	Phenylalanyl-arginine Arginyl-phenylalanine	Incomplete breakdown product of protein digestion or protein catabolism with cell signalling effects[a]	x	x	✓
14	2-[Octahydro-4,7-dimethyl-1-oxocyclopenta[c]pyran-3-yl] nepetalactam	—	x	✓	x
15	Prostaglandin D2 Prostaglandin E2 Prostaglandin H2 Prostaglandin I2 Thromboxane A2 Lipoxin A4 Lipoxin B4 20-Hydroxy-leukotriene B4 15-Keto-prostaglandin F2a	Arachidonic acid metabolism[a]/neuroactive ligand-receptor interaction[a]/serotonergic synapse[a]/cAMP signalling pathway[a]/oxytocin signalling pathway[a]/inflammatory mediator regulation of TRP channels[a]/retrograde endocannabinoid signalling[a]/VEGF signalling pathway[a]	✓	x	✓
16	18-Hydroxycorticosterone Cortisol	Steroid hormone biosynthesis	x	✓	x

(*continued*)

Table 6.1 (*continued*)

ID	Metabolite	Metabolic pathway	PLS-DA wt/HD total	PLS-DA wt aging	PLS-DA HD progression
17	alpha-peroxyachifolide Diosbulbin F Diosbulbin A Carissanol	—	✓	x	x
18	PGD2-Dihydroxypropanylamine PGE2-Dihydroxypropanylamine	—	✓	x	x
19	Artonol E Cycloartomunoxanthone	—	x	✓	✓
20	6-Deoxohomodolichosterone	—	x	✓	x
21	Limocitrin 3-glucoside	—	x	✓	x
22	Biochanin A 7-(6-malonylglucoside)	—	x	✓	x
23	Maclurin 3-C-(6″-*p*-hydroxybenzoyl-glucoside)	—	x	✓	x
24	L-Urobilinogen	Porphyrin metabolism	x	x	✓
25	Tsugarioside B Oryzanol C	—	✓	x	x
26	PC(14 : 1(9Z)/14 : 1(9Z))	Arachidonic acid metabolism[a]/retrograde endocannabinoid signalling[a]/glycerophospholipid metabolism/linoleic acid metabolism	x	✓	✓
27	Nostoxanthin sulfate	—	✓	x	x
28	PI(13 : 0/12 : 0)	—	x	x	✓
29	Ganglioside GD1b (d18 : 1/12 : 0) Ganglioside GD1a (d18 : 1/12 : 0)	Signal transduction[a]	x	✓	✓

[a]These metabolic pathways are included in the signalling pathway.

advanced chemometric data analysis tools to provide informative metabolite profiles of biological samples, which could be useful to elucidate novel biomarkers and gain insight into disease pathogenesis and progression, as well as to design novel therapeutic treatments. There is still a long way to go in the near future to exploit the high sensitivity of SPE-CE-MS in metabolomics. The following developments will probably drive the progress in the field: high capacity sorbents with appropriate selectivity and minimal non-specific retention based on particle or monolithic supports, combination with on-line preconcentration electrophoretic techniques, easy-to build and easy-to-use microcartridge arrays for high throughput parallel analysis in capillary or microchip format, and robust and affordable nanoESI sheathless CE-MS interfaces.

The authors have declared no conflict of interest.

Acknowledgements

This study was supported by a grant from the Spanish Ministry of Economy and Competitiveness (CTQ2014-56777-R) and the Cathedra UB Rector Francisco Buscarons Ubeda (Forensic Chemistry and Chemical Engineering). Laura Pont acknowledges the same Ministry for a FPI fellowship.

References

1. M. S. Monteiro, M. Carvalho, M. L. Bastos and P. Guedes de Pinho, *Curr. Med. Chem.*, 2013, **20**, 257.
2. S. J. Kim, S. H. Kim, J. H. Kim, S. Hwang and H. J. Yoo, *Endocrinol. Metab.*, 2016, **31**, 7.
3. E. Dudley, M. Yousef, Y. Wang and W. J. Griffiths, in *Advances in Protein Chemistry and Structural Biology*, ed. R. Donev, Elsevier, 2010, p. 45.
4. M. Yuan, S. B. Breitkopf, X. Yang and J. M. Asara, *Nat. Protoc.*, 2012, **7**, 872.
5. G. J. Patti, *J. Sep. Sci.*, 2011, **34**, 3460.
6. N. Okuma, M. Saita, N. Hoshi, T. Soga, M. Tomita, M. Sugimoto and K. Kimoto, *PLoS One*, 2017, **12**, 1.
7. O. Beckonert, H. C. Keun, T. M. D. Ebbels, J. Bundy, E. Holmes, J. C. Lindon and J. K. Nicholson, *Nat. Protoc.*, 2007, **2**, 2692.
8. A. Garcia and C. Barbas, *Methods Mol. Biol.*, 2011, **708**, 191.
9. B. Zhou, J. F. Xiao, L. Tuli and H. W. Ressom, *Mol. BioSyst.*, 2012, **8**, 470.
10. R. Ramautar, G. W. Somsen and G. J. de Jong, *Electrophoresis*, 2015, **36**, 212.
11. R. Ramautar, G. W. Somsen and G. J. de Jong, *Electrophoresis*, 2017, **38**, 190.
12. M. C. Breadmore, A. Wuethrich, F. Li, S. C. Phung, U. Kalsoom, J. M. Cabot, M. Tehranirokh, A. I. Shallan, A. S. A. Keyon, H. H. See and M. Dawod, *Electrophoresis*, 2017, **38**, 33.

13. R. Ramautar, G. W. Somsen and G. J. de Jong, *Electrophoresis*, 2014, **35**, 128.
14. R. Ramautar, G. W. Somsen and G. J. de Jong, *Electrophoresis*, 2016, **37**, 35.
15. F. Benavente, M. C. Vescina, E. Hernández, V. Sanz-Nebot, J. Barbosa and N. A. Guzman, *J. Chromatogr. A*, 2007, **1140**, 205.
16. S. Medina-Casanellas, F. Benavente, J. Barbosa and V. Sanz-Nebot, *Electrophoresis*, 2011, **32**, 1750.
17. Z. Malá, P. Gebauer and P. Boček, *Electrophoresis*, 2017, **38**, 9.
18. N. A. Guzman, M. A. Trebilcock and J. P. Advis, *J. Liq. Chromatogr.*, 1991, **14**, 997.
19. N. A. Guzman and T. M. Phillips, *Electrophoresis*, 2011, **32**, 1565.
20. N. A. Guzman, T. Blanc and T. M. Phillips, *Electrophoresis*, 2008, **29**, 3259.
21. N. A. Guzman and D. E. Guzman, *J. Chromatogr. B: Anal. Technol. Biomed. Life Sci.*, 2016, **1021**, 14.
22. E. Hernández, F. Benavente, V. Sanz-Nebot and J. Barbosa, *Electrophoresis*, 2008, **29**, 3366.
23. F. Benavente, S. Medina-Casanellas, J. Barbosa and V. Sanz-Nebot, *J. Sep. Sci.*, 2010, **33**, 1294.
24. S. Medina-Casanellas, E. Domínguez-Vega, F. Benavente, V. Sanz-Nebot, G. W. Somsen and G. J. de Jong, *J. Chromatogr. A*, 2014, **1328**, 1.
25. I. Botello, F. Borrull, M. Calull, C. Aguilar, G. W. Somsen and G. J. De Jong, *Anal. Bioanal. Chem.*, 2012, **403**, 777.
26. Y. H. Tak, J. S. Toraño, G. W. Somsen and G. J. de Jong, *J. Chromatogr. A*, 2012, **1267**, 138.
27. T. Baciu, F. Borrull, M. Calull and C. Aguilar, *Bioanalysis*, 2015, **7**, 437.
28. L. Pont, F. Benavente, J. Jaumot, R. Tauler, J. Alberch, S. Ginés, J. Barbosa and V. Sanz-Nebot, *Electrophoresis*, 2016, **37**, 795.
29. C. A. Ross and I. Shoulson, *Parkinsonism Relat. Disord.*, 2009, **15**(suppl. 3), S135.
30. S. J. A. van den Bogaard, E. M. Dumas, W. M. Teeuwisse, H. E. Kan, A. Webb, M. A. van Buchem, R. A. C. Roos and J. van der Grond, *J. Huntington's Dis.*, 2014, **3**, 377.
31. F. J. Lara, A. M. García-Campaña, F. Alés-Barrero and J. M. Bosque-Sendra, *Electrophoresis*, 2008, **29**, 2117.
32. K. Jooß, J. Sommer, S. C. Bunz and C. Neusüß, *Electrophoresis*, 2014, **35**, 1236.
33. L. Pont, F. Benavente, J. Barbosa and V. Sanz-Nebot, *J. Sep. Sci.*, 2013, **36**, 3896.
34. F. Benavente, S. Medina-casanellas, E. Giménez and V. Sanz-nebot, *Methods Mol. Biol.*, 2016, **1466**, 67.
35. D. Heiger, in *High Performance Capillary Electrophoresis*, Agilent Technologies, 2010, p. 84.
36. J. Jaumot, R. Gargallo, A. De Juan and R. Tauler, *Chemom. Intell. Lab. Syst.*, 2005, **76**, 101.
37. J. Jaumot, A. de Juan and R. Tauler, *Chemom. Intell. Lab. Syst.*, 2015, **140**, 1.

38. G. Golub, K. Solna and P. van Dooren, *SIAM J. Matrix. Anal. Appl.*, 2000, **22**, 1.
39. W. Windig and J. Guilment, *Anal. Chem.*, 1991, **63**, 1425.
40. R. Tauler, A. Smilde and B. Kowalski, *J. Chemom.*, 1995, **9**, 31.
41. M. Barker and W. Rayens, *J. Chemom.*, 2003, **17**, 166.
42. S. Wold, M. Sjöström and L. Eriksson, *Chemom. Intell. Lab. Syst.*, 2001, **58**, 109.
43. R. Madsen, T. Lundstedt and J. Trygg, *Anal. Chim. Acta*, 2010, **659**, 23.
44. E. Gorrochategui, J. Jaumot, S. Lacorte and R. Tauler, *Trends Anal. Chem.*, 2016, **82**, 425.
45. J. Saurina, in *Chemometric Methods in Capillary Electrophoresis*, ed. G. Hanrahan and F. A. Gomez, A John Wiley & Sons Inc., 2009, p. 199.
46. E. Ortiz-Villanueva, J. Jaumot, F. Benavente, B. Piña, V. Sanz-Nebot and R. Tauler, *Electrophoresis*, 2015, **36**, 2324.
47. I. Joliffle, in *Principal Component Analysis*, Springer, 2002, p. 150.
48. C. A. Smith, G. O. Maille, E. J. Want, C. Qin, S. A. Trauger, T. R. Brandon, D. E. Custodio, R. Abagyan and G. Siuzdak, *Ther. Drug Monit.*, 2005, **27**, 747.
49. D. S. Wishart, T. Jewison, A. C. Guo, M. Wilson, C. Knox, Y. Liu, Y. Djoumbou, R. Mandal, F. Aziat, E. Dong, S. Bouatra, I. Sinelnikov, D. Arndt, J. Xia, P. Liu, F. Yallou, T. Bjorndahl, R. Perez-Pineiro, R. Eisner, F. Allen, V. Neveu, R. Greiner and A. Scalbert, *Nucleic Acids Res.*, 2013, **41**, D801.
50. F. Benavente, E. Balaguer, J. Barbosa and V. Sanz-Nebot, *J. Chromatogr. A*, 2006, **1117**, 94.
51. M. Kanehisa, S. Goto, Y. Sato, M. Furumichi and M. Tanabe, *Nucleic Acids Res.*, 2012, **40**, 109.
52. C. J. Krieger, P. Zhang, L. A. Mueller, A. Wang, S. Paley, M. Arnaud, J. Pick, S. Y. Rhee and P. D. Karp, *Nucleic Acids Res.*, 2004, **32**, D438.
53. R. Caspi, R. Billington, L. Ferrer, H. Foerster, C. A. Fulcher, I. M. Keseler, A. Kothari, M. Krummenacker, M. Latendresse, L. A. Mueller, Q. Ong, S. Paley, P. Subhraveti, D. S. Weaver and P. D. Karp, *Nucleic Acids Res.*, 2008, **36**, D623.

CHAPTER 7

CE-MS for Anionic and Cationic Metabolic Profiling: System Optimization and Applications

AKIYOSHI HIRAYAMA* AND TOMOYOSHI SOGA

Institute for Advanced Biosciences, Keio University, 246-2 Mizukami, Kakuganji, Tsuruoka, Yamagata 997-0052, Japan
*E-mail: hirayama@ttck.keio.ac.jp

7.1 Introduction

The metabolome is defined as the total set of low molecular weight compounds (generally less than 1.5 kDa) that exist in biological samples. Metabolomics (or metabolome analysis), together with genomics, transcriptomics and proteomics, is one of the main "omics" technologies in systems biology. Therefore, metabolomics has become an important research field for gaining functional insight into cell biology. In particular, metabolites are the final products in biological systems, and their variations are thought to reflect the responses caused by a number of metabolic perturbations. Currently, the human metabolome database lists 74 461 metabolites (as of July 1, 2017),[1] but the actual number of metabolites that are still unknown could be much higher.

The ultimate goal of metabolomics is to develop a single platform that can analyze all metabolites present in a given sample. Moreover, unlike for other

New Developments in Mass Spectrometry No. 6
Capillary Electrophoresis–Mass Spectrometry for Metabolomics
Edited by Rawi Ramautar

Published by the Royal Society of Chemistry, www.rsc.org

omics approaches, metabolites are not species-specific and theoretically a single analytical platform could be adapted for multiple species. However, from an analytical chemistry point of view, it is quite difficult to simultaneously analyze all metabolites with a single analytical platform. This is because the physical and chemical properties of the metabolites are quite varied, and their concentrations can extend over an estimated seven to nine orders of magnitude (pM to mM).[2] In addition, variations in extraction efficiencies for different classes of metabolites also complicate quantification. Therefore, researchers currently determine the priority of metabolites for quantification, and then select an appropriate analytical approach that can detect as many target metabolites as possible. Two approaches are generally employed in metabolomics; namely, targeted analysis and non-targeted analysis. Targeted analysis involves identification and quantification of only pre-selected metabolites using a suitable analytical method for these metabolites. In contrast, non-targeted analysis involves comprehensive (and sometimes rapid) sample analysis, generally for observing signal patterns (*i.e.* fingerprints) without a set target; however, the detected metabolites are often identified and quantified.

To date, both nuclear magnetic resonance (NMR) spectroscopy and mass spectrometry (MS)-based methods have been developed for metabolomics. NMR has several advantages over other techniques. For instance, NMR is non-destructive and features high throughput, high reproducibility and simple sample preparation; direct analysis of solid samples with no preparation is also possible.[3] Furthermore, NMR offers advantages for compounds that are difficult to ionize or require derivatization for MS-based analysis, in addition to allowing the identification of metabolites with identical masses, including those with different isotopomer distributions.[4] However, NMR has relatively low sensitivity and, therefore, requires large sample volumes for analysis.

By comparison, MS has high selectivity and sensitivity. Both NMR and MS can be used as stand-alone techniques, but MS can be coupled with separation analyses, such as gas chromatography (GC), liquid chromatography (LC) and capillary electrophoresis (CE); this enhances the performance of MS for complex biological samples.

Among MS-based analyses, GC-MS offers high resolution, selectivity and sensitive analyses of volatile metabolites such as fatty acids and organic acids. Because GC separation is performed in an oven at high temperatures (>200 °C), the compounds must be volatile and thermally stable. For non-volatile compounds, it is necessary to derivatize the samples before analysis. However, derivatization involves additional sample processing, which is generally time-consuming and can result in metabolite loss. Electron ionization (EI) is frequently used as the ionization method for GC-MS, and provides universal, reproducible, and characteristic spectra even among different mass spectrometers. Therefore, EI enables reliable matching of experimental data to publicly or commercially available databases. The major disadvantage of GC-MS in metabolomics is that a number of metabolites, especially phosphorylated metabolites, are still non-volatile after derivatization and thus cannot be quantified by this method.

LC is applicable to the analysis of a wide range of metabolites, with a large variety of molecular sizes. Many kinds of columns are commercially available, based on different retention mechanisms; among these, reversed-phase columns such as C8 and C18 are most commonly employed in metabolomics. However, polar metabolites do not exhibit adequate retention on these columns and often co-elute close to the column void volume, thus obstructing their detection with MS. For polar metabolites, ion-pair LC-MS and hydrophilic interaction chromatography-MS are becoming common.[5,6] Meanwhile, detailed attention should be paid to matrix effects, which may lead to quantification inaccuracy when LC is combined with MS.

CE-MS has proven to be a powerful analytical tool for charged metabolites. Most primary metabolite intermediates contain hydroxyl, amino, carboxyl and phosphate groups and, therefore, are charged, which makes them suitable for CE-MS analysis. In contrast to GC and LC, CE separates compounds according to their mass-to-charge (*m*/*z*) ratios. This orthogonality in separation principle establishes CE as a complementary tool to other chromatographic techniques. In addition, CE only requires small sample volumes and is suitable for volume-limited samples. CE-MS-based metabolomics was first applied to the comprehensive and quantitative analysis of charged metabolites from a metabolic extract of *Bacillus subtilis*, and 1692 peaks were successfully detected.[7] To date, CE-MS has been used in various research fields, including medical,[8–10] biochemical,[11,12] environmental[13] and plant metabolomics.[14–16]

Many reviews have been published on CE-MS-based metabolomics. For example, Hirayama *et al.* provided a general description of CE-MS application in metabolome analysis, including sample preparation, data processing, peak identification and data interpretation.[17] Since 2008, Ramautar *et al.* have published a biannual review and summary of the latest CE-MS-based metabolomics studies.[18–21] Barbas *et al.* have discussed the general advantages and disadvantages of both CE-UV and CE-MS for non-targeted metabolome analysis of biological samples.[2]

In this chapter, we focus on the general CE-MS conditions specifically developed for profiling cationic and anionic metabolites, and we also focus on several biological samples frequently used in medical and biochemical metabolomics, including blood (serum and plasma), urine, tissue, cultured cells and other related samples.

7.2 CE-MS for Metabolome Analysis

CE is a versatile technique that can be used to separate a wide range of analytes. Although different additives, such as micelles, chiral selectors or complexing agents, can be incorporated into the running buffer, also known as the background electrolyte (BGE), to achieve a variety of separation modes, capillary zone electrophoresis (CZE) is the main CE separation mode in

CE-MS-based metabolome analysis. This is because the buffer component is restricted when coupling CE with MS *via* electrospray ionization (ESI). In CZE, separation is achieved by the different migration velocities of analytes in an electric field based on their charge-to-size (ion radius) ratios. In addition, the presence and magnitude of the electro-osmotic flow (EOF) can ensure that all molecules, regardless of charge, migrate in the same direction and, therefore, can be analyzed simultaneously. Meanwhile, MS provides a second separation dimension for analytes that co-migrate in CE. In this marriage of techniques, CE provides rapid analysis and efficient resolution, while MS provides highly selective and sensitive detection.

For comprehensive analysis, an attractive approach is to use two or more sets of conditions, which have been optimized for different compound groups, and then combine the results. In this manner, wider sample coverage can be obtained.

7.2.1 Cationic Metabolite Analysis

In cation analysis, MS is coupled to the cathode and separation is carried out in a fused-silica capillary (Figure 7.1).[22] To analyze most of the cations simultaneously, a low pH solution (*e.g.* 1 mol L^{-1} formic acid) is used as the running buffer to confer positive charge on every cation.[23] In this manner, cationic metabolites can be efficiently separated by CE and then selectively and sensitively detected by MS. Good linearity was obtained from 1 to 1000 μM for most cations, and the limits of detection (LODs) for primary amino acids were between 0.3 and 11 μM with CE-single quadrupole MS.[23] Further sensitivity improvements were achieved by using CE coupled to time-of-flight (TOF)-MS, and the LODs obtained were in the range of 0.1–0.6 μM.[24] Figure 7.2 shows the original CE-MS results obtained for cationic metabolites in extracts from exponentially-growing *B. subtilis* cells, for a *m/z* range of 101–150.[7]

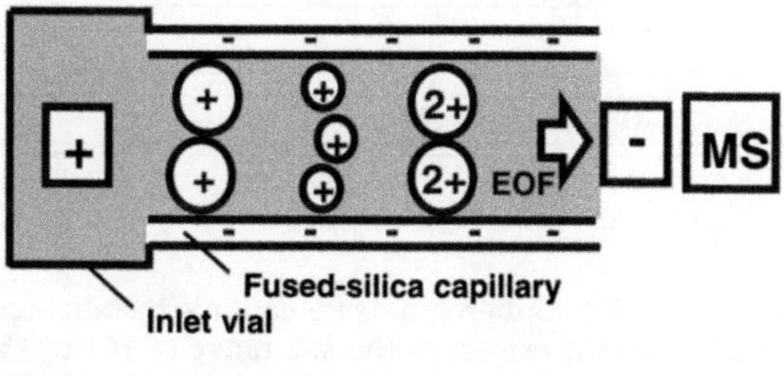

Figure 7.1 Schematic diagram of the CE-MS configuration for cationic metabolites. Both cationic metabolites and electroosmotic flow (EOF) are directed toward the cathode under this configuration. Reproduced from *Metabonomics. Methods in Molecular Biology*, Capillary Electrophoresis–Mass Spectrometry, **1277**, 2015, 113–122, M. Wakayama, A. Hirayama, T. Soga,[22] (© Springer Science + Business Media New York 2015) with permission of Springer.

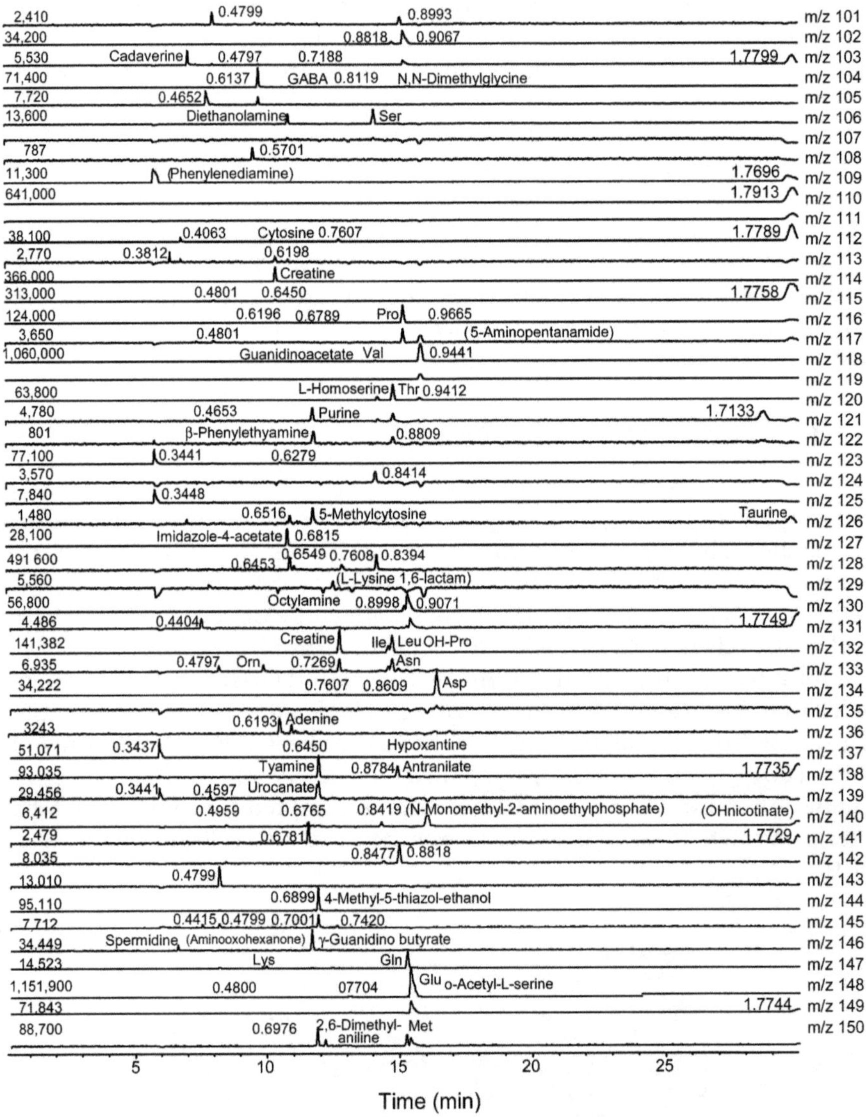

Figure 7.2 Selected ion electropherograms for cationic metabolites obtained from a *Bacillus subtillis* extract in the *m/z* range of 101 to 150. The number provided in the upper left corner of each trace is the abundance associated with the tallest peak in the electropherogram, for each *m/z*, and the peak labels are the relative migration times normalized to methionine sulfone (internal standard). Reproduced with permission from ref. 7. Copyright (2003) American Chemical Society.

The separation of biological samples by CE using fused-silica capillaries may be strongly affected by solute adsorption onto the capillary wall. In CE-MS, a constant and appreciable EOF is often essential to achieve adequate and reproducible spray conditions. However, it is impossible to remove all possible matrix components adsorbed onto the capillary during sample preparation. To reduce these problems, different coatings have been developed for CE-MS-based metabolomics studies. Details are described in Chapter 3.

7.2.2 Anionic Metabolite Analysis

The metabolites of key pathways for cellular energy production, such as glycolysis, the tricarboxylic acid (TCA) cycle and the pentose phosphate pathway, are almost entirely anionic species, *e.g.*, carboxylic acids, phosphorylated carboxylic acids and phosphorylated saccharides. An enormous number of anionic metabolites also exist in other pathways. Therefore, a methodology suitable for anionic metabolite analysis is very important. However, the analysis of anions by CE-MS using a fused-silica capillary in negative mode, where the capillary inlet is the cathode and the capillary outlet is the anode (Figure 7.3a),[22] has proved difficult. Since the CE-MS system does not contain an outlet solution vial, the EOF toward the cathode (opposite to the MS direction) generates a gap in the liquid phase at the capillary exit, resulting in a current drop (Figure 7.3b). To overcome this problem, we have designed a way to reverse the EOF toward the anode by employing a SMILE(+)™[25,26] and COSMO(+)™[27] cationic polymer-coated capillary (Figure 7.3c). This EOF reversal technique enables successive anion analysis without a deleterious current drop.

The second approach uses a running buffer with a high pH (above 7.5) to create a strong EOF. The electrophoretic polarity configuration for cationic metabolite analysis is also used for anionic analysis. In this setting, all anions migrate toward the anode. Concurrently, strong EOF movement occurs in the opposite direction (from anode to cathode), and most of the anionic metabolites are pushed back to the cathode and detected by negative-ion mode.[28] However, with this method, it is impossible to detect all anionic metabolites because faster migrating metabolites override the EOF and do not reach the MS.[26] Therefore, pressure from the capillary inlet is often applied,[29] although this can cause peak broadening because parabolic flow is generated by an external pump.

In addition, another obstacle exists when anionic metabolome analysis is performed under normal polarity with negative-ion mode detection. Alkaline aqueous ammonia solutions lead to chemical degradation of the outer polyimide capillary coating, which compromises the integrity of the fused-silica capillaries *via* aminolysis.[30] Irreversible aminolysis of the outer polyimide coating results in polymer dissolution, which occurs under aqueous conditions that are dependent on the ammonia concentration, buffer pH and exposure time. To prevent polymer dissolution, the use of weakly alkaline

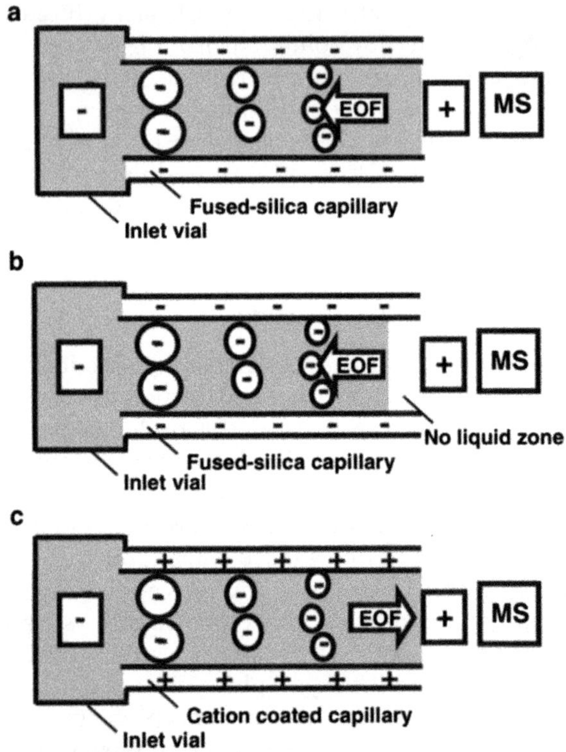

Figure 7.3 Schematic diagram of the CE-MS configuration for anionic metabolites. (a) When a fused-silica capillary is used, the anionic metabolites migrate toward the anode, while the EOF is directed toward the cathode. (b) A no-liquid zone at the capillary outlet results in an associated current drop. (c) This problem can be overcome by reversing the EOF using a cation-coated capillary. Reproduced from *Metabonomics, Methods in Molecular Biology*, Capillary Electrophoresis-Mass Spectrometry, **1277**, 2015, 113–122, M. Wakayama, A. Hirayama, T. Soga,[22] (© Springer Science + Business Media New York 2015) with permission of Springer.

ammonia-containing buffers (pH < 9), or buffers with reduced nucleophilic primary/secondary amine content, or chemically resistant coating capillaries, is recommended.

7.2.3 Interface

One soft ionization technique, ESI, has been most frequently used in CE-MS metabolomics. ESI enables liquid-phase molecules to be converted directly into gas-phase ions.[31] This enables successful coupling of MS with CE. Because CE separation is performed in the buffer, ESI is a good ionization technique for CE-MS.

A CE-ESI-MS system has to complete the electrical circuit for analyte separation, while simultaneously providing an electrical potential to the spray tip. To accomplish this, a sheath–flow interface is generally used. Smith *et al.* first developed a coaxial sheath–flow interface consisting of a triple-tube design;[32] this type of interface is most commonly adopted in commercially available instruments. In this configuration, the separation capillary is inserted coaxially into a stainless steel tube with a slightly larger diameter, and a sheath liquid is mixed with the buffer at the capillary outlet. The nebulizing gas is supplied *via* a third coaxial tube; this assists with stable spray formation and desolvation.[33]

As a sheath liquid, a mixture of an aqueous volatile solution (*e.g.* formic acid, acetic acid and its ammonium salts) and organic solvent (*e.g.* methanol, acetonitrile and 2-propanol) is often used. Optimization of the sheath liquid composition and its flow rate, in addition to the nebulizing gas flow rate, is important to create a stable spray and maintain separation efficiency and detection sensitivity.[34,35]

It should be noted that the spray needle material has a significant effect on the measurement of anionic metabolites in reverse polarity configuration.[27] Due to electrolysis, stainless steel sprayer needles are easily oxidized and generate metal ions, causing needle corrosion and capillary clogging. In addition, the generated metal ions, such as Fe^{2+} and Ni^{2+}, form complexes with many anionic metabolites, resulting in significantly decreased detection sensitivity. The use of a platinum spray needle is one way to prevent this phenomenon. Platinum is a low ionization tendency metal and can prevent both generation of metal ions and needle corrosion caused by electrolysis, potentially providing better sensitivity for anionic metabolome analysis. CE-TOF-MS with a COSMO(+)™ capillary in the negative mode using the platinum spray needle exhibited more than several-fold sensitivity improvement compared with the stainless steel spray needle.[27] The calibration curves for the tested anions were linear at 1–200 μM with correlation coefficients between 0.980 and 0.999, and the LODs were between 0.03 and 0.87 μM.

Both atmospheric pressure chemical ionization (APCI) and atmospheric pressure photoionization (APPI), known as gas-phase ionization techniques, are also available in CE-MS. APCI is a harder ionization process than ESI and is preferred over ESI for the analysis of low polarity and low molecular weight compounds.[36] APCI-MS signals are not affected by the composition of the CE buffer, even when the buffer contains a non-volatile salt or sodium dodecyl sulfate.[37] Meanwhile, it is known that APCI can only ionize thermally stable compounds, since buffer evaporation is achieved by a heated nebulizer (350–400 °C). APPI appears to be a good alternative to ESI and APCI for the coupling of low flow rate separation techniques (<50 μL min^{-1}) such as CE and capillary LC, due to the lower solvent vapor concentration in the ion source.[38–40] In addition, APPI is considered to be a compatible ionization technique for apolar compounds. While these techniques have been frequently used in LC-MS, several applications have also been reported for CE-MS.[41–43]

7.2.4 MS Instrument

As explained in the previous section, the major advantages of CE-MS are its high and efficient resolution, high selectivity and sensitivity. In addition, ideally almost any charged metabolite can be infused into the mass spectrometer. Currently, the most commonly used mass spectrometers in CE-MS-based metabolomics studies are single quadrupole, ion trap (IT), TOF, triple quadrupole, quadrupole TOF (Q-TOF) and Fourier transform ion cyclotron resonance (FT-ICR) mass spectrometers. These instruments operate under different physical principles and have different strengths and weaknesses. The general features of each MS instrument frequently used in CE-MS-based metabolomics studies are summarized in Table 7.1.

Single quadrupole instruments are one of the most commonly used mass spectrometers because they are less expensive and more robust than most other types of mass spectrometers. On the contrary, they are also considered to be low-resolution mass spectrometers, because the mass resolution of a single quadrupole is approximately 1 Da over the entire mass range. High resolution is an indispensable requirement for metabolite identification; thus, the low resolution of single quadrupole instruments represents their most significant drawback.[44]

Triple quadrupole instruments are mainly used in targeted analysis. Triple quadrupole mass spectrometers can provide high sensitivity in selective reaction monitoring mode, enabling high selectivity by monitoring a specific fragment ion of a selected precursor mass.[45] On the contrary, an important consideration of triple quadrupole linear ion trap mass spectrometers is their low mass cut-off. Notably, ion traps have a low mass cut-off that usually corresponds to about one-third of the precursor ion mass.[46]

IT instruments enable MS^n experiments without the need for multiple analyzers. In addition, ITs can accumulate specific ions by ejecting all other ions during the accumulation step, resulting in improved signal to noise for target analytes. However, a major disadvantage of IT instruments is their limited dynamic range; interferences by excess endogenous metabolites often cause space–charge effects, restricting the quantitative use of IT instruments.

TOF instruments can provide high sensitivity, high mass resolution and accuracy, with mass errors typically below 5 ppm. In addition, TOF is generally considered to be one of the fastest mass analyzers, making it ideal for highly efficient separation analysis such as CE. TOF also allows unambiguous determination of empirical formulae from elemental compositions, providing an additional analytical dimension for compound identification.[47] For these reasons, TOF instruments have been widely used in non-targeted metabolome analysis. The Q-TOF hybrid analyzer can provide accurate product ion masses, increasing the selectivity for target metabolites in complex biological matrices. While the accurate mass obtained from TOF can be used to establish the elemental composition of

Table 7.1 General features of mass spectrometers used in CE-MS-based metabolome analysis. Reproduced from *TrAC Trends in Analytical Chemistry*, **61**, A. Hirayama, M. Wakayama and T. Soga, Metabolome analysis based on capillary electrophoresis-mass spectrometry, 215.[17] Copyright 2014, with permission from Elsevier.

				Hybrid MS			
	Quad	IT	TOF	TripleQuad	Q-TOF	FT-ICR	Orbitrap
Mass resolution	Unit mass resolution	Unit mass resolution	1000–10 000	Unit mass resolution	1000–10 000	1 000 000	1 000 000
Mass accuracy	100 ppm	100 ppm	<5 ppm	100 ppm	<5 ppm	<1 ppm	<3 ppm
Scan speed	Low	Medium	High	Medium	High	Low	Medium
Sensitivity	Medium	Medium	Medium	High	Medium	Medium	High
Cost	Low	Medium	Medium	Medium	High	Very high	Very high

a particular compound, QTOF enables the elemental compositions of all product ions to be obtained, which is helpful when attempting to elucidate unknowns.[48]

FT-ICR instruments provide excellent mass accuracy (<1 ppm) and resolution (>10^6). For example, Que *et al.* analyzed complex mixtures of O-linked oligosaccharides isolated from bovine mucin and bile salt-stimulated lipase using capillary electrochromatography-FT-ICR-MS.[49] However, the slow data acquisition speed (1–2 s $scan^{-1}$) of FT-ICR instruments limits their suitability for direct combination with fast separation techniques such as CE, where the average peak width is less than 15 s.

Orbitrap, another FT mass spectrometer, is one of the most advanced mass analyzers in terms of resolution (up to 1 000 000 full width at half maximum at *m*/*z* 200 with Orbitrap Fusion Lumos from Thermo Fisher Scientific, Waltham, MA) and mass accuracy (typically sub 1 ppm accuracy for injected analytes). The Orbitrap mass analyzer employs orbital trapping of ions in its static electrostatic fields, in which the ions orbit around a central electrode and oscillate in an axial direction.[50–52] The ion oscillation frequencies depend on their *m*/*z* ratios; the Orbitrap uses a fast Fourier transform algorithm to convert time-domain signals into mass-to-charge spectra.[53] The ability to analyze small molecules with high accuracy is helpful to identify metabolites by allowing MS data to be filtered with respect to mass defect. Although the hyphenation of Orbitrap with CE is currently more common in proteomics, this technique may become a promising choice for metabolomics in the near future.[54,55]

7.3 Application

In metabolomics, two alternative approaches can be adopted; targeted analysis (metabolic profiling) and non-targeted analysis (metabolic fingerprinting).[56,57] However, in this section, detailed classification of approaches for each application into either targeted or non-targeted analysis has not been provided.

7.3.1 Serum/Plasma

Serum and plasma are widely used for metabolome analysis to detect characteristic metabolites of certain diseases. For example, Alzheimer's disease (AD) is the most common cause of dementia, but there are still no effective treatments for treating or slowing disease progression in advanced stages. For this reason, biomarkers that can diagnose early-stage AD are strongly desired. To address this need, serum metabolome analysis by CE-MS has been applied to obtain representative metabolites in order to distinguish between patients with AD and mild cognitive impairment, and healthy controls.[58] Cationic metabolites were analyzed with a fused-silica capillary using 0.8 M formic acid in 10% methanol as BGE, and 535 metabolic features were obtained. In AD patients, large increases have been observed in the levels of choline, creatinine, asymmetric dimethylarginine, homocysteine-cysteine disulfide, phenylalanyl-phenylalanine and different medium-chain acylcarnitines. By contrast, decreased levels of asparagine, methionine, histidine, carnitine, acetylspermidine and C5-carnitine have been observed in AD patients. These metabolites are considered to be related to oxidative stress, defects in energy metabolism and vascular risk factors associated with AD.

Chronic fatigue syndrome (CFS) is a persistent and unexplained pathological state characterized by exertional and severely debilitating fatigue lasting at least six consecutive months; however, its pathogenesis is not fully understood. Yamano and co-workers performed non-targeted CE-MS metabolome analysis of 133 plasma samples obtained from CFS patients and healthy controls to establish an objective diagnosis of CFS.[59] Cationic metabolites were analyzed with a fused-silica capillary under low-pH conditions; meanwhile, anionic metabolites were analyzed under high-pH conditions using a coated capillary with reversed polarity. CFS patients exhibit decreased activity in the TCA and urea cycles compared with healthy controls, and a combination of ornithine/citrulline and pyruvate/iso-citrate ratios discriminated CFS patients from healthy controls (Figure 7.4). Although further validation studies are needed, these markers might contribute not only to objective diagnosis of CFS, but also to the treatment of CFS patients by indicating appropriate nutrients to be ingested in food or supplements.

CE-MS-based metabolome analysis is also used as a biomarker discovery technique for cancer. Zeng *et al.* employed CE-TOF-MS to discover novel biomarkers for hepatocellular carcinoma (HCC).[60] Cation and anion analysis was performed with low-pH and high-pH conditions, respectively, using a fused-silica capillary under normal CE polarity. They analyzed 183 human serum samples (77 in the discovery study and 106 in the validation study) and finally developed a serum biomarker model, with components including tryptophan, glutamine and 2-hydroxybutyric acid. For distinguishing HCC from non-HCC, the area under the curve (AUC) values of receiver operating characteristic (ROC) analysis were 0.969 and 0.99 for the discovery and validation datasets, respectively. Also, this model could discriminate small HCC from pre-cancer cirrhosis with high accuracy. These results indicate that the development of a multiple biomarker model has potential as an effective clinical diagnosis tool in the future.

More recently, a population-based large-scale cohort study has been performed using CE-MS. To identify potential biomarkers of alcohol intake and alcohol-induced liver injury, Harada *et al.* analyzed 896 serum samples of Japanese men using CE-MS.[61] Cationic metabolites were analyzed with a fused-silica capillary under low-pH conditions; meanwhile, anionic metabolites were analyzed under high-pH conditions using a coated capillary with reversed polarity. The associations between ethanol intake and

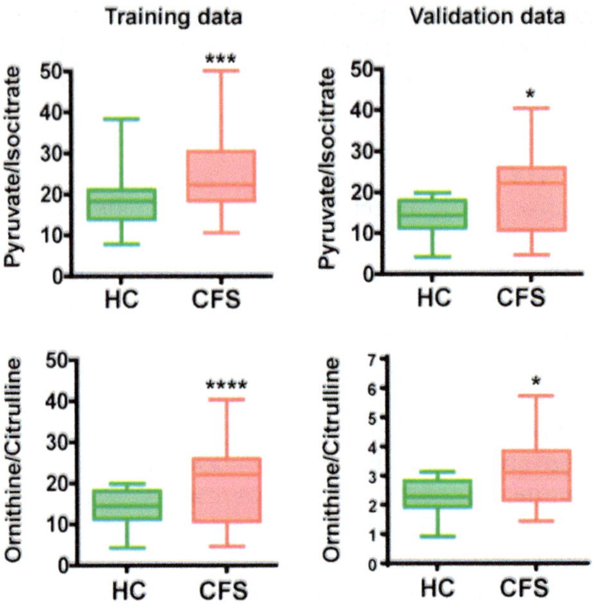

Figure 7.4 Box and whisker plots of the ratios of pyruvate/iso-citrate and ornithine/citrulline in training (left) and validation (right) datasets. HC denotes the healthy control and CFS denotes chronic fatigue syndrome. The horizontal lines indicate the minimum, maximum, median, and first and third quartile. *$P < 0.05$; ***$P < 0.001$; ****$P < 0.0001$ (Mann–Whitney *U*-test). Reproduced with permission from ref. 59. Reprinted under a Creative Commons Attribution 4.0 International License. Copyright © 2016, The Author(s).

plasma metabolite concentrations were examined. In addition, the relationships between plasma concentration and serum liver enzyme levels in the highest ethanol intake group were also examined by linear regression analysis. As a result, 19 metabolites have been found to be associated with daily alcohol consumption, and three metabolites, including threonine, glutamine, and guanidinosuccinate, as well as the ratio of glutamate/glutamine, have been selected as potential biomarker candidates of alcohol-induced liver injury.

Iida *et al.* investigated the associations of hydrophilic metabolites with metabolic syndrome (MetS) in postmenopausal women in lean Asian populations.[62] CE-MS analysis was performed on samples from 877 (594 in the original population and 283 in the replication population) naturally postmenopausal women. The analytical conditions of CE-MS were identical to those in the work by Harada *et al.* Finally, thirteen metabolites significantly related to MetS have been identified. The levels of branched-chain amino acids and their metabolism intermediates, alanine, glutamate, proline and alpha-aminoadipate, generated by lysine degradation, are significantly higher in women with MetS, while the level of 3-hydroxybutyrate is

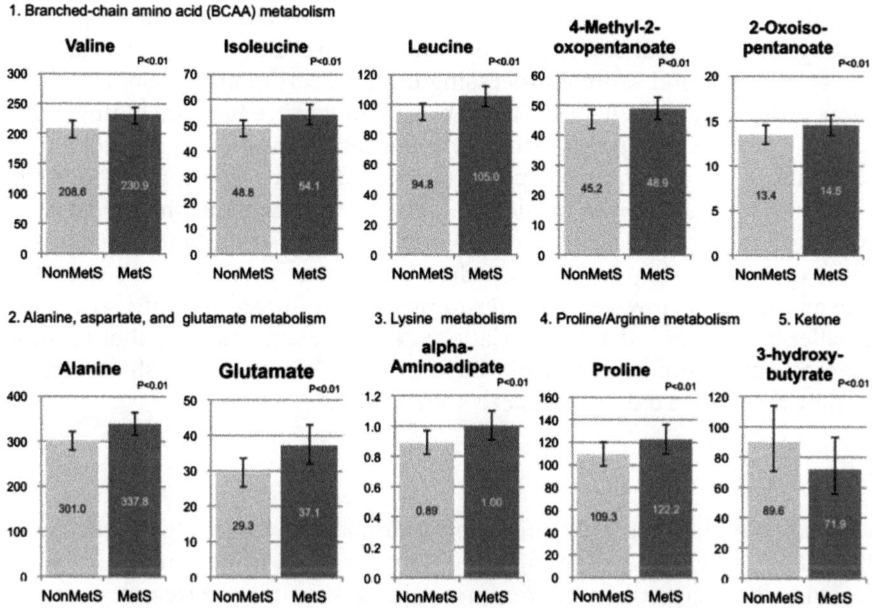

Figure 7.5 Comparison of plasma metabolite concentrations in women with and without metabolic syndrome, denoted by MetS and NonMetS, respectively. Metabolite concentrations are in µmol L^{-1} and the bar labels are the adjusted mean concentrations. The error bars reflect the 95% confidence intervals. False discovery rate *P* values are shown. Reproduced from ref. 62 with permission. Copyright © 2016, by The North American Menopause Society.

significantly lower (Figure 7.5). These results may help to understand metabolic disturbance and have clinical utility in the future.

In addition, Fukai *et al.* investigated the relationship between daily physical activity level and plasma metabolites in a population-based study.[63] A total of 1193 male participants aged from 35 to 74 years were selected from the baseline survey of a cohort study. Information on daily total physical activity, classified into metabolic equivalent score quartiles, and sedentary behavior, defined as hours of sitting per day, were collected through a self-administered questionnaire. Plasma samples were analyzed by CE-MS (the separation conditions were the same as Harada *et al.*) and linear regression analysis models with multivariable adjustments and corrected *p*-values for multiple testing were developed. Higher levels of total physical activity are associated with various metabolites, including lower concentrations of amino acids and their derivatives, and higher concentrations of pipecolic acid. Furthermore, lower levels of iso-leucine, leucine, valine, 4-methyl-2-oxoisopentanoic acid, 2-oxoisopentanoic acid, alanine, and proline are associated with longer sitting times. These metabolites might play a key role in the protective effect of higher physical activity and/or less sedentary behavior.

Standard operation protocols are necessary to obtain unbiased profiles through multiple runs over long periods of time. Hirayama *et al.* evaluated the effects of blood sample processing on both serum and plasma metabolome profiles by CE-TOF-MS.[64] They investigated procedures including: incubation duration and temperature (plasma) or clotting (serum) after blood sampling; storage conditions prior to metabolite extraction; and, the number of freeze–thaw cycles of plasma, serum, and their metabolite extracts. Although good correlations are observed between serum and plasma profiles, the processing procedures alter the serum profiles more than the plasma profiles. However, changes derived from procedural differences are smaller than inter-individual variations. These results indicate that optimal metabolomics profiles can be obtained from plasma stored for short periods of time and subjected to few freeze–thaw cycles.

7.3.2 Urine

Compared with other body fluids, urine can be non-invasively collected and has a low associated risk of infection. Furthermore, urine can be easily collected in large quantities and requires minimal sample pre-treatment due to its low protein content. Metabolites in urine, as end products of normal and pathologic cellular processes, are closely linked to phenotypes.[65] Because of this, considerable effort has been devoted to the study of urinary metabolome analysis as a diagnostic tool.

Gastric cancer remains the second leading cause of cancer death in the world, but its early detection is difficult because early-stage gastric cancer is usually asymptomatic. Chen *et al.* applied moving reaction boundary (MRB)-CE-MS to the discovery of potential urinary tumor markers in patients with gastric cancer.[66] MRB is one of the on-line sample stacking techniques for CE-MS. Separations were carried out using 1 M formic acid as BGE. Urine samples from 26 gastric cancer patients and 14 controls were subjected to MRB-CE-MS, and the levels of lactic acid, arginine, leucine, iso-leucine and valine were found to be much higher in gastric cancer patients compared with controls, while the levels of citric acid, histidine, methionine, serine, aspartic acid, malic acid and succinic acid were lower. In addition, Chen *et al.* performed ROC analysis and obtained AUC values between gastric cancer patients and controls (AUC = 1.0), as well as between early- and advanced-stage patients (AUC = 0.847). In conclusion, several amino acids are recognized as potential tumor markers for gastric cancer.

Alberice *et al.* performed non-targeted urinary biomarker discovery for bladder cancer recurrence using LC-MS and CE-MS.[67] Pre-treatment urine samples were collected from 48 patients diagnosed with urothelial bladder cancer, classified as high or low risk and as stable or recurrent according to follow-up observation. As a result, a total of 27 metabolites were highlighted as significantly different between the patient groups. Among these, potential markers of disease stability are betaine, cysteine, histidine, tyrosine, carnosine, decanoylcarnitine and uric acid, and potential markers associated with recurrence are *N*ε,*N*ε,*N*ε-trimethyllysine, *N*-acetyltryptophan, dopaquinone, leucine and hypoxanthine. These candidate biomarkers could be useful for clinical diagnosis and/or prognosis of bladder cancer.

A considerable drawback to the use of urine in metabolomics is that urine volumes can vary widely depending on water uptake and other pathophysiological factors, such as kidney disease and diabetes. To tackle this problem, several normalization procedures have been reported. Both 24 h urine collection and spot urine normalized by creatinine are most frequently used in metabolomics. However, these techniques have their own merits and limitations. For example, 24 h urine collection is cumbersome, while, on the contrary, the creatinine-based normalization method is affected by a multitude of donor factors, such as sex, age, physical activity and muscle mass.[68,69] It is important to select an appropriate normalization method suited to the purpose of the study.

In the case of urine, additional metabolism caused by bacteria should also be considered during sample storage. Saude and Sykes tested the effects of different sample preparation methods, *i.e.* centrifugation, filtration or preservative addition, as well as sample storage temperature, on changes of urinary metabolome profiles over time.[70] In conclusion, they recommended that the bacteriostatic preservative sodium azide should be added to all urine samples, and the samples should be stored in a deep freeze at −80 °C, avoiding sample freeze–thaw cycles whenever possible. Urine filtration could also help maintain the original metabolite concentrations due to the removal of bacterial contamination.

7.3.3 Tissue

Tissue-based metabolome analysis can be applied across multiple research fields, such as medicine, toxicology and environmental sciences.[71] In particular, tissues obtained from model animals can eliminate unexpected external factors like food intake and medication. Thus, these tissue samples are considered suitable for metabolome analysis.

Dilated cardiomyopathy (DCM) is a general cause of heart failure, characterized by cardiac dilation and reduced left ventricular ejection fraction; however, its detailed mechanism remains unclear. To investigate the mechanistic basis of DCM, Maekawa *et al.* performed global metabolomics analysis of myocardial tissue using both CE-MS and LC-MS.[72] Myocardial tissues were obtained from the left ventricles of J2N-k cardiomyopathic hamsters, which exhibit symptoms similar to those of human DCM patients, and their metabolomics profiles were compared with those of J2N-k healthy controls at 4 (presymptomatic phase) and 16 (symptomatic phase) weeks of age. Both cationic and anionic CE-MS analysis was performed. Cationic metabolites were analyzed with a fused-silica capillary under low-pH conditions; meanwhile, anionic metabolites were analyzed under high-pH conditions using a coated capillary with reversed polarity. A total of 180 charged metabolites were identified and quantified by CE-TOF-MS. Reduced levels of metabolite intermediates for glycolysis, the pentose phosphate pathway and the TCA cycle, combined with drastic decreases in the levels of triacylglycerols detected by LC-MS, suggest that decreased energy production leads to cardiac contractile dysfunction in the symptomatic phase. Furthermore, a mild reduction in glutathione concentration and a compensatory increase in ophthalmic acid concentration imply increased oxidative stress in the diseased tissue. These observations provide insight into the mechanisms involved in DCM pathogenesis and may lead to the identification of new targets for therapeutic intervention or diagnosis.

In the case of metabolome analysis of human tissues, especially in cancer studies, both tumor and surrounding non-tumor tissues are simultaneously analyzed.[73] This enables clarification of the metabolites that are specific to tumor tissues, and reduces inter-individual bias.

Patients with advanced adenomas are thought to be at higher risk of colorectal cancer (CRC). Endoscopy-based adenoma removal strategies greatly contribute to reducing the progression of adenoma to CRC, but show limited ability to predict submucosal invasion risk. Gao *et al.* employed CE-MS-based non-targeted analysis to detect metabolite changes related to CRC malignancy.[74] Cation and anion analysis were performed under low-pH and high-pH conditions, respectively, using a fused-silica capillary under normal CE polarity. They analyzed 11 paired CRC and paracancerous tissues, and found that 152 of 203 identified metabolites significantly changed between the CRC and paracancerous tissues. The intermediates of one-carbon metabolism are altered in CRC tissues (Figure 7.6). Gao *et al.* also focused on amino acids and performed targeted analysis of validation samples using ultra-performance LC-MS. As a result, the levels of nine amino acids, including hydroxyproline, histamine, proline, valine, methionine, tyrosine, isoleucine, leucine, and phenylalanine, were found to be conspicuously elevated with CRC onset. Furthermore, Gao *et al.* developed a regression model to diagnose adenoma and CRC differentiation; the final model includes four amino acids (methionine, valine, iso-leucine, and tyrosine) and shows a diagnosis sensitivity of 90.9% and a specificity of 100%.

Kami *et al.* compared the metabolomics of tumor and surrounding normal tissues surgically resected pairwise from nine lung and seven prostate cancer patients using CE-TOF-MS.[75] Analyses were conducted in the same way as performed by Gao *et al.* In addition, Kami *et al.* quantified the phosphorylation levels of enzymes involved in central carbon metabolism by nanoLC-tandem mass spectrometry (MS/MS). Kami *et al.* quantified 114 and 86 metabolites from lung and prostate tissues, respectively, and the profiles not only clearly distinguish tumor from non-tumor tissues, but also squamous cell carcinoma from other tumor types in lung cancer, and poorly differentiated tumors from moderately differentiated tumors in prostate cancer. The concentrations of most amino acids are significantly higher in both tumor

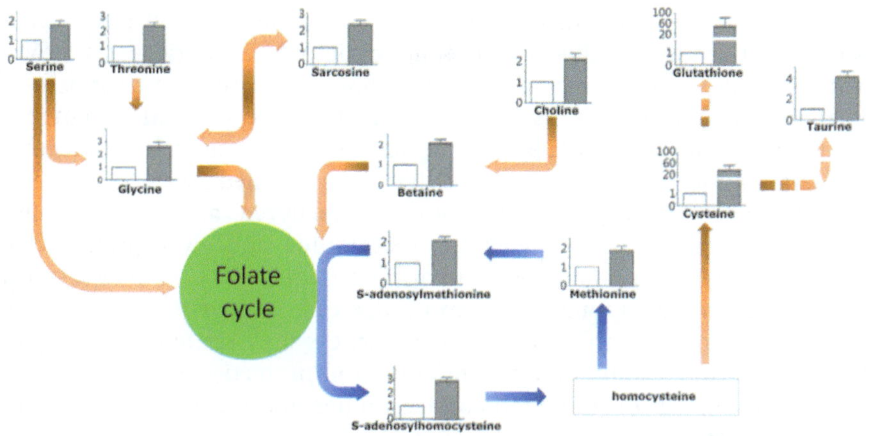

Figure 7.6 Schematic representation of the one-carbon metabolic pathways. The metabolite content difference between the paracancerous (white columns) and colorectal cancer (grey columns) tissues is shown by histograms. The *y* axis represents the relative fold change of the corresponding metabolite between the two groups. Reproduced from *Journal of Pharmaceutical and Biomedical Analysis*, **118**, P. Gao, C. J. Zhou, L. Zhao, G. H. Zhang and Y. Zhang, Tissue amino acid profile could be used to differentiate advanced adenoma from colorectal cancer, 349.[74] Copyright 2015 with permission from Elsevier.

tissues, and essential amino acid concentrations are particularly higher in prostate cancer tumors with poor differentiation compared to moderately differentiated tumors. High lactate concentrations and elevated activating phosphorylation levels of phosphofructokinase and pyruvate kinase in lung tumors are indicators of hyperactive glycolysis. The integration of metabolomics and phosphoproteomics could improve understanding of tumor-specific metabolism.

7.3.4 Cultured Cells

Metabolome analysis of cultured cells has emerged as an important approach for studying the cellular biochemistry and pathophysiology of some specific diseases. The major advantage of cell-based experiments is their ability to replicate discrete stages of physiological development. Cell lines grown in identical conditions and population contexts should display lower levels of biological variability than animal models and human subjects, which include inter-individual variation and confounding factors such as age, sex, overall health and environmental exposures.[76] Most metabolic intermediates, including those from glycolysis, the pentose phosphate pathway, the TCA cycle, and amino acid metabolism, are highly polar and charged; therefore, CE-MS is frequently used in energy metabolism studies.

Multiple sclerosis is a chronic, inflammatory, autoimmune disease, which is characterized by multifocal demyelinating lesions exhibiting as sclerotic plaques in the central nervous system. However, no specific treatment is currently available for this disease. Experimentally, administration of methionine enkephalin (MENK) partially improves disease manifestation, and the detailed mechanism of this effect has been clarified by using CE-MS.[77] Five astrocyte-related glioma cell lines were analyzed, and the intracellular levels of tyrosine, phenylalanine, methionine and glycylglycine were observed to increase after MENK treatment (Figure 7.7). Moreover, extra glycylglycine could suppress cell growth, decrease endogenous nitric oxide release and enhance the expression of some genes related to remyelination. This study exhibits the potential neuron protective effects of MENK and the promising role of metabolomics studies in therapeutic research on multiple sclerosis.

In dentistry, sodium fluoride is used as a preventive agent for dental caries because of its ability to remineralize the tooth surface and its antibacterial

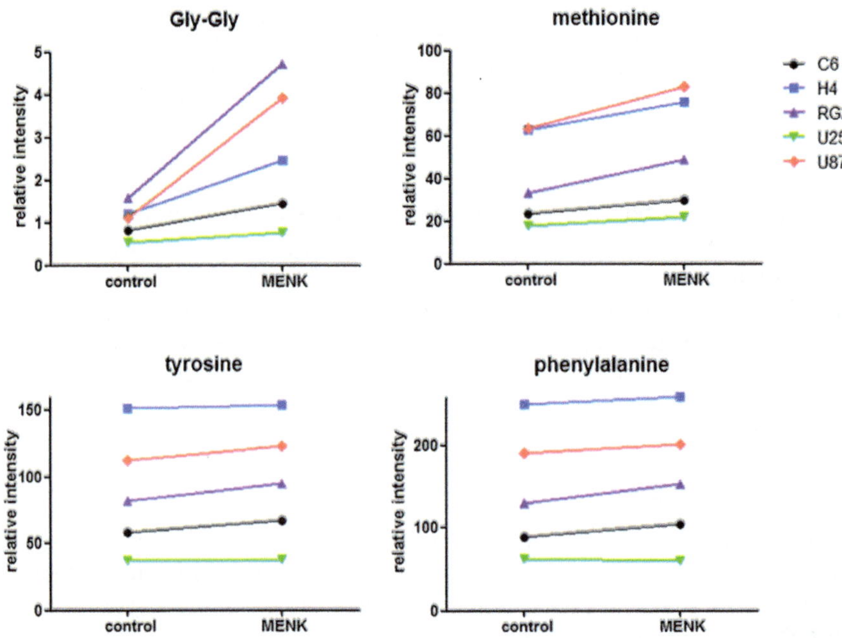

Figure 7.7 Alterations of four intracellular metabolites in cells after methionine encephalin (MENK) administration. "Relative intensity" has been normalized by cell protein dry weight. Reproduced from *Journal of Pharmaceutical and Biomedical Analysis*, **115**, C. G. Zhao, H. J. Du, L. Xu, J. Wang, L. Tang, Y. F. Cao, C. Li, Q. J. Wang, Y. Liu, F. P. Shan, J. Feng, F. Xu and P. Gao, Metabolomic analysis revealed glycylglycine accumulation in astrocytes after methionine enkephalin administration exhibiting neuron protective effects, 48.[77] Copyright 2016, with permission from Elsevier.

effect. One of the bacterial targets of fluoride is enolase, which is known as a glycolytic enzyme; however, the global impact of fluoride on metabolic pathways in oral cells has not been investigated. Sakagami *et al.* applied CE-MS metabolomics to investigate the target metabolites or metabolic pathways that are affected by NaF administration in cultured human oral squamous cell carcinoma cell lines.[78] Cationic metabolites were separated in a low-pH BGE with a fused-silica capillary, whereas anionic metabolites were separated in a high-pH BGE using a COSMO(+)™ capillary. In the early stages after NaF administration, inhibition of enolase activity is followed by rapid inhibition of the TCA cycle. On the contrary, in the later stages, gradual increases in the AMP/ATP ratio, known as a putative marker of apoptosis, and the levels of oxidized products, such as oxidized glutathione and methionine sulfoxide, are observed. In addition, the levels of polyamines (putative markers of necrosis), including putrescine, spermidine and spermine, decrease slightly. These results provide new insight into the global impact of sodium fluoride on metabolic pathways in human oral squamous cell carcinoma cells.

Although metabolic content must be representative of the metabolic state of the cell, the effectiveness of the different sample preparation protocols described in the literature is still under discussion. Ibáñez *et al.* described an effective sample preparation procedure for intracellular metabolite extraction using HT-29 cells, an adherent human colon cancer cell line.[79] They first tested four different extraction solvents; namely, water, water/formic acid (95/5, v/v), acetonitrile and isopropanol/acetonitrile/water (3/3/1, v/v/v), detecting 105, 86, 44 and 61 metabolites, respectively. Furthermore, only 27 out of a total of 145 detected metabolites are common to the tested four extracts, indicating that highly complementary information is obtained from these solvents. Finally, Ibáñez *et al.* selected water as the best extraction solvent. The suitability of this methodology was evaluated by examination of metabolic changes in the polyamine pathway caused by the potent ornithine decarboxylase inhibitor, difluoromethylornithine. Ten metabolites, including putrescine, ornithine, γ-aminobutyric acid, oxidized and reduced forms of glutathione, 5′-deoxy-5′-(methylthio)adenosine, *N*-acetylputrescine, cysteinyl-glycine, spermidine and one unknown compound, were found to be significantly altered ($p < 0.05$) by difluoromethylornithine administration. In addition, minor modifications of metabolic pathways related to the intracellular thiol redox state were also observed.

7.3.5 Other Samples

In addition to blood and urine, other biofluids can also be used in CE-MS metabolomics. For example, cerebrospinal fluid (CSF) is the preferred biofluid for the discovery of new biomarkers for central nervous system disorders, because the metabolite concentrations in CSF are considered to be directly correlated to their brain contents.[80] One of the central nervous system disorders, AD, is characterized by progressive loss of memory and other cognitive functions leading to dementia. Non-targeted CE-MS metabolomics has been applied to investigate metabolite differences in CSF from patients with different cognitive statuses related to AD progression.[81] Eighty-five CSF samples from patients with subjective cognitive impairment (control), mild cognitive impairment (MCI) that remained stable after a follow-up period of 2 years, MCI that progressed to AD within 2 years after initial MCI diagnosis and AD were subjected to CE-TOF-MS analysis. A prediction model for AD progression using multivariate statistical analysis was developed, which could correctly classify 97–100% of the tested samples into their diagnosis group. Furthermore, choline, dimethylarginine (symmetric- or asymmetric-dimethylarginine was not specified in the manuscript), arginine, valine, proline, serine, histidine, creatine, carnitine and suberylglycine were identified as possible disease progression markers. These results suggest that CSF is a promising biofluid to predict AD progression.

Saliva can also be collected non-invasively, like urine. Sugimoto *et al.* performed non-targeted metabolome analysis of saliva obtained from 215 individuals, including 69 oral, 18 pancreatic and 30 breast cancer patients, 11 periodontal disease patients and 87 healthy controls using CE-TOFMS.[82] As a result, they identified 57 principal metabolites that can be used as predictive biomarkers for each individual disease. In addition, Sugimoto *et al.* developed multiple logistic regression models and obtained high AUC values to discriminate healthy controls from each disease (the AUC values are 0.865 for oral cancer, 0.973 for breast cancer, 0.993 for pancreatic cancer and 0.969 for periodontal disease). These values remained high with cross-validation study. These results indicate that CE-MS can be readily applied to salivary metabolomics and is a promising tool for biomarker discovery for medical screening.

Gut microbiota is involved in uremic solute metabolism. To investigate the influence of gut microbiota on the retention of uremic solutes in chronic kidney disease (CKD), Mishima *et al.* compared feces metabolite profiles, in addition to plasma and urine metabolite profiles, of adenine-induced renal failure and control mice under germ-free or specific pathogen-free conditions by CE-TOF-MS.[83] Of the 183 detected metabolites, 11 metabolites, including indoxyl sulfate, *p*-cresyl sulfate, phenyl sulfate, cholate, hippurate, dimethylglycine, γ-guanidinobutyrate, glutarate, 2-hydroxypentanoate, trimethylamine *N*-oxide and phenaceturate, are considered to be microbiota-derived uremic solutes. In particular, indoxyl sulfate, *p*-cresyl sulfate, phenyl sulfate, cholate and hippurate were found to be completely derived from microbiota (Figure 7.8). In addition, the disappearance of colonic short-chain fatty acids and decreased utilization of intestinal amino acids were also shown in germ-free renal failure mice. Microbiota contributes substantially to the production of harmful uremic solutes; meanwhile, a lack of microbiota also has harmful effects on CKD progression.

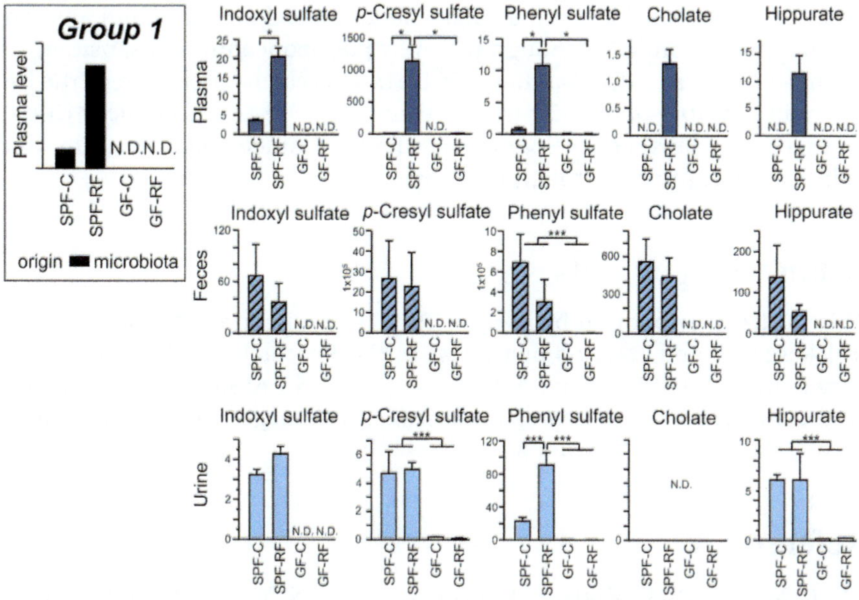

Figure 7.8 Profiles of microbiota-derived uremic solutes. GF-C, germ-free control; GF-RF, germ-free renal failure; SPF-C, specific pathogen-free control; SPF-RF, specific pathogen-free renal failure. Plasma concentrations are in μmol L^{-1}. Fecal and urinary levels are corrected for fecal weight (nmol g^{-1}) and urinary creatinine content (μmol mg^{-1} urinary creatinine), respectively. The *p*-cresyl sulfate and phenyl sulfate levels are not concentrations, but rather relative intensities estimated from the peak areas. *$P < 0.05$, ***$P < 0.001$ compared between indicated groups (analysis of variance). N.D. denotes Not Detected. Reproduced from *Kidney International*, **92**, E. Mishima, S. Fukuda, C. Mukawa, A. Yuri, Y. Kanemitsu, Y. Matsumoto, Y. Akiyama, N. N. Fukuda, H. Tsukamoto and K. Asaji, Evaluation of the impact of gut microbiota on uremic solute accumulation by a CE-TOFMS–based metabolomics approach, 634–645.[83] Copyright 2017 International Society of Nephrology, with permission from Elsevier.

7.4 Conclusions

Currently, several MS-based analytical methods, such as GC-MS, LC-MS and CE-MS, are widely employed in metabolomics. Both the physical and chemical properties of metabolites are varied; therefore, the choice of separation platform is important. Different methodologies have distinct advantages that can be exploited in investigating different metabolite classes. Among the existing methods, CE-MS is particularly suited for the analysis of polar and charged metabolites.

In this chapter, we focused on general CE-MS conditions frequently used in metabolomics and provided definite examples of medical and biochemical applications. Of course, CE-MS-based metabolome analysis can be applied to other research fields, such as plant, food, biotechnological and environmental metabolomics. Although relatively poor concentration sensitivity is often described as a disadvantage of CE-MS, this can be overcome by the use of on-line sample pre-concentration techniques[84,85] and sheathless CE-MS interfaces.[86–89] Further development of these techniques adapted to any kind of biological sample is strongly expected.

Acknowledgements

This work was supported by AMED-CREST from the Japan Agency for Medical Research and Development, AMED, the Research on Development of New Drugs (GAPFREE) from AMED, and the JSPS KAKENHI Grant Number JP15K21364, as well as research funds from the Yamagata prefectural government and the City of Tsuruoka.

References

1. D. S. Wishart, C. Knox, A. C. Guo, R. Eisner, N. Young, B. Gautam, D. D. Hau, N. Psychogios, E. Dong, S. Bouatra, R. Mandal, I. Sinelnikov, J. G. Xia, L. Jia, J. A. Cruz, E. Lim, C. A. Sobsey, S. Shrivastava, P. Huang, P. Liu, L. Fang, J. Peng, R. Fradette, D. Cheng, D. Tzur, M. Clements, A. Lewis, A. De Souza, A. Zuniga, M. Dawe, Y. P. Xiong, D. Clive, R. Greiner, A. Nazyrova, R. Shaykhutdinov, L. Li, H. J. Vogel and I. Forsythe, *Nucleic Acids Res.*, 2009, **37**, D603.
2. C. Barbas, E. P. Moraes and A. Villasenor, *J. Pharm. Biomed. Anal.*, 2011, **55**, 823.
3. H. K. Kim, Y. H. Choi and R. Verpoorte, *Trends Biotechnol.*, 2011, **29**, 267.
4. J. L. Markley, R. Bruschweiler, A. S. Edison, H. R. Eghbalnia, R. Powers, D. Raftery and D. S. Wishart, *Curr. Opin. Biotechnol.*, 2017, **43**, 34.
5. H. Kato, Y. Izumi, T. Hasunuma, F. Matsuda and A. Kondo, *J. Biosci. Bioeng.*, 2012, **113**, 665.
6. D. Q. Tang, L. Zou, X. X. Yin and C. N. Ong, *Mass Spectrom. Rev.*, 2016, **35**, 574.
7. T. Soga, Y. Ohashi, Y. Ueno, H. Naraoka, M. Tomita and T. Nishioka, *J. Proteome Res.*, 2003, **2**, 488.
8. I. Garcia-Perez, J. C. Lindon and E. Minet, *Bioanalysis*, 2014, **6**, 2733.
9. S. Koike, M. Bundo, K. Iwamoto, M. Suga, H. Kuwabara, Y. Ohashi, K. Shinoda, Y. Takano, N. Iwashiro, Y. Satomura, T. Nagai, T. Natsubori, M. Tada, H. Yamasue and K. Kasai, *Transl. Psychiatry*, 2014, **4**, e379.
10. L. C. Soliman, Y. Hui, A. K. Hewavitharana and D. D. Y. Chen, *J. Chromatogr. A*, 2012, **1267**, 162.
11. D. Kim, B. P. Fiske, K. Birsoy, E. Freinkman, K. Kami, R. L. Possemato, Y. Chudnovsky, M. E. Pacold, W. W. Chen, J. R. Cantor, L. M. Shelton, D. Y. Gui, M. Kwon, S. H. Ramkissoon, K. L. Ligon, S. W. Kang, M. Snuderl, M. G. Vander Heiden and D. M. Sabatini, *Nature*, 2015, **520**, 363.

12. S. Tohyama, J. Fujita, T. Hishiki, T. Matsuura, F. Hattori, R. Ohno, H. Kanazawa, T. Seki, K. Nakajima, Y. Kishino, M. Okada, A. Hirano, T. Kuroda, S. Yasuda, Y. Sato, S. Yuasa, M. Sano, M. Suematsu and K. Fukuda, *Cell Metab.*, 2016, **23**, 663.
13. V. R. Robledo and W. F. Smyth, *Electrophoresis*, 2009, **30**, 1647.
14. A. Anegawa, M. Ohnishi, D. Takagi, C. Miyake, C. Shichijo, K. Ishizaki, H. Fukaki and T. Mimura, *Plant Biotechnol.*, 2015, **32**, 65.
15. K. Maruyama, K. Urano, K. Yoshiwara, Y. Morishita, N. Sakurai, H. Suzuki, M. Kojima, H. Sakakibara, D. Shibata, K. Saito, K. Shinozaki and K. Yamaguchi-Shinozaki, *Plant Physiol.*, 2014, **164**, 1759.
16. W. S. Suharti, A. Nose and S. H. Zheng, *J. Plant Physiol.*, 2016, **206**, 13.
17. A. Hirayama, M. Wakayama and T. Soga, *TrAC, Trends Anal. Chem.*, 2014, **61**, 215.
18. R. Ramautar, O. A. Mayboroda, G. W. Somsen and G. J. de Jong, *Electrophoresis*, 2011, **32**, 52.
19. R. Ramautar, G. W. Somsen and G. J. de Jong, *Electrophoresis*, 2013, **34**, 86.
20. R. Ramautar, G. W. Somsen and G. J. de Jong, *Electrophoresis*, 2015, **36**, 212.
21. R. Ramautar, G. W. Somsen and G. J. de Jong, *Electrophoresis*, 2017, **38**, 190.
22. M. Wakayama, A. Hirayama and T. Soga, *Metabonomics: Methods and Protocols*, ed. J. T. Bjerrum, Springer New York, New York, 2015, pp. 113–122.
23. T. Soga and D. N. Heiger, *Anal. Chem.*, 2000, **72**, 1236.
24. T. Soga, R. Baran, M. Suematsu, Y. Ueno, S. Ikeda, T. Sakurakawa, Y. Kakazu, T. Ishikawa, M. Robert, T. Nishioka and M. Tomita, *J. Biol. Chem.*, 2006, **281**, 16768.
25. H. Katayama, Y. Ishihama and N. Asakawa, *Anal. Chem.*, 1998, **70**, 5272.
26. T. Soga, Y. Ueno, H. Naraoka, Y. Ohashi, M. Tomita and T. Nishioka, *Anal. Chem.*, 2002, **74**, 2233.
27. T. Soga, K. Igarashi, C. Ito, K. Mizobuchi, H. P. Zimmermann and M. Tomita, *Anal. Chem.*, 2009, **81**, 6165.
28. H. Sawada and C. Nogami, *Anal. Chim. Acta*, 2004, **507**, 191.
29. K. Harada, E. Fukusaki and A. Kobayashi, *J. Biosci. Bioeng.*, 2006, **101**, 403.
30. M. Yamamoto, R. Ly, B. Gill, Y. J. Zhu, J. Moran-Mirabal and P. Britz-McKibbin, *Anal. Chem.*, 2016, **88**, 10710.
31. J. M. Ding and P. Vouros, *Anal. Chem.*, 1999, **71**, 378a.
32. R. D. Smith, C. J. Barinaga and H. R. Udseth, *Anal. Chem.*, 1988, **60**, 1948.
33. V. Sanz-Nebot, F. Benavente, E. Balaguer and J. Barbosa, *Electrophoresis*, 2003, **24**, 883.
34. J. Krenkova and F. Foret, *Proteomics*, 2012, **12**, 2978.
35. M. Mokaddem, P. Gareil, J. E. Belgaied and A. Varenne, *Electrophoresis*, 2009, **30**, 1692.
36. B. Desmazieres, W. Buchmann, P. Terrier and J. Tortajada, *Anal. Chem.*, 2008, **80**, 783.
37. P. Hommerson, A. M. Khan, G. J. de Jong and G. W. Somsen, *Mass Spectrom. Rev.*, 2011, **30**, 1096.
38. S. J. Bos, S. M. van Leeuwen and U. Karst, *Anal. Bioanal. Chem.*, 2006, **384**, 85.

39. K. A. Hanold, S. M. Fischer, P. H. Cormia, C. E. Miller and J. A. Syage, *Anal. Chem.*, 2004, **76**, 2842.
40. I. Marchi, S. Rudaz and J. L. Veuthey, *Talanta*, 2009, **78**, 1.
41. J. Schappler, D. Guillarme, J. Prat, J. L. Veuthey and S. Rudaz, *Electrophoresis*, 2007, **28**, 3078.
42. Y. Takada, M. Sakairi and H. Koizumi, *Anal. Chem.*, 1995, **67**, 1474.
43. Y. Tanaka, K. Otsuka and S. Terabe, *J. Pharm. Biomed. Anal.*, 2003, **30**, 1889.
44. V. R. Robledo and W. F. Smyth, *Electrophoresis*, 2014, **35**, 2292.
45. J. Ohnesorge, C. S. V. de Griend and H. Watzig, *Electrophoresis*, 2005, **26**, 2360.
46. G. Hopfgartner, E. Varesio, V. Tschappat, C. Grivet, E. Bourgogne and L. A. Leuthold, *J. Mass Spectrom.*, 2004, **39**, 845.
47. G. Bringmann, I. Kajahn, C. Neususs, M. Pelzing, S. Laug, M. Unger and U. Holzgrabe, *Electrophoresis*, 2005, **26**, 1513.
48. J. V. Sancho, O. J. Pozo, M. Ibanez and F. Hernandez, *Anal. Bioanal. Chem.*, 2006, **386**, 987.
49. A. H. Que, Y. Mechref, Y. P. Huang, J. A. Taraszka, D. E. Clemmer and M. V. Novotny, *Anal. Chem.*, 2003, **75**, 1684.
50. S. Forcisi, F. Moritz, B. Kanawati, D. Tziotis, R. Lehmann and P. Schmitt-Kopplin, *J. Chromatogr. A*, 2013, **1292**, 51.
51. M. Himmelsbach, *J. Chromatogr. B*, 2012, **883**, 3.
52. Q. Z. Hu, R. J. Noll, H. Y. Li, A. Makarov, M. Hardman and R. G. Cooks, *J. Mass Spectrom.*, 2005, **40**, 430.
53. M. W. Senko, J. D. Canterbury, S. H. Guan and A. G. Marshall, *Rapid Commun. Mass Spectrom.*, 1996, **10**, 1839.
54. K. Faserl, B. Sarg, V. Maurer and H. H. Lindner, *J. Chromatogr. A*, 2017, **1498**, 215.
55. Y. M. Zhao, L. L. Sun, G. J. Zhu and N. J. Dovichi, *J. Proteome Res.*, 2016, **15**, 3679.
56. K. Dettmer, P. A. Aronov and B. D. Hammock, *Mass Spectrom. Rev.*, 2007, **26**, 51.
57. S. G. Villas-Boas, S. Mas, M. Akesson, J. Smedsgaard and J. Nielsen, *Mass Spectrom. Rev.*, 2005, **24**, 613.
58. R. Gonzalez-Dominguez, A. Garcia, T. Garcia-Barrera, C. Barbas and J. L. Gomez-Ariza, *Electrophoresis*, 2014, **35**, 3321.
59. E. Yamano, M. Sugimoto, A. Hirayama, S. Kume, M. Yamato, G. H. Jin, S. Tajima, N. Goda, K. Iwai, S. Fukuda, K. Yamaguti, H. Kuratsune, T. Soga, Y. Watanabe and Y. Kataoka, *Sci. Rep.*, 2016, **6**, 34990.
60. J. Zeng, P. Y. Yin, Y. X. Tan, L. W. Dong, C. X. Hu, Q. Huang, X. Lu, H. Y. Wang and G. W. Xu, *J. Proteome Res.*, 2014, **13**, 3420.
61. S. Harada, T. Takebayashi, A. Kurihara, M. Akiyama, A. Suzuki, Y. Hatakeyama, D. Sugiyama, K. Kuwabara, A. Takeuchi, T. Okamura, Y. Nishiwaki, T. Tanaka, A. Hirayama, M. Sugimoto, T. Soga and M. Tomita, *Environ. Health Prev. Med.*, 2016, **21**, 18.

62. M. Iida, S. Harada, A. Kurihara, K. Fukai, K. Kuwabara, D. Sugiyama, A. Takeuchi, T. Okamura, M. Akiyama, Y. Nishiwaki, A. Suzuki, A. Hirayama, M. Sugimoto, T. Soga, M. Tomita, K. Banno, D. Aoki and T. Takebayashi, *Menopause*, 2016, **23**, 749.
63. K. Fukai, S. Harada, M. Iida, A. Kurihara, A. Takeuchi, K. Kuwabara, D. Sugiyama, T. Okamura, M. Akiyama, Y. Nishiwaki, Y. Oguma, A. Suzuki, C. Suzuki, A. Hirayama, M. Sugimoto, T. Soga, M. Tomita and T. Takebayashi, *PLoS One*, 2016, **11**, e0164877.
64. A. Hirayama, M. Sugimoto, A. Suzuki, Y. Hatakeyama, A. Enomoto, S. Harada, T. Soga, M. Tomita and T. Takebayashi, *Electrophoresis*, 2015, **36**, 2148.
65. A. H. Zhang, H. Sun, X. H. Wu and X. J. Wang, *Clin. Chim. Acta*, 2012, **414**, 65.
66. J. L. Chen, J. Fan and X. J. Lu, *Electrophoresis*, 2014, **35**, 1032.
67. J. V. Alberice, A. F. S. Amaral, E. G. Armitage, J. A. Lorente, F. Algaba, E. Carrilho, M. Marquez, A. Garcia, N. Malats and C. Barbas, *J. Chromatogr. A*, 2013, **1318**, 163.
68. S. B. Heymsfield, C. Arteaga, C. Mcmanus, J. Smith and S. Moffitt, *Am. J. Clin. Nutr.*, 1983, **37**, 478.
69. J. C. Verhave, P. Fesler, J. Ribstein, G. du Cailar and A. Mimran, *Am. J. Kidney Dis.*, 2005, **46**, 233.
70. E. J. Saude and B. D. Sykes, *Metabolomics*, 2007, **3**, 19.
71. C. Y. Lin, H. F. Wu, R. S. Tjeerdema and M. R. Viant, *Metabolomics*, 2007, **3**, 55.
72. K. Maekawa, A. Hirayama, Y. Iwata, Y. Tajima, T. Nishimaki-Mogami, S. Sugawara, N. Ueno, H. Abe, M. Ishikawa, M. Murayama, Y. Matsuzawa, H. Nakanishi, K. Ikeda, M. Arita, R. Taguchi, N. Minamino, S. Wakabayashi, T. Soga and Y. Saito, *J. Mol. Cell. Cardiol.*, 2013, **59**, 76.
73. A. Hirayama, K. Kami, M. Sugimoto, M. Sugawara, N. Toki, H. Onozuka, T. Kinoshita, N. Saito, A. Ochiai, M. Tomita, H. Esumi and T. Soga, *Cancer Res.*, 2009, **69**, 4918.
74. P. Gao, C. J. Zhou, L. Zhao, G. H. Zhang and Y. Zhang, *J. Pharm. Biomed. Anal.*, 2016, **118**, 349.
75. K. Kami, T. Fujimori, H. Sato, M. Sato, H. Yamamoto, Y. Ohashi, N. Sugiyama, Y. Ishihama, H. Onozuka, A. Ochiai, H. Esumi, T. Soga and M. Tomita, *Metabolomics*, 2013, **9**, 444.
76. B. M. Hounoum, H. Blasco, P. Emond and S. Mavel, *TrAC, Trends Anal. Chem.*, 2016, **75**, 118.
77. C. G. Zhao, H. J. Du, L. Xu, J. Wang, L. Tang, Y. F. Cao, C. Li, Q. J. Wang, Y. Liu, F. P. Shan, J. Feng, F. Xu and P. Gao, *J. Pharm. Biomed. Anal.*, 2015, **115**, 48.
78. H. Sakagami, M. Sugimoto, S. Tanaka, H. Onuma, S. Ota, M. Kaneko, T. Soga and M. Tomita, *Metabolomics*, 2014, **10**, 270.
79. C. Ibáñez, C. Simo, A. Valdes, L. Campone, A. L. Piccinelli, V. Garcia-Canas and A. Cifuentes, *J. Pharm. Biomed. Anal.*, 2015, **110**, 83.

80. C. E. Johanson, J. A. Duncan, P. M. Klinge, T. Brinker, E. G. Stopa and G. D. Silverberg, *Cerebrospinal Fluid Res.*, 2008, **5**, 10.
81. C. Ibanez, C. Simo, P. J. Martin-Alvarez, M. Kivipelto, B. Winblad, A. Cedazo-Minguez and A. Cifuentes, *Anal. Chem.*, 2012, **84**, 8532.
82. M. Sugimoto, D. T. Wong, A. Hirayama, T. Soga and M. Tomita, *Metabolomics*, 2010, **6**, 78.
83. E. Mishima, S. Fukuda, C. Mukawa, A. Yuri, Y. Kanemitsu, Y. Matsumoto, Y. Akiyama, N. N. Fukuda, H. Tsukamoto and K. Asaji, *Kidney Int.*, 2016, **92**, 634.
84. F. Kitagawa and K. Otsuka, *J. Chromatogr. A*, 2014, **1335**, 43.
85. A. Slampova, Z. Mala, P. Gebauer and P. Bocek, *Electrophoresis*, 2017, **38**, 20.
86. A. Hirayama, M. Tomita and T. Soga, *Analyst*, 2012, **137**, 5026.
87. M. Moini, *Anal. Chem.*, 2007, **79**, 4241.
88. R. Ramautar, J. M. Busnel, A. M. Deelder and O. A. Mayboroda, *Anal. Chem.*, 2012, **84**, 885.
89. L. H. Shi, Y. X. Jin, D. C. Moon, S. K. Kim and S. R. Park, *Electrophoresis*, 2009, **30**, 1661.

CHAPTER 8

CE-MS for Metabolomics: A Comparison with Other Techniques

JOANNA GODZIEN, ANTONIA GARCIA, ANGELES LÓPEZ-GONZALVEZ AND CORAL BARBAS*

Centre for Metabolomics and Bioanalysis (CEMBIO), Facultad de Farmacia, Universidad San Pablo CEU, Boadilla del Monte 28668, Madrid, Spain
*E-mail: cbarbas@ceu.es

8.1 Introduction

Untargeted metabolomics aims to measure the whole set of metabolites within a given biological material and to compare the levels of the small molecules present between different conditions, *e.g.*, healthy *vs.* disease, treated *vs.* untreated *etc.*, providing potential markers of the situation under study. The novelty of this global approach is that it involves working without *a priori* hypotheses, often just assuming that the conditions under comparison are different. In order to reach sound conclusions, as many metabolites as possible have to be measured, and advanced statistical analysis has to be performed to reveal those changes that will be later used to build new hypotheses about the mechanism of action of a drug or the pathophysiological changes caused by a disease or any other condition. This profoundly changes the dogma of research from hypothesis rejection/confirmation to hypothesis generation.[1]

New Developments in Mass Spectrometry No. 6
Capillary Electrophoresis–Mass Spectrometry for Metabolomics
Edited by Rawi Ramautar

Published by the Royal Society of Chemistry, www.rsc.org

The diversity in terms of mass, concentration, polarity, volatility, solubility, pK_a and charge of metabolites in a biological sample makes their simultaneous analysis difficult, since the analytical platforms used to detect these compounds tend to exploit specific chemical characteristics of different classes of metabolites.[2] The use of different high-resolution separation techniques such as gas chromatography (GC), liquid chromatography (LC) or capillary electrophoresis (CE) coupled to mass spectrometry (MS) in a multiplatform approach provides broader metabolite coverage. GC-MS was introduced in the late sixties and is the oldest and most robust coupling between a separation technique and mass spectrometry, and is also the most reproducible in relation to retention times and mass spectra. LC-MS was not commercialised until the seventies. Its long development time was related to the interphase limitations and the need to design new ionisation methods suitable for liquid samples. CE-MS was introduced in the late eighties and although it is the newest coupling approach, it is also the least robust and, moreover, the least reproducible in relation to migration time (MT). That said, coupled to an accurate mass analyser, it can provide crucial information in metabolomics.

GC-MS and LC-MS are clearly complementary analytical tools, as the mechanism of the separation is based on the principles of chromatography in both. However, GC can only be applied to compounds that are volatile or can be made volatile by derivatisation, meaning that its use is limited to small metabolites, while LC, in the most common mechanism, that is reversed phase, is mainly devoted to medium to non-polar compounds soluble in organic solvents with no weight limitation. Furthermore, GC is pressure/temperature-driven, while LC is flow-driven. CE is, as an orthogonal technique based on electro-driven separations, mainly focused on polar or ionic compounds.

Regarding LC, different mechanisms applicable to different types of compounds can also be considered, along with their respective strengths and weaknesses: (*i*) reversed phase liquid chromatography (RPLC), which mainly uses C18 and C8 stationary phases, is the most common type of LC and the one preferred for medium to non-polar compounds, mostly lipids; (*ii*) HILIC is a modification of normal phase liquid chromatography which uses hydrophilic stationary phases with reversed-phase type eluents. Analytes elute with decreasing polarity for RP and increasing polarity for HILIC, which makes HILIC more suitable for polar metabolites; (*iii*) ion chromatography (IC), devoted to ionic compounds or compounds that can be ionised through pH, which includes (*iii.a*) ion pairing (IP) — a modification of RPLC, where the mobile phase is supplemented with an ion-pairing reagent. Ion-pairing reagents consist of large ionic molecules that have a charge opposite to that of the analyte of interest, as well as a substantial hydrophobic region that allows interaction with the stationary phase, plus associated counter-ions; (*iii.b*) ion-exchange chromatography (IEC), which depends upon the reversible adsorption of charged solute molecules to immobilized ion exchange groups of opposite charge based on coulombic (ionic) interactions. IC-MS is more robust and reproducible than CE-MS; however, it is limited to ionic metabolites and because of this, it is considered mainly for target or semi-target approaches rather than for global profiling. Given the small number of applications in metabolomics studies,[3–6] this platform will be omitted from further comparisons.

A top-down approach opens up research to new, unexpected findings but also provides new hypotheses for other research areas. As a result, the metabolomics workflow can be represented as a cycle instead of a linear scheme (Figure 8.1).

It is a cascade of successive stages including: design of the experiment, sampling and sample preparation, analysis, data reprocessing and treatment, identification, and finally data interpretation. It is important to mention that this scheme is true for LC-MS- and CE-MS-based metabolomics, where—due to metabolite identification issues—identification is usually performed after statistical analysis to reduce the number of signals requiring identification. Meanwhile, in the case of GC-MS analysis, the high inter-laboratory reproducibility of retention time (RT) and fragmentation patterns allows quick and reliable identification by using libraries. In this way, most measured signals can be identified after deconvolution, and identification comes together with data reprocessing and is followed by statistical treatment and interpretation.

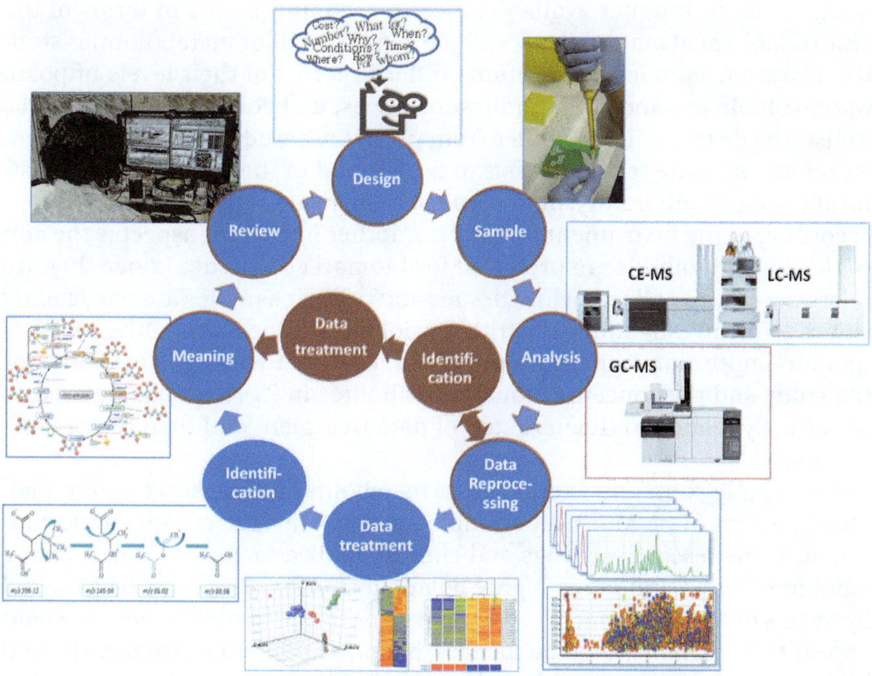

Figure 8.1 Scheme of the metabolomics workflow, highlighting differences between CE-MS and LC-MS and GC-MS platforms.

We will now focus on each step of the metabolomics process, highlighting the main advantages and disadvantages of CE-MS in comparison with other separation techniques including GC-MS and LC-MS, with LC-MS further divided into RP-LC-MS and HILIC-LC-MS.

8.2 Design of the Experiment

The very first step in any analytical process is the definition of the question to be answered by the experiment. A clearly defined problem, in terms of both biological and analytical aspects, is necessary to plan the entire process properly, designing each phase of work appropriately to match the aims of the study. This seems trivial, and indeed it is often neglected due to its obviousness, resulting in the failure of the project. This step is the same for all analytical platforms and performing it is equally important for each technique, describing in detail the requirements for all steps of the workflow through standard operation procedures (SOPs).

8.3 Sample Selection and Collection

Sample selection must be performed considering mainly the aim of the study and the analytical platforms available. Due to the continuous improvement of the techniques available, there are no limitations in terms of the type of biological material; any sample can be used for metabolomics studies. However, sample compositions differ in terms of their levels of polar, water-soluble metabolites, lipidic substances, and volatile compounds, as well as the degree of interference from proteins. Based on this information, either an adequate technique has to be selected in the case of a multi-platform equipped laboratory, or the type of biological sample has to be chosen according to the instrument available. Another important aspect is the aim of the study: biofluids are preferred for biomarker searches (since they are relatively easy to collect), while tissues and cellular samples are particularly suitable for investigation of pathophysiological processes.[7] Other aspects that are important to mention are the amount of the sample available for the study and the concentration of metabolites in it. These characteristics are directly related to the next step of data treatment and then to the analysis itself.

Any sample containing small, polar metabolites can be successfully analysed by means of CE-MS. A key example is urine, which plays a very important role in many research settings and clinical applications. It contains a large amount of water and many highly polar substances. Urine can be analysed by means of any platform; however, CE seems to be the best choice, as would appear to be demonstrated by the fact that urine is the most frequently used biological sample in CE-MS-based metabolomics.[8] For GC-MS analysis, the high content of urea affects the profiles and therefore requires the use of urease, which in turn increases the variability during sample preparation. HILIC analysis can be performed with urine; however, the high content of salts affects the chromatographic resolution, leading to broadening of the peaks, which is especially problematic in the case of large-scale studies. As already mentioned, salts are also problematic for samples such as cells and cell culture media, which are a valuable source of information regarding mechanisms of action. Usually the simultaneous analysis of intra- (fingerprinting)[9] and extracellular content (footprinting)[10] is required. Other samples containing high levels of salts such as saline solution, tears or sweat can be better analysed by means of CE than with other techniques. Apart from the content of the sample itself, the amount of the sample needed is also very important. This is a crucial consideration for multiplatform approaches where a single amount of biological material needs to be used for different analyses. Volume-limited samples such as those coming from small experimental animals are well suited for CE-MS, because only a few nanolitres are used for the analysis.

Hyphenation of capillary electrophoresis to mass spectrometry is generally performed *via* a sheath–liquid interface, which dilutes the analytes. Recently, some groups have paid attention to the sheathless porous tip interface,[11] which could increase the sensitivity, although robustness remains a challenge.

CE is not suitable for lipidic metabolites; therefore, information about almost the entire lipidome cannot be obtained by means of this platform. This constraint is mostly due to the solubility of lipids; only lipids soluble in water can be analysed, which limits them to small and relatively polar ones, containing some polar groups such as nitriles or alcohols. Therefore, although very lipidic samples such as brain, liver and breast milk can be analysed by CE-MS, this technique will not provide all information about the content of the sample and it should therefore be analysed by RP-LC-MS.

8.4 Sample Preparation

Sample preparation aims to extract as many metabolites as possible from the biological material, eliminating at the same time all potential interferences such as proteins, DNA, particles, salts, *etc.* To achieve this, mixtures of different solvents are usually used, allowing the extraction of the metabolites of interest and precipitation of proteins. Depending on the sample type and selected platform, either more polar or nonpolar metabolites are required and, therefore, the proportions between water and organic solvents will change. The most common solvents include methanol, ethanol, acetonitrile, chloroform and methyl-*tert*-butylether (MTBE) alone or in combination at different proportions, together with water.[12] The choice should reflect a compromise between efficient recovery of polar and non-polar metabolites. Protein precipitation is usually achieved with ice-cold organic solvents. Another strategy involves the use of filtration devices with a particular cut-off to separate particles and proteins from low molecular mass substances.

The great advantage of sample preparation for CE-MS together with LC-MS is its simplicity in comparison to GC-MS. The reason for this is that sample preparation for GC analysis includes long derivatisation processes[13] aiming to transform non-volatile metabolites into their volatile forms. This increases the time needed for sample preparation, as well as the variability of the obtained data, and requires very good sample manipulation skills. Furthermore, the derivatisation step can introduce a number of biases to the GC-MS metabolomic data, which are not encountered in chemical analyses that do not require derivatisation and will be explained in detail below in Table 8.1. Moreover, biases can affect each metabolite to a different extent.[14]

HILIC-LC-MS is particularly suitable for polar molecules; however, ironically it requires some acetonitrile (usually around 50%) in the sample to perform chromatography successfully. This causes problems with the solubility of very polar metabolites, which precipitate resulting in the loss of information and also often causing system clogging.

Sample stability is also important, especially in the case of very big batches or multi-injections. From this perspective, GC-MS samples are the most problematic since derivatised metabolites can easily evaporate, changing the content of the sample in an uncontrolled way. Similar problems can be noted in relation to LC-MS samples prepared with chloroform or MTBE. However, here the loss of solvent instead of metabolites is observed, and this can be partially overcome with internal standards. In this regard, the limited amount of organic solvent in CE-MS samples makes them more resistant to solvent evaporation but more prone to degradation, which can be limited by the addition of formic acid.

A clear drawback of sample preparation for CE-MS is in regard to protein precipitation. As mentioned, this is usually obtained by means of cold organic solvents, which cannot be infused into the CE system, because their low conductivity affects the current stability and therefore the electrophoretic separation. To avoid this, organic solvents have to be evaporated and the dried pellet has to be resuspended in a water-based buffer. Such sample manipulation increases time and variability. This can be overcome by using ultrafiltration devices, which avoid most of these problems, although they do increase the cost of the sample treatment.

8.5 Analysis

Extracted samples are subjected to analysis, which usually includes two steps: separation and detection. Separation techniques are preferred due to the large sample complexity, which results in the need to separate metabolites prior to their measurement. In this way, matrix effects and ion suppression can be reduced, providing higher quality information. Detection involves the use of mass spectrometers, which—in most cases—provide a highly

Table 8.1 Metabolites with controversial identification by GC-MS and description of the conditions used with CE-MS to address and solve the problem.

Compound	Potential reason	Sample	Conditions	CE-MS reference
Amino acids				
Arginine - Ornithine	Conversion of arginine into the TMS[a] derivative of ornithine Halket 2005[37]	Rat urine	Cap: fused-silica. *V*: +30 kV BGE: 1 M formic acid ESI: 4.0 kV SL: methanol/water (50/50)	Zeng 2013[38]
Asparagine	Reduced derivatisation efficiency Koek 2006[13]	Rat urine Rat plasma Mouse lung tissue	Cap: fused-silica. *V*: +30 kV BGE: 0.8 M formic acid in 10% methanol ESI: 3.5 kV SL: 1 mM formic acid in methanol/water (50/50)	Godzien 2013[39] Shama 2013[40] Shama 2013[41]
Glutamic acid-pyroglutamic	Reduced derivatisation efficiency and conversions Koek 2006[13]	Human plasma and serum	Cap: fused-silica. *V*: +30 kV BGE: 1 M formic acid ESI: 4.0 kV. SL: methanol/water (50/50)	Hirayama 2015[42]
Glutamine-glutamic acid	Reduced derivatisation efficiency and conversions Koek 2006[13]	Human urine from children and adults	Cap: fused-silica. *V*: +30 kV. BGE: 0.8 M formic acid in 10% methanol ESI: 3.5 kV SL: 1 mM formic acid in methanol/water (50/50)	Balderas 2013[43] Alberice 2013[44]
Lysine	Reduced derivatisation efficiency Koek 2006[13]	Human saliva samples	Cap: fused-silica. *V*: +30 kV BGE: 1 M formic acid ESI: 4.0 kV SL: methanol/water (50/50 v/v)	Sugimoto 2010[45]

(*continued*)

Table 8.1 (*continued*)

Compound	Potential reason	Sample	Conditions	CE-MS reference
Methionine	Reduced derivatisation efficiency Koek 2006 [13]	24 h human breast milk	Cap: fused-silica. *V*: +25 kV BGE: 0.1 M formic acid ESI: 3.5 kV SL: 1 mM formic acid in methanol/water (70/30)	Andreas 2015 [46]
Tryptophan	Reduced derivatisation efficiency	Human aqueous humour	Cap: fused-silica. *V*: +30 kV BGE: 0.8 M formic acid ESI: 3.5 kV SL: 1 mM formic acid in methanol/water (50/50)	Barbas-Bernandos 2016 [47]
	Koek 2006 [13]	Myocardial tissue	Cap: fused-silica. *V*: +30 kV BGE: 1 M formic acid ESI: 4.0 kV SL: methanol/water (50/50)	Maekawa 2013 [32]
		Colon cancer cells	Cap: fused-silica. *V*: +30 kV BGE: 1 M formic acid ESI: 3.5 kV SL: isopropanol/water (50/50)	Ibáñez 2012 [48]
Short chain organic acid				
Glycolic acid	High volatility Nielsen 2007 [10]	Human serum	Cap: fused-silica. *V*: +30 kV BGE: 1 M formic acid ESI: 3.5 kV SL: isopropanol/water (50/50)	Yoshimi 2016 [49]
Malic acid	Reduced derivatisation efficiency Koek 2006 [13]	Rat urine and plasma	Cap: FunCap-CE type S. *V*: −30 kV BGE: 50 mM ammonium acetate solution pH 9.0. ESI: 3.5 kV. Platinum ESI needle SL: 5 mM ammonium formate in acetonitrile/water (50/50)	Shima 2011 [50]

Citric/Isocitric	Isomers	Human plasma and serum	Cap: cationic polymer coated COSMO (+). *V*: −30 kV BGE: 50 mM ammonium acetate solution pH 8.5 SL: 5 mM ammonium acetate in methanol/water (50/50) ESI: 3.5 kV. Platinum ESI needle	Hirayama 2015[42]
		Human U-87 MG glioblastoma cell line	Cap: fused-silica. *V*: −30 kV BGE: 10% acetic acid pH 2.2 Sheathless MS: −1.5 kV	Gulersonmez 2016[52]
		Rat plasma	Cap: cationic polymer coated COSMO (+). *V*: −30 kV BGE: 50 mM ammonium acetate solution pH 8.5 ESI: 3.5 kV. Platinum ESI needle	Kume 2015[53]
		Bacillus subtilis	Cap: cationic polymer coated SMILE (+). *V*: −30 kV BGE: 50 mM ammonium acetate solution pH 8.5 ESI: 3.5 kV. Platinum ESI needle	Soga 2003[54]

(*continued*)

Table 8.1 (*continued*)

Compound	Potential reason	Sample	Conditions	CE-MS reference
Sugar phosphates				
Fructose-6-phosphate Glucose-6-phosphate Glyceraldehyde-3-phosphate Glycerol-3-phosphate	Reduced derivatisation efficiency Koek 2006[13] Low recovery (lost into the Speedvac) Nielsen 2007[10]	Human U-87 MG glioblastoma cell line	Cap: fused-silica. *V*: −30 kV BGE: 10% acetic acid pH 2.2 Sheathless MS[18] MS: −1.5 kV	Gulersonmez 2016[52]
		Human urine	Cap: fused-silica. *V*: +30 kV BGE: 1 M formic acid. ESI: 4.0 kV. SL: 0.1% formic acid in methanol/water (50/50)	Chen 2014[55]
		Rat urine	Cap: fused-silica. *V*: +30 kV BGE: 25 mM pH 11.7 ESI: 4.0 kV SL: 5 mM TEA in methanol/water (50/50)	Kok 2013[56]
		Human serum	Cap: cationic polymer coated SMILE (+). *V*: −30 kV BGE: 50 mM ammonium acetate solution pH 8.5 ESI: 3.5 kV. Platinum ESI needle	Yoshimi 2016[49]
		Escherichia coli	Cap: fused-silica coated with PolyE. *V*: −30 kV BGE: 50 mM ammonium acetate pH 8.7 in water/methanol (95/5) SL: 20 mM ammonium hydroxide in isopropanol/water (50/50) ESI: 3.5 kV. Platinum ESI needle	Timischl 2008[57]

Nucleosides and nucleotides				
Adenosine 5′-phosphosulfate	Degradation Koek 2006[13]	Tumour and normal human tissue (colon and stomach)	Cap: fused-silica. *V*: +30 kV BGE: 50 mM ammonium acetate solution pH 7.5 ESI: 3.5 kV SL: 5 mM ammonium acetate solution in methanol/water (50/50)	Hirayama 2009[58]
		Bacillus subtilis	Cap: poly-dimethylsiloxane (DB-1). *V*: −30 kV BGE: 50 mM ammonium acetate solution pH 7.5 ESI: 3.5 kV. Platinum ESI needle	Soga 2003[54]
		Escherichia coli	Cap: fused-silica coated with PolyE-323. *V*: −30 kV BGE: 50 mM ammonium acetate pH 8.7 in water/methanol (95/5) SL: 20 mM ammonium hydroxide in isopropanol/water (50/50) ESI: 3.5 kV. Platinum ESI needle	Timischl 2008[57]
Uridine 5′-monophosphate	Poor recovery Koek 2006[13]	Human erythrocytes	Cap: poly-dimethylsiloxane (DB-1). *V*: −30 kV BGE: 50 mM ammonium acetate solution pH 7.5 ESI: 3.5 kV. Platinum ESI needle	Kinoshita 2007[59]
		Mouse heart tissue	Cap: fused silica. *V*: −30 kV BGE: 50 mM ammonium acetate solution pH 7.5 ESI: 3.5 kV. Platinum ESI needle	Shimura 2013[60]

[a]TMS: trimethylsilyl, Cap: CE capillary; *V*: capillary voltage; BGE: background electrolyte; SL: sheath liquid; ESI: MS capillary voltage.

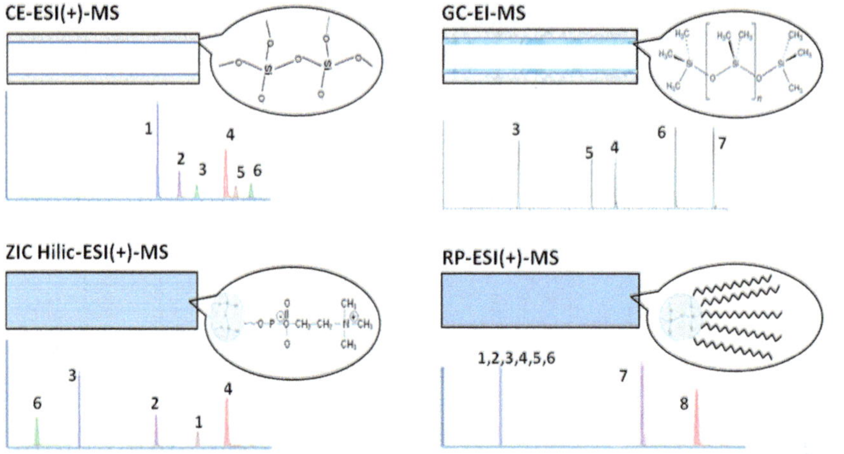

Figure 8.2 Scheme of analytical platforms used in MS-based metabolomics covering CE-MS, GC-MS, HILIC-LC-MS and RP-LC-MS, including capillary/column packing and examples of separation of several compounds: 1-carnitine; 2-propionylcarnitine; 3-nicotinamide; 4-glutamic acid; 5-aspartic acid; 6-hypoxanthine; 7-arachidonic acid; 8-LPC(16 : 1).

accurate mass. Different types of separations with characteristic metabolites are illustrated in Figure 8.2.

There are no clear differences between platforms in terms of the time needed for the analysis of a single sample. This time can vary from several minutes to as much as one hour. However, on average, global profiling takes around 20–40 minutes in all three platforms, with the shorter one probably being LC when using UPLC conditions. When starting and finishing a batch, analysis of blanks is necessary to control carryover (mainly a problem in LC-MS) and different sources of contamination in the three platforms. This analysis is crucial for non-targeted metabolomics and is directly linked to the identification power.

The quality of the hundreds or thousands of unknown signals that are measured is controlled by means of quality control samples (QCs).[15,16] Briefly, a QC is a sample that is representative of all the samples being investigated. Ideally, such a sample is a pool of aliquots of all the samples to be analysed. QCs are injected in replicates to equilibrate the system and to control its stability and performance. Any change in their signal illustrates analytical variation providing information about the system suitability. The small amount of sample required for every injection in CE-MS results in the possibility of using one single QC sample for an entire batch. On the contrary, GC-MS requires preparation of many QCs since the high volatility of both the solvent and metabolites precludes the use of the same sample. LC-MS requirements are in between CE-MS and GC-MS; however, the final amount of QC needed depends on the injection volume, which usually varies between 0.5 μL and 20 μL.

The physical contact between the instrument and the biological material causes changes, which are particularly noticeable at the beginning of the batch. For this reason, it is recommended to perform a series of injections of QC samples at the beginning of the run to equilibrate the system. The time needed to obtain this equilibrium, expressed as the number of subsequent injections, varies greatly and depends on the platform: CE-MS usually requires 5–10 injections, while GC-MS only requires 5. This number can be very different for LC-MS and it is often related to the time of the single run: longer methods require fewer injections, while shorter methods require more. Another difference between RP and HILIC platforms is that HILIC requires a very long equilibration time, often exceeding 10 or 20 injections, while RP usually requires 5–10 injections. This aspect of the analysis is important for two main reasons: the amount of QC sample available and the total time of the analysis for the whole batch.

Another factor to be considered is the volume of solvent required and waste disposal. In CE-MS, the amount of buffer needed for a single analysis is very small — usually around 1 mL every 5 injections. In fact, not all of this volume is consumed, but after a certain number of injections, the buffer is replaced with a new, fresh portion to avoid changes in pH due to the electrolysis process. By contrast, chromatographic separations in LC-MS occur with a gradient of two solvents (water and organic solvent) and the total amount of solvents used ranges from 10 to as high as 50 mL (depending on the flow rate and total length of the run). Therefore, taking into account green chemistry and economic considerations, CE-MS is favourable mainly for large batches of samples.

Regarding the cleaning of the system, in GC-MS the column is cleaned by increasing the oven temperature. This can be performed in any run as part of the method. The capillary in CE-MS is easily cleaned by a rinsing step before or after the separation process — using a buffer or other rinsing solutions to avoid carry-over. LC-MS columns are cleaned by a properly designed gradient, which includes a sufficient volume of strong solvent passing through the column for long enough to ensure elution of all highly retained metabolites and requires posterior re-equilibration. Consequently, this technique is more prone to carry-over.

The use of MS as a detector requires the ionisation of metabolites, since only charged molecules can be measured. Although different ion sources and analysers are available, the most common configurations are GC-EI-Q-MS, CE-ESI-TOF-MS and LC-ESI-QTOF. This is a simplification since examples of all possible couplings can be found in the literature. The electron ionization (EI) ion source in GC-MS produces high fragmentation of the analytes according to some specific rules due to the 70 eV applied by the incandescent filament. It is important to highlight that this value is constant and produces a fragmentation pattern that is reproducible, regardless of the instrument used, which is of great value for identifications based on libraries. By contrast, the ionisation voltages used in LC and CE are not standardised at all and are widely modified. The voltages used in CE oscillate around 100–125 eV and usually do not exceed 150 eV to avoid in-source metabolite fragmentation. In the case of LC-MS, similar voltages are applied for HILIC analysis, while for RP they shift up to 200–250 eV to increase the ionisability of bigger and more neutral lipids. However, different degrees of in-source fragmentation, depending on the metabolite, are unavoidable and this is one of the reasons why chromatographic resolution is required for reliable identification of those metabolites.

In liquid phase separation techniques, the most common ionisation method is electrospray ionisation (ESI) — a soft method, which induces ionisation of molecules and provides information about the molecular ion. However, the ionisation efficiency can be affected by the chemical conditions. This is particularly true for LC-MS where a gradient of mobile phases is used. The change in the proportion of two solvents over the course of the analysis causes changes in the ionisation of metabolites eluting at different times. This does not prevent comparisons between samples; however, it does affect comparisons of signals along a series in the same sample. This limitation is partially overcome in CE-MS where the composition of both the buffer and the sheath liquid is constant, thereby assuring constant ionisation efficiency.

There are three different sources of variability: pre-analytical, intra-analytical and post-analytical.[10] Pre-analytical variability includes the variability introduced during the collection, storage and preparation of samples; this is higher in GC-MS, while the analytical variation during the sequence analysis is the highest in CE-MS, especially in terms of MT. The main characteristics of each platform are summarised in Table 8.2.

Table 8.2 Main characteristics of different platforms with emphasis on analytical aspects.

Technique	CE-ESI-MS-TOF	GC-EI-MS-Q	RP-LC-ESI-MS-QTOF	HILIC-LC-ESI-MS-QTOF
Disadvantage	Multi-signals: Yes	Multi-derivatives: Yes	Multi-signals: Yes	Multi-signals: Yes
	Ghost peaks: No	Ghost peaks: Yes	Ghost peaks: Yes	Ghost peaks: Yes
	Bleeding: No	Bleeding: Yes	Bleeding: Yes	Bleeding: Yes
Time of sample preparation	Low	High	Low	Low
Metabolite coverage	Medium	Medium	Large	Large
Efficiency	High	High	High	High
Sensitivity	ppm	<ppm	≥ppm	≥ppm
Injection volume of sample	nl	µl	µl	µl
Identification by comparison of	Relative migration time + accurate mass + in-source fragmentation	Retention index + mass spectrum	Retention time + MS/MS spectrum	Retention time + MS/MS spectrum
Commercial spectral libraries	No	Yes	No	No
RT reproducibility	Low	High	Medium	Medium
Consumables	Low	Low	High	High

Regarding sensitivity, this is directly related to the separation efficiency, ionisation efficiency and amount of sample introduced into the analyser, among other factors. In this regard, it should be remembered that the injection volume is more than 100 times lower in CE than in LC. Therefore, sensitivity is not the main value of CE.

There are few studies providing a fair comparison of the different platforms. In this regard, it is interesting to mention the research by Kok *et al.*,[17] in which the CE-MS and HILIC-MS methods were compared in terms of reproducibility, sensitivity, selectivity and coverage of the anionic urinary metabolome. Rat urine analysis by HILIC-MS provided detection of 1360 molecular features compared to 347 molecular features revealed with CE-MS. Of these, 144 molecular features were found with both HILIC-MS and CE-MS, which showed on average 10 times higher peak areas in HILIC-MS. The HILIC retention and CE migration times of the common features were clearly not correlated, indicating a markedly different separation selectivity, and thus a significant degree of orthogonality of HILIC and CE.

8.6 Data Reprocessing

The acquired data comprise two main parts: the chromatogram/electropherogram and the mass spectrum. This information needs to be converted to three-dimensional structures including measured mass, RT/MT and abundance/intensity of the signal. To achieve this, several different software packages and tools are available including commercial and open-source ones. Data reprocessing can be considered for LC-MS and CE-MS together and separately for GC-MS. This is mostly because of the type of ionisation method: as already mentioned, LC-MS and CE-MS both use ESI, while most GC-MS applications in metabolomics are based on electron impact ionisation (EI), which causes significant fragmentation of ionised molecules. For this reason, data reprocessing for GC-MS data requires effective deconvolution with assignment of fragments to their precursors according to known fragmentation rules under an EI source, removing noise and non-related ions and extracting the pure component spectra to be searched against target or non-target libraries (commercial or homemade ones).[18] Regarding this task, the retention index based on the alkane scale as well as accurate mass analysers enhance the capability for identification purposes. It is important to highlight that fragmentation is not the only source of multi-signals corresponding to a single molecule. The derivatisation process often leads to formation of more than one derivative, which in turn produces more than one signal. For this reason, in GC, the signals corresponding to the same molecule can appear in different parts of the chromatogram. The data reprocessing in LC-MS and CE-MS is very similar and aims to remove noise and unrelated signals while, at the same time, finding co-eluting related ions corresponding to the same molecule. Such multi-signals arise from the ionisation and refer to different adduct charges, aggregation and neutral losses. As mentioned, CE-MS samples often contain a high concentration of salts, which causes formation of many different adducts, such as Na^+, K^+, and NH_4^+. This can be problematic during data reprocessing and requires a clear evaluation of all possible adducts and neutral losses that might occur in the source. In addition, although ESI is assumed to be a soft ionisation method, very fragile metabolites can be easily fragmented, further increasing the complexity of the signal due to the presence of such fragments. This is a problem for data reprocessing because, in contrast to adducts, the list of possible fragments for each metabolite is not defined and therefore an automatic assignment of them can be performed only for a very limited number of metabolites. Regardless of the origin, adduct formation or fragmentation, all produced multi-signals have exactly the same RT, which partly helps to identify them. This problem is common for LC-MS, however, its impact can be larger or smaller depending on the application. Considering the HILIC mode, small molecule fragmentation is equally problematic as it is for CE-MS; however, in RP analysis of lipids, the fragmentation phenomenon is not so important. In conclusion, each platform produces data containing multi-signals, but their origin is different and they might have different impacts on further data analysis.

Signals have to be matched afterwards to assign the same metabolites across different samples. This step, called alignment, is crucial and especially problematic for CE-MS data. The reason is the relatively big shift in the MT between different runs, which is commonly observed for CE-MS. By contrast, this shift is much smaller for LC and is almost negligible in the case of GC-MS. For this reason, tools exclusively devoted to CE data include a step for MT correction in the data reprocessing. This step is usually performed either using internal standards or probabilistic methods. One of the solutions is to use a series of reference molecules allocated along the entire electropherogram, which can be used to adjust the MT. Another practical option, especially in the case of isobaric compounds, is to visualise compounds independently, inspecting the quality of feature assignment and modifying if necessary.[8] More advanced methods are based on genetic algorithms and breakpoints of the piece-wise function.[19] After correction, alignment, based on MT and mass matching, can be performed much more effectively, minimising the risk of misalignment of isobaric compounds such as betaine and valine or alanine and sarcosine. Due to the similar properties of the data, the majority of software designed for LC-MS data can also be used for CE-MS. However, there is a list of software programs and tools devoted exclusively to CE-MS data to solve problems with MT shift. This includes the above-mentioned correction of the MT, as well as sophisticated methods for data alignment itself. The most popular — and so far most successful — method is based on dynamic time warping, where the corresponding (matched) features among different samples have to be found to produce the wrapping functions.[20,21]

8.7 Data Treatment

Statistical analysis of reprocessed data aims to find those signals responsible for the differences between the conditions under comparison. There are already many different tools and tests available and, indeed, the number continues to grow, offering new and better methods to treat such complex data. Prior to the statistical analysis, the data have to be evaluated in terms of quality. For this purpose, a quality assurance (QA) procedure has to be performed based on the information contained in QCs. Only reproducible signals can be considered reliable. The reproducibility is assessed by the coefficient of variation (CV) or relative standard deviation (RSD) calculated for each signal across all QCs. Depending on the analytical platform, slightly different criteria are used:[22] for GC-MS, the CV has to be below 30%, while for LC-MS usually this value is decreased to 20%. This difference is mostly related to the derivatisation process for GC-MS, which — as already mentioned — is a source of variation. In the literature there are no clear requirements for CE-MS data. However, as a general criterion, only signals with a CV below 30% should be accepted.

Data analysis *sensu stricto* is very similar for all platforms and it is selected considering biological rather than analytical aspects. Uni- or multivariate analysis can be performed for selection of statistically significant signals. The analysis itself is the same regardless of the platform, although some aspects of data scaling should be noted. Metabolomics data have very specific properties that do not fit classical statistical definitions. Within the same dataset, signals with large abundances/intensities are present that require data scaling prior to the statistical analysis; otherwise such signals will contribute more to the models than the ones that are less abundant. The two most common scaling methods are unit variance (UV) and Pareto (Par).[23,24] UV attributes the same importance to all signals in terms of their contribution to the model, whereas Par attributes more importance to less abundant signals, while also reducing the impact of highly abundant ones. UV scaling can be 'dangerous' in the case of multi-adducts and in-source fragments, which are of rather low abundance and after scaling will contribute strongly to the plots. This can be the case for both LC-MS and CE-MS data; however, it might be more problematic for CE-MS since LC-MS covers polar metabolites, but also a large number of nonpolar lipids, which are less prone to fragmentation.

As mentioned previously, the constant composition of the separation buffer along the separation makes CE-MS data especially suitable for ratio analysis of different metabolites within the same sample, which is useful for biomarker searches — especially in the case of samples with different physiological concentrations of metabolites where absolute abundances cannot be compared.

8.8 Identification

Once statistically significant signals have been selected, they have to be identified and assigned to metabolites. For this purpose, the most reliable method is comparison with authentic standards. However, the number of such standards is limited and often other approaches have to be used. The most common one is tandem mass analysis, where selected signals are isolated and fragmented providing information about the metabolite structure.

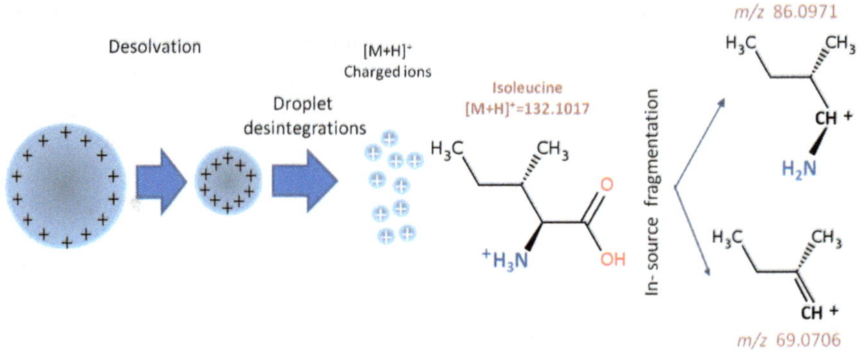

Figure 8.3 Scheme of the mechanism of in-source fragmentation of molecules.

The spectrum can be either used for comparison with databases or for structural elucidation of unknown metabolites. This methodology is very powerful, but is limited to certain types of mass analysers such as QTOF or Orbitrap.[25–27] These types of analysers are very commonly combined with LC, whereas their use with CE is limited, probably because the small amount of sample injected makes MS/MS challenging. Although there are studies performed with QTOF, TOF is undeniably the most commonly used mass analyser. TOF provides mass accuracy, but does not perform MS/MS analysis. For this reason, identification, especially of unknown metabolites, is challenging. However, recently, in-source fragmentation has been used as a source of information for identification purposes. Although this phenomenon is undesirable in regular analysis, it can be very useful for confirmation of the identity of metabolites. Intentionally increasing the fragmentor voltage causes fragmentation of molecules (Figure 8.3). In contrast to MS/MS analysis, there is no selection and isolation of a particular ion; instead, all molecules present in the source undergo fragmentation. For this reason, special data analysis is needed to assign fragments to their precursors. This can be performed by means of correlation analysis assuming that very high positive correlation reflects analytical correlation. It has been proved that fragments correlate very strongly with their precursors with a critical value of 0.9.[28] This method is limited to confirmation of some candidates and it is not suitable for structural elucidation of unknown compounds.

Similar to RT, MT also provides information about the chemical properties of molecules, which can be used for identification purposes. With CE-MS, this can be used in two ways: for prediction of the MT of new compounds based on the behaviour of other metabolites or for standardisation of the absolute MT to the relative one. MT depends on the charge-to-size ratio of the molecule. Taking into consideration information such as the pH of the sample and buffer, and the chemical formula and pK_a of the metabolite and its mass, charge formation can be predicted and therefore migration order can be established. However, to ensure the usefulness of such tools in different laboratories, a number of standards have to be analysed, applying particular conditions, to obtain a dataset for algorithm learning. Thanks to this tool, a tentative annotation has been performed in many studies. This has been done by either 'manually' inspecting each case or using automated tools to predict the MT. These tools usually use algorithms such as artificial neural networks[29] or support vector regression,[30] where prediction of the behaviour of analyte electromigration is achieved using molecular descriptors, net charge and absolute mobility.[31] Similar tools can be found for LC-MS but the reliability is not as high as in CE-MS, since there are many more factors affecting the separation.

Any change in the capillary and its length significantly affects the MT of metabolites. Even within the same batch, the variation of the MT is relatively high in comparison to other platforms, thus making the comparison of the MT between batches even more challenging. For this reason, instead of using absolute values, the relative migration time (RMT), calculated with an internal standard, is frequently used.[32–34] In this way, all absolute values are standardised to the same substance, allowing their comparison in different analyses. This strategy can be successfully used only when analysing samples under the same conditions. Any changes in the length of the capillary, buffer composition or applied voltage might cause changes in the electromigration of metabolites. This migration can also be affected by drastic changes in the composition of the sample. Despite all these alterations, the effectiveness of identification using the RMT is always higher than the predicted MT and, therefore, using a reliable internal standard is highly recommended. In the case of GC-MS, a similar strategy is used, although it works in a different way: the adjustment of the RT is performed prior to the analysis. For this purpose, a series of target compounds is used with very well-known RTs.[35] A series of injections with different pressures are then performed in order to obtain the desired RT. This process, called RT locking, has to be repeated whenever any changes in chromatography are observed. In this way, very reproducible separations are achieved, allowing the use of standardised libraries for the identification of compounds. It is important to emphasize that GC-MS is the quickest and easiest technique in terms of identification, where using RT locking and libraries is sufficient for robust and convenient identification. The main drawback is that the number of metabolites is limited to those already described.

8.9 Pathway Analysis and Data Interpretation

Although over recent decades a large advance in the understanding of global metabolism has been achieved, there is still a large lack of knowledge especially in terms of disorders and pathological processes. Identified significant metabolites have to be placed within metabolic processes and connected to the stated biological question. In this way, the obtained results can be interpreted, bringing biological meaning to the work as a whole. This step is very important but also extremely challenging. This step can be performed in two ways, including pathway analysis and/or literature revision.[36] Both of these approaches can be performed either 'manually' or using some automated tools, and there are no differences related to the source of data, since — at this stage — the data is already converted to particular metabolites, which need to be characterised by their name and identifiers, *e.g.*, CAS, and KEGG ID. Only metabolites with these descriptors can be successfully forwarded for pathway analysis. In the case of GC-MS and CE-MS data, this is not very problematic thanks to the nature of the detected metabolites. Small molecules such as amino acids, organic acids, sugars, carnitines, *etc.* are quite well characterised and labelled with several identifiers. Consequently, pathway analysis of GC-MS and CE-MS data can be very successful, covering the most important metabolic processes such as glycolysis and gluconeogenesis, the TCA or Krebs cycle, the urea cycle, oxidative phosphorylation or the pentose phosphate pathway. By contrast, most lipids do not have any identifier related to their number and they cannot be placed in metabolic pathways, which is a problem for LC-MS data interpretation.

8.10 Conclusions

CE-MS, with all its pros and cons, is a technique that has already found a place in the metabolomics field. The fact that it requires personnel trained in its use, knowledge of the fundamentals and careful handling of the instrument and data obtained has prevented its more widespread, general use. However, users consider that it provides information that cannot be easily obtained with other analytical platforms; its separation mechanism is orthogonal; it requires minimum sample pre-treatment; and it consumes small volumes of sample and solvents. For all these very reasons, until a unique platform for metabolome studies appears, CE-MS is a useful approach to increase metabolite coverage and to obtain broad metabolic phenotyping.

Given the rapid evolution of instrumental techniques nowadays, making predictions about trends and future developments in a given field is difficult. However, CE-MS seems especially well suited for one of the current challenges, that is "omics" single cell analysis, if the sensitivity can be improved through the development of robust sheathless interfaces. Moreover, with appropriate capillaries and methodologies, proteins, nucleic acids and metabolites could be measured in a single run, which would be ideal for a proper multi-omic approach.

References

1. T. Cakır and M. J. Khatibipour, *Front. Bioeng. Biotechnol.*, 2014, **2**, 62.
2. J. Haggarty and K. E. Burgess, *Curr. Opin. Biotechnol.*, 2017, **43**, 77–85.
3. M. Schwaiger, E. Rampler, G. Hermann, W. Miklos, W. Berger and G. Koellensperger, *Anal. Chem.*, 2017, **89**, 7667–7674.

4. J. Wang, T. T. Christison, K. Misuno, L. Lopez, A. F. Huhmer, Y. Huang and S. Hu, *Anal. Chem.*, 2014, **86**, 5116–5124.
5. K. Burgess, D. Creek, P. Dewsbury, K. Cook and M. P. Barrett, *Rapid Commun. Mass Spectrom.*, 2011, **25**, 3447–3452.
6. C. Petucci, A. Zelenin, J. A. Culver, M. Gabriel, K. Kirkbride, T. T. Christison and S. J. Gardell, *Anal. Chem.*, 2016, **88**, 11799–11803.
7. A. J. Chetwynd, W. B. Dunn and G. Rodriguez-Blanco, *Adv. Exp. Med. Biol.*, 2017, **965**, 19–44.
8. A. García, J. Godzien, Á. López-Gonzálvez and C. Barbas, *Bioanalysis*, 2017, **9**, 99–130.
9. K. Hollywood, D. R. Brison and R. Goodacre, *Proteomics*, 2006, **6**, 4716–4723.
10. J. Nielsen and M. C. Jewett, *Metabolomics. A Powerful Tool in Systems Biology*, Springer, 2007.
11. W. Zhang, T. Hankemeier and R. Ramautar, *Curr. Opin. Biotechnol.*, 2017, **43**, 1–7.
12. A. Mastrangelo, A. Ferrarini, F. Rey-Stolle, A. García and C. Barbas, *Anal. Chim. Acta*, 2015, **900**, 21–35.
13. M. M. Koek, B. Muilwijk, M. J. van der Werf and T. Hankemeier, *Anal. Chem.*, 2006, **78**, 1272–1281.
14. H. Kanani, P. K. Chrysanthopoulos and M. I. Klapa, *J. Chromatogr. B: Anal. Technol. Biomed. Life Sci.*, 2008, **871**, 191–201.
15. H. G. Gika, G. A. Theodoridis, M. Earll and I. D. Wilson, *Bioanalysis*, 2012, **4**, 2239–2247.
16. J. Godzien, V. Alonso-Herranz, C. Barbas and E. G. Armitage, *Metabolomics*, 2014, 1–11.
17. M. G. Kok, G. W. Somsen and G. J. de Jong, *Talanta*, 2015, **132**, 1–7.
18. L. G. Johnsen, P. B. Skou, B. Khakimov and R. Bro, *J. Chromatogr. A*, 2017, **1503**, 57–64.
19. E. Nevedomskaya, R. Derks, A. M. Deelder, O. A. Mayboroda and M. Palmblad, *Anal. Bioanal. Chem.*, 2009, **395**, 2527–2533.
20. G. Jeelani, D. Sato, A. Husain, A. Escueta-de Cadiz, M. Sugimoto, T. Soga, M. Suematsu and T. Nozaki, *PLoS One*, 2012, **7**, e37740.
21. M. Sugimoto, A. Hirayama, T. Ishikawa, M. Robert, R. Baran, K. Uehara, K. Kawai, T. Soga and M. Tomita, *Metabolomics*, 2010, **6**, 27–41.
22. W. B. Dunn, I. D. Wilson, A. W. Nicholls and D. Broadhurst, *Bioanalysis*, 2012, **4**, 2249–2264.
23. J. Godzien, M. Ciborowski, S. Angulo and C. Barbas, *Electrophoresis*, 2013, **34**, 2812–2826.
24. R. A. van den Berg, H. C. Hoefsloot, J. A. Westerhuis, A. K. Smilde and M. J. van der Werf, *BMC Genomics*, 2006, **7**, 142.
25. A. M. Knolhoff, K. M. Nautiyal, P. Nemes, S. Kalachikov, I. Morozova, R. Silver and J. V. Sweedler, *Anal. Chem.*, 2013, **85**, 3136–3143.
26. J. T. Aerts, K. R. Louis, S. R. Crandall, G. Govindaiah, C. L. Cox and J. V. Sweedler, *Anal. Chem.*, 2014, **86**, 3203–3208.

27. P. Nemes, S. S. Rubakhin, J. T. Aerts and J. V. Sweedler, *Nat. Protoc.*, 2013, **8**, 783–799.
28. J. Godzien, E. G. Armitage, S. Angulo, M. P. Martinez-Alcazar, V. Alonso-Herranz, A. Otero, A. Lopez-Gonzalvez and C. Barbas, *Electrophoresis*, 2015, **36**, 2188–2195.
29. M. Sugimoto, S. Kikuchi, M. Arita, T. Soga, T. Nishioka and M. Tomita, *Anal. Chem.*, 2005, **77**, 78–84.
30. M. Sugimoto, A. Hirayama, T. Ishikawa, M. Robert, R. Baran, K. Uehara, K. Kawai, T. Soga and M. Tomita, *Metabolomics*, 2010, **6**, 27–41.
31. R. Lee, A. S. Ptolemy, L. Niewczas and P. Britz-McKibbin, *Anal. Chem.*, 2007, **79**, 403–415.
32. K. Maekawa, A. Hirayama, Y. Iwata, Y. Tajima, T. Nishimaki-Mogami, S. Sugawara, N. Ueno, H. Abe, M. Ishikawa, M. Murayama, Y. Matsuzawa, H. Nakanishi, K. Ikeda, M. Arita, R. Taguchi, N. Minamino, S. Wakabayashi, T. Soga and Y. Saito, *J. Mol. Cell. Cardiol.*, 2013, **59**, 76–85.
33. X. Y. Qin, F. Wei, M. Tanokura, N. Ishibashi, M. Shimizu, H. Moriwaki and S. Kojima, *PLoS One*, 2013, **8**, e82860.
34. C. Zhao, H. Du, L. Xu, J. Wang, L. Tang, Y. Cao, C. Li, Q. Wang, Y. Liu, F. Shan, J. Feng, F. Xu and P. Gao, *J. Pharm. Biomed. Anal.*, 2015, **115**, 48–54.
35. N. Etxebarria, O. Zuloaga, M. Olivares, L. J. Bartolomé and P. Navarro, *J. Chromatogr. A*, 2009, **1216**, 1624–1629.
36. J. Xia and D. S. Wishart, *Nat. Protoc.*, 2011, **6**, 743–760.
37. J. M. Halket, D. Waterman, A. M. Przyborowska, R. K. Patel, P. D. Fraser and P. M. Bramley, *J. Exp. Bot.*, 2005, **56**, 219–243.
38. J. Zeng, H. Kuang, C. Hu, X. Shi, M. Yan, L. Xu, L. Wang, C. Xu and G. Xu, *Environ. Sci. Technol.*, 2013, **47**(13), 7457–7465.
39. J. Godzien, D. García-Martínez, P. Martinez-Alcazar, F. Ruperez and C. Barbas, *Metabolomics*, 2013, **9**, 188–202.
40. S. Naz, A. García and C. Barbas, *Anal. Chem.*, 2013, **85**, 10941–10948.
41. S. Naz, A. Garcia, M. Rusak and C. Barbas, *Anal. Bioanal. Chem.*, 2013, **405**, 4849–4858.
42. A. Hirayama, M. Sugimoto, A. Suzuki, Y. Hatakeyama, A. Enomoto, S. Harada, T. Soga, M. Tomita and T. Takebayashi, *Electrophoresis*, 2015, **36**(18), 2148–2155.
43. C. Balderas, F. J. Rupérez, E. Ibañez, J. Señorans, J. Guerrero-Fernández, I. G. Casado, R. Gracia-Bouthelier, A. García and C. Barbas, *Electrophoresis*, 2013, **34**, 2882–2890.
44. J. V. Alberice, A. F. Amaral, E. G. Armitage, J. A. Lorente, F. Algaba, E. Carrilho, M. Márquez, A. García, N. Malats and C. Barbas, *J. Chromatogr. A*, 2013, **1318**, 163–170.
45. M. Sugimoto, D. T. Wong, A. Hirayama, T. Soga and M. Tomita, *Metabolomics*, 2010, **6**, 78–95.
46. N. J. Andreas, M. J. Hyde, M. Gomez-Romero, M. A. Lopez-Gonzalvez, A. Villaseñor, A. Wijeyesekera, C. Barbas, N. Modi, E. Holmes and I. Garcia-Perez, *Electrophoresis*, 2015, **36**(18), 2269–2285.

47. C. Barbas-Bernardos, E. G. Armitage, A. García, S. Mérida, A. Navea, F. Bosch-Morell and C. Barbas, *J. Pharm. Biomed. Anal.*, 2016, **127**, 18–25.
48. C. Ibáñez, C. Simó, V. García-Cañas, A. Gómez-Martínez, J. A. Ferragut and A. Cifuentes, *Electrophoresis*, 2012, **33**, 2328–2336.
49. N. Yoshimi, T. Futamura, K. Kakumoto, A. M. Salehi, C. M. Sellgren, J. Holmén-Larsson, J. Jakobsson, E. Pålsson, M. Landén and K. Hashimoto, *BBA Clin.*, 2016, **5**, 151–158.
50. N. Shima, I. Miyawaki, K. Bando, H. Horie, K. Zaitsu, M. Katagi, T. Bamba, H. Tsuchihashi and E. Fukusaki, *Toxicology*, 2011, **287**, 29–37.
51. T. Kind, V. Tolstikov, O. Fiehn and R. H. Weiss, *Anal. Biochem.*, 2007, **363**, 185–195.
52. M. C. Gulersonmez, S. Lock, T. Hankemeier and R. Ramautar, *Electrophoresis*, 2016, **37**, 1007–1014.
53. S. Kume, M. Yamato, Y. Tamura, G. Jin, M. Nakano, Y. Miyashige, A. Eguchi, Y. Ogata, N. Goda, K. Iwai, E. Yamano, Y. Watanabe, T. Soga and Y. Kataoka, *PLoS One*, 2015, **10**, e0120106.
54. T. Soga, Y. Ohashi, Y. Ueno, H. Naraoka, M. Tomita and T. Nishioka, *J. Proteome Res.*, 2003, **2**, 488–494.
55. J. L. Chen, J. Fan and X. J. Lu, *Electrophoresis*, 2014, **35**, 1032–1039.
56. M. G. Kok, M. M. Ruijken, J. R. Swann, I. D. Wilson, G. W. Somsen and G. J. de Jong, *Anal. Bioanal. Chem.*, 2013, **405**, 2585–2594.
57. B. Timischl, K. Dettmer, H. Kaspar, M. Thieme and P. J. Oefner, *Electrophoresis*, 2008, **29**, 2203–2214.
58. A. Hirayama, K. Kami, M. Sugimoto, M. Sugawara, N. Toki, H. Onozuka, T. Kinoshita, N. Saito, A. Ochiai, M. Tomita, H. Esumi and T. Soga, *Cancer Res.*, 2009, **69**, 4918–4925.
59. A. Kinoshita, K. Tsukada, T. Soga, T. Hishiki, Y. Ueno, Y. Nakayama, M. Tomita and M. Suematsu, *J. Biol. Chem.*, 2007, **282**, 10731–10741.
60. D. Shimura, G. Nakai, Q. Jiao, K. Osanai, K. Kashikura, K. Endo, T. Soga, N. Goda and S. Minamisawa, *Am. J. Physiol.: Heart Circ. Physiol.*, 2013, **305**, H494–H505.

CHAPTER 9

Potential of CE-MS for Chiral Metabolic Profiling

ELENA SÁNCHEZ-LÓPEZ, MARÍA LUISA MARINA AND ANTONIO L. CREGO*

Department of Analytical Chemistry, Physical Chemistry and Chemical Engineering, University of Alcalá, Ctra. Madrid-Barcelona Km. 33.600, 28871 Alcalá de Henares, Madrid, Spain
*E-mail: antonio.crego@uah.es

9.1 Introduction to Chiral Analysis

Chirality is an inherent property of every biological system, including the basic building blocks of life (amino acids and sugars), and larger molecules derived from them, such as peptides, proteins, and polysaccharides.[1] Therefore, it is of utmost importance to ensure adequate chiral separation in order to be able to deepen the existing knowledge on chiral systems. This chapter describes the potential of the coupling of CE to MS to carry out chiral metabolic profiling. With this aim, the interest in chirality and its relevance, chiral separation in CE, the specific strategies to be successful in chiral CE-MS coupling, and the most relevant applications developed in various representative fields are discussed. The chapter ends with concluding remarks and the future trends of chiral metabolic profiling in CE-MS taking into account the present state-of-the-art.

New Developments in Mass Spectrometry No. 6
Capillary Electrophoresis–Mass Spectrometry for Metabolomics
Edited by Rawi Ramautar

Published by the Royal Society of Chemistry, www.rsc.org

9.1.1 Chirality and Its Relevance

Chirality is understood as the geometric property of some molecules that makes them non-superimposable with their mirror images. These mirror images are known as enantiomers. Enantiomers are molecules that have identical atomic compositions and bonds, but are different in the three-dimensional arrangement of their atoms around a central element (*e.g.* a carbon atom), known as the chiral center. Thus, a pair of enantiomers share both physical and chemical properties, except those acquired in a chiral environment. When two enantiomers are present in equal amounts, the resulting mixture is known as a racemate. In this sense, according to the IUPAC systematic nomenclature, enantiomers are *R* or *S* following the priority rules of the Cahn–Ingold–Prelog system.[2] Regarding biomolecules, *e.g.* amino acids and sugars, these are commonly denoted as D and L following the Latin words *dexter* and *laevus*, right and left, respectively. For instance, L-amino acids have an *S* configuration, whereas D-sugars have an *R* configuration. Other nomenclature for enantiomers is based on how they rotate the plane of polarized light, (+) if light rotates clockwise and (−) if it is rotated anticlockwise.

Nowadays, there is clear evidence that many endogenous molecules exist as both enantiomers. Therefore, chirality will determine how a chiral molecule interacts with its environment, a process known as enantioselectivity. Probably the first time that enantioselectivity was demonstrated in a biological context is attributed to Louis Pasteur in 1858.[3] In his research, Pasteur found that microorganisms consumed (+)-tartaric acid in a more rapid manner than its (−)-enantiomeric counterpart.[3] This is nowadays explained as the enantioselective metabolism that occurs in living organisms. As a consequence of the enantioselective metabolism, a wide array of metabolic processes are mediated by stereospecific interactions, which might lead to differences in the activity of a pair of enantiomers. Thus, enantiomers are considered to be different compounds because, in a biological context, they may have different properties.

Stereoselective interactions between macromolecules, such as enzymes, and endogenous chiral molecules are involved in many biological processes, for example, metabolic activities.[1] Amino acids are paramount small molecules known to be chiral. Despite being considered as unnatural at first, over the past few years, D-amino acids have been demonstrated to be endogenous in numerous organisms.[4–7] This has resulted in an increase in the research carried out to improve the existing knowledge on the role of D-amino acids.[8] As a consequence, the physiological functions of the D-proteinogenic amino acids and/or their role as potential biomarkers have been demonstrated in different publications. Figure 9.1 displays the number of publications reported in the Web of Science database when searching for the terms biomarker and each of the selected D-amino acids. As can be seen, D-serine, D-aspartic acid, and D-alanine are the most popular ones. D-serine is a co-agonist of the *N*-methyl-D-aspartate-type glutamate receptors in the brain, and

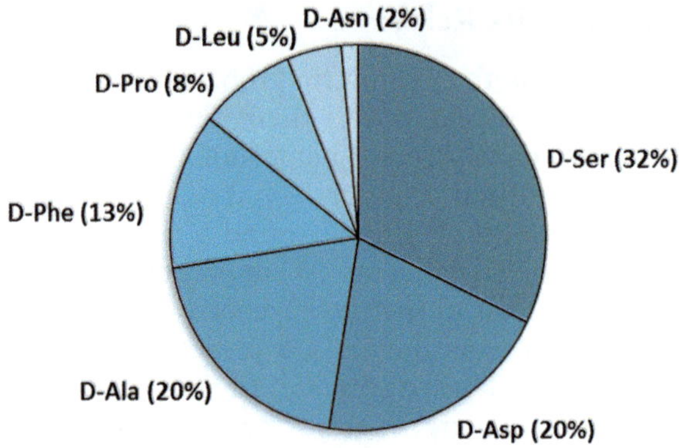

Figure 9.1 Graphic displaying the percentage of publications when looking up the terms "biomarker" and the corresponding D-amino acid in the Web of Science (October 2017). Legend: D-Ser: D-serine; D-Asp: D-aspartic acid; D-Ala: D-alanine; D-Phe: D-phenylalanine; D-Pro: D-proline; D-Leu: D-leucine; D-Asn: D-asparagine.

it is involved in different neuronal functions, such as memory, neuroplasticity and learning,[9–11] as well as in the function of the kidneys.[12] In addition, D-serine, D-aspartic acid, and D-alanine have been related to schizophrenia, depression and Alzheimer's disease,[9,10,13] among other disorders. Another set of interesting molecules whose D-enantiomer has proven to have biological function includes non-proteinogenic amino acids such as D-ornithine or D-citrulline,[14] as well as other enantiomeric compounds, such as D-lactic acid.[15]

All of the above-mentioned examples clearly indicate the high relevance in studying the role of enantiomers. Many fascinating questions regarding the roles of enantiomers still need to be answered.[16] Thus, strategies need to be developed to separate these interesting chiral molecules, which might help in improving the existing knowledge of enantioselective biological systems. In this sense, current state-of-the-art analytical techniques demonstrate a good capability to perform this well-known challenging task. Decades after the first chiral separation, research on this matter is still being actively carried out, in which new approaches are focused on enhancing enantioselectivity and sensitivity.

9.1.2 Chiral Recognition by Separation Techniques

Among the different analytical techniques that have been employed for chiral analysis, CE, liquid chromatography (LC), gas chromatography (GC), and supercritical fluid chromatography (SFC) are the most widely used. Figure 9.2 illustrates the evolution of the application of these techniques over the past 30 years for chiral analysis. As can be seen, LC is the technique of choice in chiral separations. Even though GC was also popular in the beginning, CE soon took over, being still nowadays more used than GC in chiral analysis.

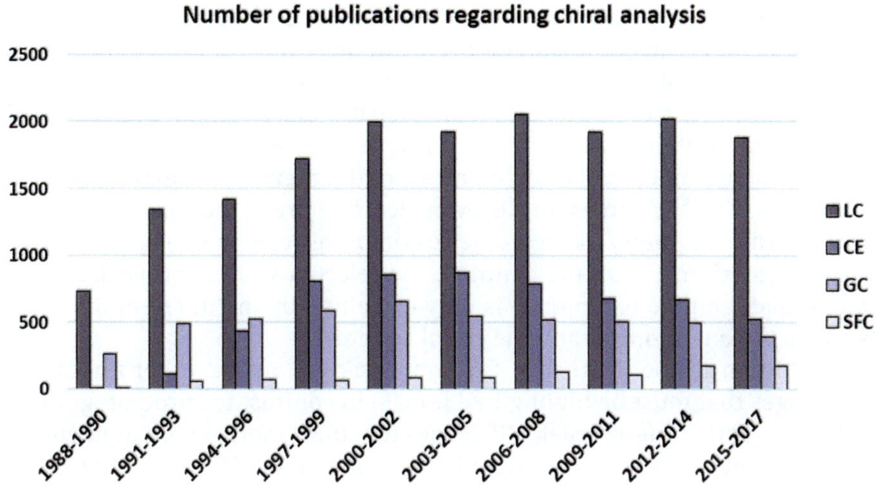

Figure 9.2 Bar plot illustrating the number of publications related to chiral analysis by each technique over the past three decades. Search carried out using the Web of Science (October 2017) by looking up "chiral" or "enantio*" and each of the corresponding analytical techniques.

SFC is undergoing constant growth but, still, has not reached the popularity of the rest of the analytical techniques regarding enantioseparations.

Regardless of the analytical technique employed, discrimination of enantiomers requires the presence of a chiral environment, also known as a chiral selector, *i.e.* a chiral reagent involved in the formation of the corresponding diastereomeric species formed between the enantiomeric forms of the analyte of interest and the corresponding enantiomer of the chiral selector employed. This involves an energetic difference between them, enabling their enantioseparation, in the best case.

Also, unrelated to the analytical technique used, there are two approaches to carry out a chiral separation: indirect or direct methods. In the indirect method, the enantiomers are derivatized with a chiral reagent in order to form the corresponding pair of diastereomers, which will later be separated based on their different physicochemical properties (*i.e.* without the need of a chiral selector). The main drawback of the indirect approach is that the chiral derivatization reagent must be used as a very pure enantiomer, which in some cases might not be possible to acquire either due to its limited commercial availability, or if it is accessible, it might have a prohibitive cost. Thus, the direct approach is the most commonly used. In the direct method, a chiral selector is used to establish noncovalent and temporary diastereomeric complexes between the enantiomers of the analyte and the chiral selector, which can be either in solution or static in the stationary phase.

According to the direct approach, the largest difference between the enantioseparations carried out by chromatographic techniques and CE is the effect of the electrophoretic mobility in CE, which causes species in the same phase to migrate at different velocities, thus separating one from another. However, in chromatographic techniques, pressure as the migration force does not enable different velocities of the species residing in the same phase. This is the reason why the enantioseparations in CE can be performed based on the different association constants, but also due to the different electrophoretic mobilities of their corresponding temporary diastereomeric complexes.[17–20] Oftentimes, in CE, even electrophoretic mobility differences between the temporary diastereomeric complexes can be enough to enable chiral separation.[21,22] On the contrary, the selectivity of enantioseparations in chromatographic techniques is only explained by the different affinities between the enantiomers and the chiral selector.

In addition to the distinctive features described above, other important advantages that must be highlighted for CE, in contrast to chromatographic techniques, are: a higher peak efficiency enabling a more potent adjustment of the separations, so chiral recognition is based on small energy differences between the temporary diastereomers (a separation factor of 1.01–1.02 is usually enough to achieve baseline enantioseparation); a very reduced use of solvents and reagents, including chiral selectors, and smaller sample volumes due to the inherent small dimensions of the capillary, what makes it a green and cost-saving technique; and, finally, an impressive flexibility and versatility due to the wide variety of chiral selectors that can be added in different concentrations, or even using combinations of different chiral selectors, facilitating method development.[23] In this context, there is no doubt that CE plays a very important role nowadays due to its inherent attractive advantages. Therefore, it should be noted that LC remains the dominant technique for chiral analysis only for reasons of robustness and familiarity of researchers with this technique and not because of its greater performance in achieving success in a chiral separation.

9.1.3 Possibilities of CE in Chiral Separations

About thirty years after the first time a chiral separation was carried out by CE,[24] chiral CE appears to be one of the most powerful techniques for analytical-scale enantioseparations, and its use has shown impressive possibilities. Today, the flexibility and versatility of chiral CE are unrivalled and by using the same instrumentation it is possible to separate a diverse range of analytes, including large or small molecules, that are either charged or uncharged.

The two main modes of CE to perform chiral separations are electrokinetic chromatography (EKC) and capillary electrochromatography (CEC).[25] The main difference between these two modes is that, in EKC, the chiral selector is dissolved in the BGE acting as a pseudo-stationary phase, while in the CEC mode the chiral selector is immobilized in the capillary forming a truly stationary phase. As described above, the EKC mode has important advantages over the CEC mode, which shares the limitations of any chromatographic technique. Thus, the EKC mode continues to be the favorite one in the field of chiral separations.

In chiral analysis, both charged and neutral molecules can be separated if the experimental setup is properly established. When neutral chiral selectors are employed, only analytes that are charged will be separated unless a second migration principle is incorporated, such as a charged pseudo-phase (*e.g.* a charged micelle), which will enable such interaction with a neutral analyte. This CE mode is known as micellar electrokinetic chromatography (MEKC). Charged selectors are also useful when having a charge opposite to that of the analyte, as a means to improve the enantioresolution due to the large mobility difference between the analyte and the selector, and to enhance the interaction between the analyte and the selector *via* electrostatic forces, it being possible in some cases to achieve an enantioseparation using very low concentrations of chiral selector.

There is a myriad of chiral selectors commonly used in EKC, such as cyclodextrins (CDs), monomeric and polymeric surfactants, antibiotics, chiral crown ethers, linear mono-, oligo-, and polysaccharides, peptides, and proteins.[26–29] In certain scenarios when baseline resolution cannot be achieved by using only one chiral selector, it might be required to employ a combination of different chiral selectors. This is known as a dual system. Although some research studies have employed systems based on a combination of polysaccharides or polymeric surfactants,[30,31] the most frequently used dual selector systems include different CDs. CDs are the most widely used chiral selectors due to their excellent properties associated with their availability, diversity, universality, cost, and safety.[32] The most accepted interaction mechanism for the chiral separation of an analyte with CDs is the inclusion of the enantiomers into the cavity of the CD so that secondary interactions (hydrogen bonds and dipole–dipole interactions) with the hydroxyl groups situated on the CD rim are established. However, it is not required to achieve a complete inclusion; a partial one can also lead to chiral recognition and subsequent enantioseparation.

Along with CDs, surfactants are also very frequently employed as selectors in EKC. Polymeric surfactants (also called micellar polymers) have received great attention in the past few years due to their high compatibility with MS detection.[29] Among their excellent properties, their zero critical micelle concentration, high stability (which prevents dissociation of the monomers during the electrospray process), significant decrease in ion suppression (due to the low surface activity), low background noise, and more stable electrospray can be highlighted.[33,34]

Despite its numerous advantages, CE is not free of drawbacks. Probably the main limitation of this technique is its low concentration sensitivity when using UV-Vis detection, related to the inherent low sample volumes injected. Other drawbacks include the not-so-straightforward coupling to MS detection, and the relatively poor migration time and peak area repeatability. Regardless of these issues, there have been attempts to solve them in the past few years. For instance, as a means to enhance the sensitivity, three approaches are mainly selected: sample treatment, in-capillary sample preconcentration techniques and highly sensitive detectors. These aspects are discussed in detail in Chapters 4 and 5, and they have also been reported elsewhere, paying special attention to chiral issues.[35,36] On the other hand, the poor repeatability in the migration times often observed in CE-MS can be resolved by using an internal standard and/or an appropriate alignment software. Chapter 5 describes in detail these possibilities, so they will not be further discussed.

9.1.4 Chiral CE-MS

MS detection offers several improvements regarding selectivity and sensitivity. In addition, analyte derivatization is not a mandatory step as in other sensitive detectors, such as fluorescence detection. Another benefit of MS detection is that it enables an analyte to be unequivocally identified by its mass-to-charge ratio (*m*/*z*) and its fragmentation mass spectrum upon MS/MS experiments. Thanks to its outstanding advantages, providing universality and selectivity, high sensitivity, and rich information, MS can be considered as the most powerful detection system to achieve sensitive enantioseparations by CE.[33,37–41] However, as commented on in the previous section, the coupling of CE to MS is not always an easy task. The vast majority of the chiral selectors most commonly used in EKC are non-volatile, which will most likely result in ion suppression and contamination of the ionization source and optics causing a detrimental effect on the sensitivity. This is the reason why CEC is the most suitable CE mode to be coupled to MS detection to carry out a chiral separation, since there is no contamination caused by the entry of the chiral selector into the MS system as it is (statically) attached to the capillary. As a means to mitigate these issues, two approaches are mainly being used: the counter migration technique (CMT) and the partial filling technique (PFT).[33] In the CMT approach, the charge of the chiral selector must be opposite to that of the analyte so that the chiral selector migrates away from the MS detector, but still interacts with the analytes to enable their enantioseparation. In the PFT, the capillary is partially filled with a band of BGE containing the chiral selector; thus, the separated enantiomers will leave the chiral selector band reaching the detector before the chiral selector. The PFT is the approach more widely used when coupling EKC to MS detection. Regardless of the approach used, a loss in enantioresolution is expected as a consequence of the lower separation length. Thus, the best option would be to use strategies that allow the entry of chiral selectors into the MS system. Yet, just a few of these approaches have been reported up to now, such as the use of low-concentrated chiral selectors[42,43] or the use of micelle polymers.[33,34] Another approach that can be used is the one reported by Iwata *et al.*[44] in which the chiral selector, a highly sulfated γ-CD, not only does not ionize in ESI positive mode, but also breaks down in the ionization source. Another approach is the one developed by Moini *et al.* based on complexation of amino acids with a chiral crown ether.[45] This interesting strategy will be briefly discussed in the next sections.

9.2 Chiral Metabolic Profiling

It is clear that it is of utmost importance to study chiral processes in which proper chiral separation is required using the strategies mentioned in the previous section. In this sense, chiral metabolomics or chiral metabolic profiling is a recent trend that focuses its attention on the study of chiral molecules that were not considered at first glance. The first time the term "chiral metabonomics" was mentioned in the literature was in an NMR study published by Nicholson and collaborators in 2012.[46] Therein, using β-CD as a chiral cosolvating agent, the authors resolved the enantiomers of ibuprofen and the diastereomers of its main metabolite in intact urine samples, demonstrating the possibilities of chiral metabolic profiling. It is nowadays well-understood that many endogenous molecules are chiral; thus, chiral metabolic profiling can be considered a promising tool to monitor metabolites, including their enantiomeric counterparts, in order to provide a more detailed picture of the metabolic scenario. In the following sections, the current state-of-the-art of chiral metabolic profiling will be discussed, paying special attention to the role of chiral CE-MS on the basis of the already discussed strategies, as well as reported methodologies with a promising future in chiral metabolic profiling.

9.2.1 State-of-the-Art of Chiral Metabolic Profiling

From an experimental point of view, chiral metabolic profiling is highly challenging since a chiral environment needs to be established to enable the corresponding enantioseparation, as was discussed in earlier sections. Chiral untargeted metabolomics approaches have not been popular, being applied mainly by LC-MS. In LC-MS, indirect approaches are more widely used. In this regard, Toyo'oka and collaborators designed a strategy based on chiral derivatization using both enantiomers of the labelling reagent to identify those enantiomeric biomarkers that displayed correlation with diseases. This strategy can be seen in detail elsewhere.[47,48] Nevertheless, the untargeted approach is not the most common one, possibly due to the fact that chiral separation is a very selective strategy and, at least with the current state-of-the-art, it would be challenging to conduct an untargeted chiral metabolomic approach. Thus, research has been focused on a more controlled or targeted method, meaning that it has been dedicated only to specific groups of metabolites, primarily using LC-MS and GC-MS and to a less extent, CE-MS.

Literature reports based on enantioselective profiling by LC-MS include studies of amino acids and carboxylic acids, among others. Kimura *et al.* determined a subset of 16 D-amino acids in plasma from more than 100 patients suffering from chronic kidney disease (CKD).[12] This study revealed a correlation between CKD and the levels of D-serine, D-proline, and D-asparagine, which thereby indicates the importance of D-amino acids as potential biomarkers for renal disorders. In another contribution, researchers profiled chiral carboxylic acids in saliva from patients with diabetes, proposing the D/L-lactate ratio as a biomarker to diagnose diabetes using LC-MS.[15] The importance of studying enantiomeric profiles not only applies to the bioanalysis field. As an example, Miyoshi *et al.* employed two-dimensional LC-MS to analyze Kurozu vinegars, finding that vinegars obtained during traditional earthenware jar fermentation had larger quantities of D-amino acids.[49] GC-MS has also been employed in enantiomeric profiling, as can be seen in the work by Waldhier and collaborators, where a GCxGC-MS setup using a combination of an Rt-γDEXsa chiral column and a polar achiral column enabled the enantioseparation of 10 pairs of amino acid enantiomers.[50] When comparing the chiral amino acid profile in serum samples from patients with liver cirrhosis and healthy patients, a significant increase in the level of D-amino acids was found in the cirrhosis group. These contributions highlight the importance and utility of chiral metabolomics to obtain further knowledge on enantioselective systems. Regarding CE-MS, chiral metabolomics has not yet been fully exploited and only a limited number of studies that can be included in this theme have been reported. In the next section, the current state-of-the-art of CE-MS in chiral metabolic profiling will be discussed.

9.2.2 Applications of Chiral Metabolic Profiling by CE-MS

The advantages of CE-MS for chiral analysis have been made clear in Section 9.1.4 of this chapter. Nevertheless, to date, only a few studies devoted to studying chiral metabolic profiles have been reported by this technique. In this section, the most relevant and noteworthy uses of chiral CE-MS, in which the profile of enantiomeric species has led to and/or helped in the discovery of novel diagnostics tools, or the study of the metabolism of drugs, among others, are discussed. These applications are also included in Table 9.1, along with detailed experimental aspects regarding the analysis method.

9.2.2.1 Study of the Enantioselective Metabolism of Drugs

As in the pioneering work by Nicholson and collaborators,[46] an interesting application is the study of metabolism of drugs, in which the metabolism of drugs enantiomers, as well as their metabolites, is key in determining their pharmacologic and therapeutic effects. In recent years, CE-MS has been used to study the metabolic profile of drugs and their metabolite enantiomers, and will be herein discussed.

By means of a MEKC-MS system based on the use of polysodium *N*-undecanoyl-L-leucylvalinate (poly-L,L-SULV) polymeric micelles, fully compatible with MS detection, Hou *et al.* studied the metabolism of warfarin enantiomers in patients undergoing warfarin therapy.[51] In healthy patients, the levels of *R*-warfarin in plasma were higher than the ones for *S*-warfarin, as expected from its stereoselective metabolism (the *S*-enantiomer is more potent and it is metabolized much faster). However, in plasma from a patient with mutant genes (CYP2C9*2 or *3), the *S*-enantiomer was more abundant than its *R*-counterpart, meaning that *S*-warfarin was metabolized in a much lower manner than its enantiomer in the patient with these mutated genes.[51] A more sensitive methodology using a triple quadrupole (QqQ) as the MS analyzer was developed by the same research group in which the chiral profile of warfarin and hydroxywarfarins was evaluated in 55 human plasma samples (see Figure 9.3).[34] The authors corroborated, once again, the output found in their previous paper regarding differences in enantiomeric profile for warfarin.[51]

Another contribution based on the study of enantiomeric metabolism employing polymeric micelles was carried out by Liu *et al.*, in which enantiomers of the drug venlafaxine (VX) and its active metabolite O-desmethylvenlafaxine (O-DVX) were enantioseparated by poly-sodium *N*-undecenoyl-L,L-leucylalaninate (poly-L,L-SULA).[52] In this interesting paper, first, the authors found that, upon co-administration of VX and indinavir (an antiretroviral), *R*,*S*-VX metabolized faster to *R*,*S*-O-DVX. Second, differences were found in the *S*/*R* ratio of VX and O-DVX. The authors suggested that it could be due to differences in the pharmacogenetic profile from CYP2D6 metabolism or the drug transporter P-glycoprotein polymorphism.[52]

Iio *et al.* developed a chiral EKC-MS method to monitor amphetamine derivatives in urine.[53] The authors used a dual system based on a mixture of neutral β-CD and (2,6-di-*O*-methyl)-β-CD to enable the enantiomers to be fully resolved. Due to the inherent migration time shift that occurs in the analysis of biological samples by CE, the researchers used two internal standards for proper migration time and peak area correction. This method was applied to study the stereoselective metabolism of amphetamine and its derivatives in the urine of addicts of these drugs. The obtained results displayed differences in metabolism. On the one hand, in the urine of an (*S*,*R*)-methamphetamine addict, the *S*-enantiomers of methamphetamine and amphetamine were more abundant than the corresponding *R*-forms. On the other hand, the urine of an *S*-dimethylamphetamine addict only revealed the presence of the *S*-form of dimethylamphetamine, methamphetamine and amphetamine. The authors stated that the measured concentrations matched the ones quantified by a GC-MS method, highlighting the applicability of the CE-MS approach.[53]

Using an EKC-MS method based on a polyvinyl alcohol-coated capillary, Rudaz and collaborators studied the metabolism of the drug tramadol in plasma from a volunteer after intake of this drug in a racemic form.[54] The analysis concluded that both tramadol and *O*-desmethyltramadol, its active metabolite, were found in the plasma as a racemic mixture.

Table 9.1 Research articles related to chiral metabolic profiling using CE-MS.[b]

Objective	MS detection	BGE
Study of the enantioselective metabolism of drugs	Q ESI- SL: methanol : water (80 : 20, v/v) containing 5 mM NH_4OAc (pH 6.8) at 5 μL min^{-1}	25 mM poly-L,L-SULV in 25 mM NH_4OAc (pH 5.5)
	QqQ ESI+ SL: methanol : water (80 : 20, v/v) with 5 mM NH_4OAc (pH 6.8) at 5 μL min^{-1}	25 mM poly-L,L-SULV in 25 mM NH_4OAc (pH 5.5) + 15% (v/v) methanol
	QqQ ESI+ SL: 80/20 methanol : water (v/v) + 5 mM NH_4OAc (pH 6.8) at 5 μL min^{-1}	15 mM poly- L,L-SULA in 20 mM NH_4OAc, 25 mM TEA (pH 8.5)
	Q ESI+ SL: 10 mM ammonium acetate : methanol (50 : 50, v/v) at 4 μL min^{-1}	3 mM β-CD and 10 mM DM-β-CD in 1 M formic acid (pH 2.2)
	Q ESI+ SL: isopropanol : water (50 : 50, v/v) + 0.5% formic acid at 3 μL min^{-1}	2.5 mg mL^{-1} SBE-β-CD in 40 mM ammonium acetate (pH 4.0) PFT
Study of the enantioselective metabolism of amino acids	IT (MS/MS) ESI+ SL: methanol : water (50 : 50, v/v) + 0.1% formic acid at 3 μL min^{-1}	5 mM sulfated-β-CD in 0.2 M formic acid PFT
	IT (MS/MS) ESI+ Make up fluid: 15 mM ammonium acetate (pH 5.5)/ methanol (1 : 1) at 100 nL min^{-1}	15 mM sulfated-β-CD in 15 mM ammonium acetate (pH 5.5)/methanol (1 : 1) PFT
	IT (MS/MS) ESI+ Make up fluid: 15 mM ammonium acetate (pH 5.5)/methanol (1 : 1) at 100 nL min^{-1}	15 mM sulfated β-CD in 15 mM ammonium acetate/acetic acid (pH 5.5)/methanol (1 : 1) PFT

Other conditions	Analytes	Sample	LOD[a]	Analysis time	Ref.
Bare fused-silica capillary: 50 µm × 120 cm Voltage: +30 kV Temperature: 20 °C	*R*,*S*-warfarin and *R*,*S*-coumachlor	Human plasma	0.1 µM	22 min	51
Bare fused-silica capillary: 50 µm × 120 cm Voltage: +30 kV Temperature: 20 °C	*R*,*S*-warfarin and 4 hydroxy-warfarins (10-OH, 6-OH, 7-OH, and 4′-OH-warfarine)	Human plasma	0.15–0.9 ng mL^{-1}	45 min	34
Bare fused-silica capillary: 50 µm × 60 cm Voltage: +25 kV Temperature: 20 °C	*R*,*S*-venlafaxine and *R*,*S*-O-desmethyl-venlafaxine	Human plasma	9.5–13.6 ng mL^{-1}	15 min	52
Bare fused-silica capillary: 50 µm id × 100 cm Voltage: +30 kV Temperature: 20 °C	*R*,*S*-methamphet-amine, *R*,*S*-am-phetamine, *R*,*S*-dimethyl-amphetamine, and *R*,*S*-p-hydroxymeth-amphetamine	Human urine	30–50 ng mL^{-1}	33 min	53
Polyvinyl alcohol-coated capillary: 50 µm id × 70 cm Voltage: +25 kV Temperature: 20 °C	*R*,*S*-tramadol and its phase I metabolites	Human plasma	—	20 min	54
Bare fused-silica capillary: 75 µm × 80 cm Voltage: +30 kV Temperature: 20 °C	D,L-DOPA, D,L-Phe, D,L-Tyr	PC12 cells	0.48–0.51 µM	12 min	55
Glass/PMDS hybrid microchip CE Voltage: 3850 V	D,L-DOPA, D,L-Ser, D,L-Glu	SH-SY5Y cells	43–47 nM	2 min	56
Glass/PMDS hybrid microchip CE Voltage: 3950 V	D,L-Ser	PC-12 cells	—	2 min	57

(*continued*)

Table 9.1 (*continued*)

Objective	MS detection	BGE
D-amino acids as diagnostic markers of diseases	QqQ ESI- SL: methanol : water : ammonium hydroxide (50 : 49.5 : 0.5%, v/v/v) at 6 μL min^{-1}	25 mM vancomycin in 50 mM ammonium acetate (pH 4.5)
D-Amino acids as indicators of the state of a system	IT (MS/MS) ESI+ Sheathless	30 mM 18-C-6-TCA in water
	TOF ESI+ SL: water : 2-propanol (50 : 50 v/v) at 4 μL min^{-1}	0.5 mM of CD3NH2 in 50 mM ammonium hydrogen carbonate (pH 8.0)
	IT (MS/MS) ESI+ SL: isopropanol/25 mM ammonium carbonate (50/50 v/v) at 3.3 μL min^{-1}	Method 1: 5 mM γ-CD in 50 mM ammonium carbonate buffer (pH 10.0) Method 2: 5 mM β-CD in 50 mM ammonium carbonate buffer (pH 10.0)

[a]LOD calculated as $S/N = 3$.

[b]18-C-6-TCA: (+)-(18-crown-6)-2,3,11,12-tetracarboxylic acid; Ala: alanine; Arg: arginine; Asn: asparagine; CD3NH2: 3-monodeoxy-3-monoamino-β-CD; DM-β-CD: dimethyl-β-CD; DOPA: 3,4-dihydroxyphenylalanine; FITC: fluorescein isothiocyanate; Gln: glutamine; Glu: glutamic acid; His: histidine; IT: ion trap analyzer; Leu: leucine; Lys: lysine; Met: methionine; Orn: ornithine; PFT: partial filling technique; Phe: phenylalanine; poly-L,L-SULA: polysodium undecanoyl-L-leucyl-alaninate; poly-L,L-SULV: polysodium *N*-undecanoyl-L-leucylvalinate; Q: quadrupole analyzer; QqQ: triple quadrupole analyzer; SBE-β-CD: sulfobutyl ether-β-CD; Ser: serine; SL: sheath liquid; TEA: triethylamine; TOF: time of flight analyzer; Trp: tryptophan; Tyr: tyrosine.

Other conditions	Analytes	Sample	LOD[a]	Analysis time	Ref.
Polyacrylamide-coated capillary: 50 µm × 60 cm Voltage: −25 kV Temperature: 25 °C	D,L-2-hydroxyglutaric acid	Human urine	31–38 nM	9 min	58
Bare fused-silica capillary: 20 µm × 130 cm Voltage: +25 kV Temperature: —	D,L-Asp	Bones and Silks	Attomole	20 min	59 and 60
Bare fused-silica capillary: 50 µm × 85 cm Voltage: +30 kV Temperature: 25 °C	FITC-derivatized D,L-Arg, D,L-Asn, D,L-Ala, D,L-Glu, D,L-Asp	Vinegar and soybean	0.16–1.87 µM	20 min	61
Bare fused-silica capillary: 50 µm × 100 cm Voltage: +20 kV Temperature: 25 °C	Method 1: FITC-derivatized D,L-Ala, D,L-Asn, D,L-Asp, D,L-Gln, D,L-Glu, D,L-His, D,L-Leu, D,L-Met, D,L-Phe, D,L-Ser, D,L-Tyr Method 2: FITC-derivatized D,L-Lys, D,L-Orn, D,L-Trp	Fertilizers	0.02–0.8 µM	60 min (each method)	43

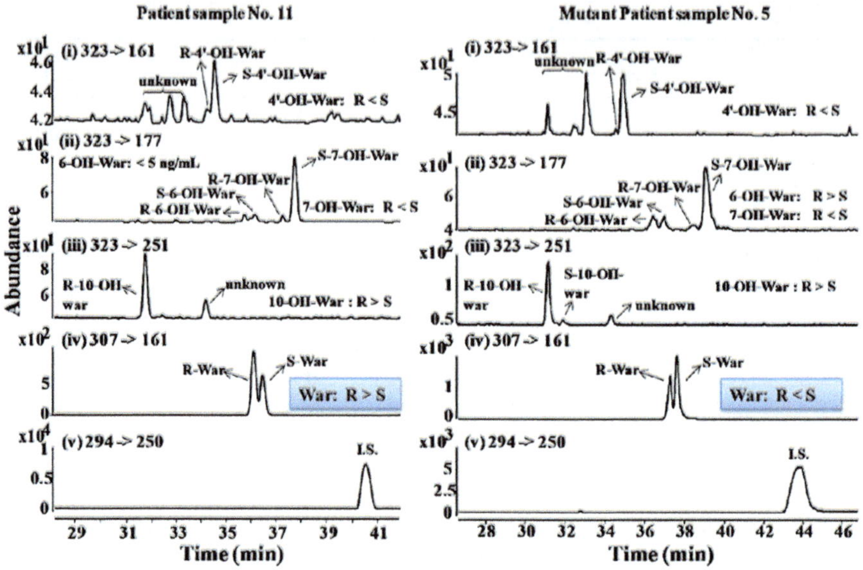

Figure 9.3 MEKC-MS/MS analysis of a plasma sample from a normal patient and a patient with mutated genes (CYP2C9*2 or *3) in which the *R*/*S*-warfarin (War) ratio is reversed. Reproduced from *Journal of Chromatography A*, **1271**, X. Wang, J. Hou, M. Jann, Y. Y. Hon and S. A. Shamsi, Development of a chiral micellar electrokinetic chromatography–tandem mass spectrometry assay for simultaneous analysis of warfarin and hydroxywarfarin metabolites: Application to the analysis of patients serum samples, 207–216. Copyright 2013, with permission from Elsevier.[34]

9.2.2.2 *Study of the Enantioselective Metabolism of Amino Acids*

As in the case of drugs, it is also interesting to address how metabolites, such as amino acids, are enantiomerically metabolized *in vivo* or *in vitro*. These studies will provide insight into the mechanisms involved in such metabolism. In this matter, Liu's research group developed a CE-MS methodology to allow enantioseparation of D,L-3,4-dihydroxyphenylalanine (DOPA), D,L-phenylalanine, and D,L-tyrosine in just 12 min using negatively charged sulfated-β-CD.[55] This method was effectively applied to monitor the enantioselective metabolism of DOPA in PC-12 cells as an *in vitro* neuronal model. When PC-12 cells were incubated with a racemic mixture of DOPA, the L-enantiomer was the only one to be metabolized by these cells, whereas D-DOPA remained intact.[55] Another contribution from Liu and collaborators was related to the investigation of amino acid metabolism using a microchip device coupled to MS to study the metabolism of DOPA in SH-SY5Y neuronal cells.[56] As in their previous work using PC12 cells,[55] upon incubation with racemic DOPA, SH-SY5Y cells only metabolized the L-enantiomer. These studies highlight the enantioselective preference of these neuronal cells when incorporating DOPA into their metabolism.

DOPA has not been the only amino acid of interest. In a recent contribution using microchip CE-MS, the authors studied both D-serine uptake and release in a PC-12 neuronal model.[57] Upon incubation of PC-12 cells with a racemic mixture of serine, the D-serine levels in the culture medium disappeared over time, whereas the L-enantiomer remained at the same level. On the other hand, the intracellular levels of D-serine increased. The authors suggested that this indicated that PC-12 cell metabolism is enantioselective, preferring D-serine over the L-enantiomer, possibly accumulating it in the cells for further use. Another finding achieved in this work was that Asc-1, a neutral amino acid transporter, participates in the uptake of D-serine in PC-12 cells. It was also pointed out that an important pathway for the release of D-serine is neuronal depolarization and it has no stereochemical preference. On the contrary, release evoked by glutamic acid was stereoselective towards D-serine.[57] All these findings clearly highlight the necessity of chiral methodologies to evaluate the enantioselective metabolism of amino acids.

9.2.2.3 D-Amino Acids as Diagnostic Markers of Diseases

It has been clearly demonstrated over the past few years how D-amino acids serve as biomarkers for the diagnosis of several diseases and disorders.[62,63] In this regard, CE-MS can be considered a useful tool, since amino acids are small chargeable molecules, which can be enantioseparated and selectively detected by this technique. An example of the application of this technique to the diagnosis of diseases is the study carried out by Svidrnoch *et al.* in which an EKC-MS/MS method based on the use of vancomycin as a chiral selector was used to determine, in just 9 min, underivatized D- and L-2-hydroxyglutaric acid (HG) in children affected with abnormal excretion of HG.[58] Urine analysis revealed that children affected with 2-hydroxyglutaric aciduria had similar levels of L-HG when compared to healthy patients, whereas the level of the D-form of HG was from 20 to 400 times higher than the levels found in healthy volunteers (see Figure 9.4). This is an interesting approach to aid in the diagnosis of different 2-hydroxyglutaric acidurias.[58]

9.2.2.4 D-Amino Acids as Indicators of the State of a System

Comparison of the chiral metabolic profile of amino acids is not only important in the diagnosis of diseases, but has also been demonstrated to provide valuable information when defining specific states and/or conditions of a particular system. This means that the presence of D-amino acids can be used to date historical artefacts,[59,60] or to classify genotypes of, for instance, vinegars,[61] or even in knowing what enzymatic treatment has been applied to fertilizers,[43] among others. These interesting applications carried out by CE-MS will be hereafter detailed.

Amino acid racemization is a post-translational modification that can be used to determine the age of different archeological artifacts. Moini *et al.*

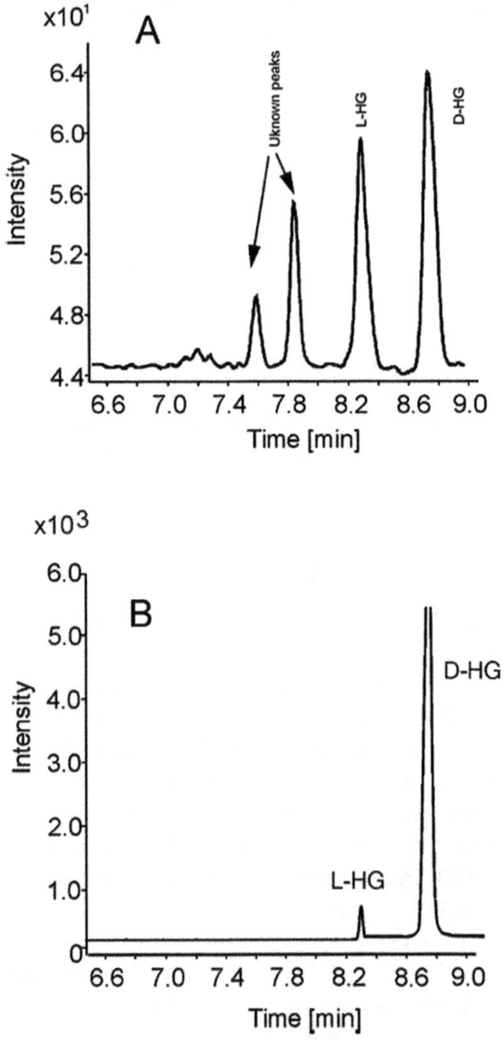

Figure 9.4 EKC-MS/MS analysis of urine from a healthy patient (A) and urine obtained from a child patient suffering from abnormal excretion of 2-HG (B). Reproduced from *Journal of Chromatography A*, **1467**, M. Svidrnoch, A. Pribylka, V. Bekarek, J. Sevcik, V. Smolka and V. Maier, Enantioseparation of D,L-2-hydroxyglutaric acid by capillary electrophoresis with tandem mass spectrometry—Fast and efficient tool for D- and L-2-hydroxyglutaracidurias diagnosis, 383–390. Copyright 2016, with permission from Elsevier.[58]

used a chiral CE-MS methodology based on amino acid complexation with the crown ether (+)-(18-crown-6)-2,3,11,12-tetracarboxylic acid (18-C-6-TCA) to assess dating of both historical silks[59] and archeological bones (see Figure 9.5)[60] using the D/L-aspartic acid ratio. According to the authors, this is the only amino acid whose conversion of L-enantiomer to D-enantiomer was

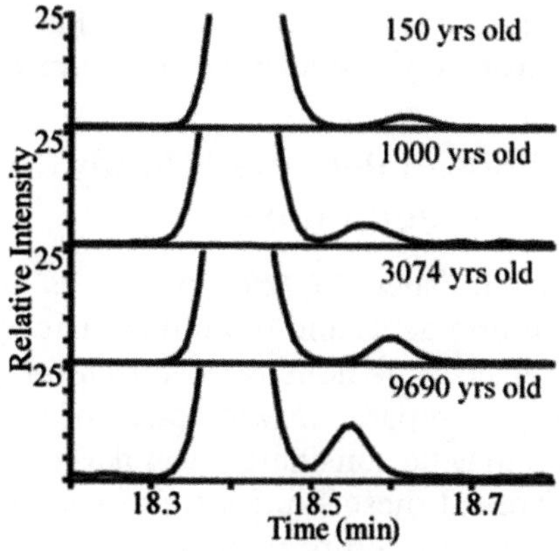

Figure 9.5 EKC-MS/MS analysis of bones with different ages, in which the large peak is the L-enantiomer of Asp and the minor peak is D-Asp. Adapted with permission from ref. 60. Copyright (2013) American Chemical Society.

quick enough to identify the aging process. In both studies, higher levels of D-aspartic acid were found in older silks[59] and older bones.[60] 18-C-6-TCA was used as the BGE, acting as a complexation reagent in the analysis of underivatized amino acids, which resulted in the formation of stable chiral selector complexes. Due to the high MS ionization yield of these complexes, careful selection of the *m/z* scan range, and the use of a sheathless ESI interface, LODs as low as at the attomole level were obtained.[45]

Even if direct approaches are commonly employed, the use of derivatization agents is popular in the determination of amino acids, since derivatization leads to a better sensitivity in MS analysis, due to the fact that larger molecules are formed and, thus, the ionization yield is higher. Moreover, as a consequence of the derivatization, the molecular mass also increases, which is beneficial as noisier backgrounds are expected when working in a lower mass range. An example of this event is the case of the work reported by Giuffrida *et al.*, in which only using 0.5 mM of a novel CD derivative, *i.e.* the cationic 3-monodeoxy-3-monoamino-β-CD (CD3NH2), it was possible to enantioseparate 10 pairs of FITC-labeled amino acid enantiomers.[61] The use of this cationic CD enabled fast chiral separation of FITC-labeled amino acids due to the extra positive charge added to the analytes. The authors applied this method to the analysis of wild and transgenic soybean. Along with the differences found in the L-amino acid profile of these samples, interestingly, differences in the D-amino acid profile were also found, with D-arginine only being present in the transgenic variety. The authors suggested that the inserted transgene could have enabled the modification of a metabolic pathway related to amino acid metabolism.[61]

Finally, Sánchez-Hernández *et al.* designed two strategies based on FITC-derivatization and native CDs (β- and γ-CD) to enantioseparate up to 13 amino acids.[43] Thanks to these two methods, the authors could study the amino acid profile of different fertilizers, finding that enzymatic hydrolysis prior to the chemical hydrolysis reduced the appearance of D-amino acids.

9.2.3 CE-MS Methods to Potentially Study Chiral Profiles for Amino Acids and/or Amines

This section outlines published CE-MS approaches specifically designed for the separation of amino acids and/or amines. The applicability of these approaches has not yet been demonstrated for chiral metabolic profiling, but due to their ability to separate a wide array of enantiomeric metabolites, these approaches may be considered useful as a starting point for this purpose. Therefore, some of these studies are discussed here with a special emphasis on the experimental setup, along with more information included in Table 9.2.

An indirect approach to enable enantioseparation of amino acids was carried out by Prior *et al.*, in which (+)-FLEC was used as a chiral derivatization reagent forming the corresponding diastereomers with the amino acids, which were then separated by the volatile negatively charged pseudostationary phase ammonium perfluorooctanoate (APFO).[64] Using this approach, 14 pairs of chiral amino acids, which were spiked into human cerebrospinal fluid (CSF), could be separated in about 45 min. Interestingly, the authors found endogenous D-serine and D-glutamine in healthy volunteers. Adaptation of this method to enable in-capillary derivatization was performed by Moldovan *et al.*, in which the authors demonstrated the applicability of the automated method in artificial CSF with good recoveries.[65] Another study by Prior *et al.* was based on the chiral separation of amino acids employing low amounts of a chiral selector that entered the MS system.[66] In this case, the authors used 10 mM of β-CD, which only resulted in moderate ion suppression of fluorenylmethyloxycarbonyl chloride (FMOC)-labeled amino acids. This method seemed to be suitable for amino acid determination in spiked human CSF samples. These studies point out the interest in using this methodology for further metabolomics studies to investigate, for example, disease mechanisms.

Among the procedures to enantioseparate many amino acids, a paper described the enantioseparation of 6 pairs of amino acid enantiomers.[67] Enantioseparation was possible with native β-CD in combination with a capillary coating based on fluorescein isothiocyanate (FITC) derivatization, which provided similar LODs (in the μM range) as compared to dansyl labelling, but the achieved enantioresolution was superior with FITC. The authors demonstrated the applicability of this method to the analysis of juices but, due to the importance of these amino acids in biological systems, this method could be further adapted to other biological matrices, such as biological fluids.

An EKC-MS/MS strategy was developed by Sánchez-Hernández *et al.* to enantioseparate 16 FMOC-derivatized amino acids.[68] This strategy is based on the use of the macrocyclic antibiotic vancomycin to enable enantioseparation

Table 9.2 Chiral CE-MS methods to potentially study profiles for amino acids and/or amines.[b]

MS detection	BGE	Other conditions	Analytes	Sample	LOD[a]	Analysis time	Ref.
IT (MS) ESI+ SL: isopropanol–water–formic acid (90:10:0.1, v/v/v) at 3 μL min^{-1}	150 mM APFO (pH 9.5)	Bare fused-silica capillary: 50 μm × 80 cm Voltage: +30 kV Temperature: 15 °C	(+)-FLEC derivatized D, L-Ala, D,L-Asn, D,L-Gln, D,L-Ile, D,L-Leu, D,L-Met, D,L-Phe, D,L-Ser, D,L-Thr, D,L-Trp, D,L-Val, D,L-Arg, D,L-His, D,L-Lys	Human CSF	0.38–4.62 μM	45 min	64
IT (MS) ESI+ SL: isopropanol–water–formic acid (90:10:0.1, v/v/v) at 2.5 μL min^{-1}	150 mM APFO (pH 9.5)	Bare fused-silica capillary: 50 μm × 80 cm Voltage: +25 kV Temperature: 15 °C	(−)-FLEC derivatized D,L-Ala, D,L-Ser, D,L-Val, D,L-Leu, D,L-Ile, D,L-Gln, D,L-Met, D,L-Phe, D,L-Tyr, D,L-Pro, D,L-Asp, D,L-Arg, D,L-His,	Artificial CSF	3–12 μM	40 min	65
IT (MS) ESI+ SL: isopropanol–water–1 M ammonium bicarbonate (50:50:1, v/v/v) at 3 μL min^{-1}	10 mM β-CD in 0.05 M ammonium bicarbonate (pH 8): isopropanol (85:15, v/v)	Bare fused-silica capillary: 50 μm × 80 cm Voltage: + 25 kV Temperature: 20 °C	FMOC-derivatized D,L-Ser, D,L-Pro, D,L-Val, D,L-Thr, D,L-Leu, D,L-Ile, D,L-Asn, D,L-Gln, D,L-Met, D,L-Phe, D,L-Trp, D,L-Cys, D,L-Asp, D,L-Glu	Human CSF	0.5–84.3 μM	52 min	66
IT (MS) ESI+ SL: methanol:water (50:50, v/v) containing 25% of running buffer at 3.5 μL min^{-1}	5 mM β-CD in 100 mM ammonium acetate (pH 6.0)	EPyM-DMA-coated capillary: 50 μm × 87 cm Voltage: −15 kV Temperature: --	FITC-derivatized D,L-Asp, D,L-Glu, D,L-Ser, D,L-Asn, D,L-Ala, D,L-Pro, D,L-Arg	Juices	2–29 μM	52 min	67

(*continued*)

Table 9.2 (*continued*)

MS detection	BGE	Other conditions	Analytes	Sample	LOD[a]	Analysis time	Ref.
IT (MS/MS) ESI+ SL: methanol : 50 mM ammonium formate (pH 7.0) (50 : 50, v/v) at 3.3 µL min^{-1}	10 mM vancomycin in 50 mM ammonium formate (pH 7.0) PFT	HDB-coated capillary: 50 µm × 100 cm Voltage: −20 kV Temperature: 25 °C	FMOC-derivatized D,L-Ala, D,L-Asn, D,L-Asp, D,L-Cit, D,L-Gln, D,L-Glu, D,L-Ile, D,L-Leu, D,L-Met, D,L-Phe, D,L-Pip, D,L-Pro, D,L-Ser, D,L-Thr, D,L-Trp, D,L-Tyr, D,L-Val	Standards	0.1–3 µM	22 min	68
IT (MS/MS) ESI+ SL: 50 : 50 (v/v) methanol/water with 0.1% (v/v) formic acid at 3.3 µL min^{-1}	180 mM M-β-CD and 40 mM HP-β-CD in 2 M formic acid (pH 1.2) PFT	Bare fused-silica capillary: 50 µm × 120 cm Voltage: +30 kV Temperature: 15 °C	D,L-Phe, D,L-Tyr, D,L-DOPA, D,L-Norepinephrine, D,L-Epinephrine	Rat plasma	40–150 nM	90 min	69

[a]LOD calculated as $S/N = 3$ except for ref. 65 where LOD was calculated as 3.3 times the standard deviation divided by the slope of the calibration curve.
[b]Ala: alanine; APFO: ammonium perfluorooctanoate; Arg: arginine; Asn: asparagine; Asp: aspartic acid; CSF: cerebrospinal fluid; Cys: cysteine; DOPA: 3,4-dihydroxyphenylalanine; EPyM-DMA: ethylpyrrolidine methacrylate-*N*,*N*-dimethylacrylamide; FITC: fluorescein isothiocyanate; FLEC: 1-(9-fluorenyl) ethyl chloroformate; FMOC: fluorenylmethyloxycarbonyl chloride; Gln: glutamine; Glu: glutamic acid; HDB: hexadimethrine bromide; His: histidine; HP-β-CD: 2-hydroxypropil-β-CD; Ile: isoleucine; IT: ion trap analyzer; Leu: leucine; Lys: lysine; Met: methionine; M-β-CD: methyl-β-CD; PFT: partial filling technique; Phe: phenylalanine; Pro: proline; Ser: serine; SL: sheath liquid; Thr: threonine; Trp: tryptophan; Tyr: tyrosine; Val: valine.

of the FMOC-labeled amino acids. Since vancomycin is a positively charged chiral selector it is necessary to coat the capillary with the cationic polymer HDB to avoid adsorption of the chiral selector onto the capillary walls. The PFT also had to be used to avoid entry of the chiral selector.[68] This methodology has potential in chiral metabolic profiling as a large number of amino acids were enantioseparated; however, readers need to bear in mind that the LODs were in the μM-range in MS^2 mode, which may limit the utility of this approach for the analysis of samples with (intrinsically) low levels of D-amino acids.

Regarding enantioseparation of amino acids and amines, recently, Sánchez-López *et al.* developed a chiral CE-MS^2 methodology that made it possible to resolve the chiral constituents involved in the phenylalanine–tyrosine metabolic pathway without the need of derivatization.[69] The PFT had to be used in order to avoid the undesired entry of the chiral selectors into the MS system. As far as we know, this is the only CE-MS methodology enabling the enantioseparation of the constituents of a metabolic pathway in a single run, although the analysis time is relatively long (90 min) and the LODs (at the nM level) were not low enough to enable the determination of all the studied compounds in the analyzed samples (rat plasma). A way to improve the sensitivity of this method could be by using a sheathless interface for coupling CE to MS.

9.3 Concluding Remarks and Future Trends

CE is a separation technique capable of successfully carrying out chiral separations thanks to its high efficiency, versatility, and low consumption of reagents, solvents, and most importantly, samples. The not-so-straightforward coupling of CE to MS detection makes its direct application to chiral metabolic profiling a challenging task. Nevertheless, in the past few years, several publications have been dedicated to studying chiral molecules by this technique, obtaining very interesting outputs. It is also important to take into account that, regardless of the analytical technique used, a chiral separation is a very selective approach; thus, with the current state-of-the-art global chiral metabolic profiling is a challenging task.

Future methodologies must be aimed at determining a larger number of molecules, ideally even mixtures of different classes of metabolites (*e.g.* amino acids, amines and carboxylic acids), in a single run. The enantioseparation of molecules belonging to the same metabolic pathway by means of sensitive methodologies should also be considered, as this would vastly improve the existing knowledge on enantioselective metabolism. Although the current state-of-the-art indicates that there is still much left to do to implement CE-MS as a technique to carry out chiral metabolic profiling, future trends should continue to provide more robust instruments (particularly to avoid migration time shifting) with higher sensitivity (by using in-capillary preconcentration approaches or using more sensitive interfaces such as the sheathless ESI interface). Another trend for the future is the development of new chiral selectors fully compatible with MS detection, which will widen the options for chiral CE-MS analysis. These improvements, will, without a doubt, increase the applicability of chiral CE-MS, expanding its use in chiral metabolomics, hopefully, in the (near) future.

Acknowledgements

E. S. L. and M. L. M. thank the Ministry of Economy and Competitiveness (Spain) for research project CTQ2016-76368 P. E. S. L. and A. L. C. thank the Government Delegation for the National Plan on Drugs of the Ministry of Health, Social Services and Equality (Spain) for the research project PNSD-2016I073. E. S. L. thanks the Comunidad Autónoma of Madrid (Spain) and European funding from the FEDER program (project S2013/ABI-3028, AVANSECAL-CM) for her contract.

References

1. N. M. Maier and W. Lindner, in *Chirality in Drug Research*, ed. E. Francotte and W. Lindner, Wiley-VCH, Weinheim, 2006, p. 189.
2. R. S. Cahn, C. K. Ingold and V. Prelog, *Angew. Chem., Int. Ed.*, 1966, **5**, 385.
3. L. Pasteur, *Compt. Rend. Acad. Sci.*, 1858, **46**, 615.
4. J. J. Corrigan and N. G. Srinivasan, *Biochemistry*, 1966, **5**, 1185.
5. J. J. Corrigan, *Science*, 1969, **164**, 142.
6. G. Kreil, *J. Biol. Chem.*, 1994, **269**, 10967.
7. R. L. Preston, *Comp. Biochem. Physiol., Part B: Biochem. Mol. Biol.*, 1987, **87**, 55.
8. A. V. Chervyakov, N. V. Gulyaeva and M. N. Zakharova, *Neurochem. J.*, 2011, **5**, 100.
9. D. L. Kirschner and T. K. Green, *J. Sep. Sci.*, 2009, **32**, 2305.
10. G. E. Tsai, P. Yang, Y. C. Chang and M. Y. Chong, *Biol. Psychiatry*, 2006, **59**, 230.
11. C. Henneberger, T. Papouin, S. H. Oliet and D. A. Rusakov, *Nature*, 2010, **463**, 232.
12. T. Kimura, K. Hamase, Y. Miyoshi, R. Yamamoto, K. Yasuda, M. Mita, H. Rakugi, T. Hayashi and Y. Isaka, *Sci. Rep.*, 2016, **6**, 26137.
13. Y. Xing, X. Li, X. Guo and Y. Cui, *Anal. Bioanal. Chem.*, 2016, **408**, 141.
14. R. Koga, Y. Miyoshi, Y. Sato, M. Mita, R. Konno, W. Lindner and K. Hamase, *J. Chromatogr. A*, 2016, **1467**, 312.
15. T. Takayama, T. Kuwabara, T. Maeda, I. Noge, Y. Kitagawa, K. Inoue, K. Todoroki, J. Z. Min and T. Toyo'oka, *Anal. Bioanal. Chem.*, 2015, **407**, 1003.
16. F. Cava, H. Lam, M. A. de Pedro and M. K. Waldor, *Cell. Mol. Life Sci.*, 2011, **68**, 817.
17. B. Chankvetadze and G. Blaschke, *J. Chromatogr. A*, 2001, **906**, 309.
18. B. Chankvetadze, W. Lindner and G. K. E. Scriba, *Anal. Chem.*, 2004, **76**, 4256.

19. B. Chankvetadze, *J. Chromatogr. A*, 2007, **1168**, 45.
20. M. Castro-Puyana, A. L. Crego, M. L. Marina and C. García-Ruiz, *Electrophoresis*, 2007, **28**, 2667.
21. K. Lomsadze, A. B. Martínez-Girón, M. Castro-Puyana, L. Chankvetadze, A. L. Crego, A. Salgado, M. L. Marina and B. Chankvetadze, *Electrophoresis*, 2009, **30**, 2803.
22. E. Sánchez-López, A. Salgado, A. L. Crego and M. L. Marina, *Electrophoresis*, 2014, **35**, 2842.
23. E. Sánchez-López, M. Castro-Puyana, M. L. Marina and A. L. Crego, in *Analytical Separation Science*, ed. J. Anderson, A. Berthod, V. Pino and A. M. Stalcup, Wiley-VCH Verlag GmbH & Co, Weinheim, 2015, vol. 2, p. 731.
24. E. Gassmann, J. E. Kuo and R. N. Zare, *Science*, 1985, **230**, 813.
25. J. Zhou, J. Tang and W. Tang, *TrAC, Trends Anal. Chem.*, 2015, **65**, 22.
26. T. J. Ward and K. D. Ward, *Anal. Chem.*, 2012, **84**, 626.
27. K. Kalíková, M. Riesová and E. Tesařová, *Cent. Eur. J. Chem.*, 2012, **10**, 450.
28. P. Jac and G. K. E. Scriba, *J. Sep. Sci.*, 2013, **36**, 52.
29. D. A. Tsioupi, R. I. Stefan-van-Staden and C. P. Kapnissi-Christodoulou, *Electrophoresis*, 2013, **34**, 178.
30. J. Chen, Y. Du, F. Zhu and B. Chen, *J. Chromatogr. A*, 2010, **1217**, 7158.
31. J. He and S. A. Shamsi, *J. Chromatogr. A*, 2009, **1216**, 845.
32. Q. Zhu and G. K. E. Scriba, *Chromatographia*, 2016, **79**, 1403.
33. C. Simó, V. García-Cañas and A. Cifuentes, *Electrophoresis*, 2010, **31**, 1442.
34. X. Wang, J. Hou, M. Jann, Y. Y. Hon and S. A. Shamsi, *J. Chromatogr. A*, 2013, **1271**, 207.
35. E. Sánchez-López, M. L. Marina and A. L. Crego, *Electrophoresis*, 2016, **37**, 19.
36. A. L. Crego, M. Mateos and L. Nozal, *Electrophoresis*, 2018, **39**, 67–81.
37. R. Ramautar, G. W. Somsen and G. J. de Jong, *Electrophoresis*, 2013, **34**, 86.
38. A. Wuethrich, P. R. Haddad and J. P. Quirino, *Electrophoresis*, 2014, **35**, 2.
39. X. F. Zhong, Z. C. Zhang, S. Jiang and L. J. Li, *Electrophoresis*, 2014, **35**, 1214.
40. K. Kleparnik, *Electrophoresis*, 2015, **36**, 159.
41. W. Zhang, T. Hankemeier and R. Ramautar, *Curr. Opin. Biotechnol.*, 2017, **43**, 1.
42. L. Sánchez-Hernández, M. Castro-Puyana, C. García-Ruíz, A. L. Crego and M. L. Marina, *Food Chem.*, 2010, **120**, 921.
43. L. Sánchez-Hernández, N. Sierras Serra, M. L. Marina and A. L. Crego, *J. Agric. Food Chem.*, 2013, **61**, 5022.
44. Y. T. Iwata, T. Kanamori, Y. Ohmae, K. Tsujikawa, H. Inoue and T. Kishi, *Electrophoresis*, 2003, **24**, 1770.
45. M. Moini, C. L. Schultz and H. Mahmood, *Anal. Chem.*, 2003, **75**, 6282.
46. M. Pérez-Trujillo, J. C. Lindon, T. Parella, H. C. Keun, J. K. Nicholson and T. J. Athersuch, *Anal. Chem.*, 2012, **84**, 2868.

47. T. Takayama, T. Mochizuki, K. Todoroki, J. Z. Min, H. Mizuno, K. Inoue, H. Akatsu, I. Noge and T. Toyo'oka, *Anal. Chim. Acta*, 2015, **898**, 73.
48. T. Toyo'oka, *Biol. Pharm. Bull.*, 2016, **39**, 1397.
49. Y. Miyoshi, M. Nagano, S. Ishigo, Y. Ito, K. Hashiguchi, N. Hishida, M. Mita, W. Lindner and K. Hamase, *J. Chromatogr. A*, 2014, **966**, 187.
50. M. C. Waldhier, M. F. Almstetter, N. Nürnberger, M. A. Gruber, K. Dettmer and P. J. Oefner, *J. Chromatogr. A*, 2011, **1218**, 4537.
51. J. Hou, J. Zheng and S. A. Shamsi, *J. Chromatogr. A*, 2007, **1159**, 208.
52. Y. Liu, M. Jann, C. Vandenberg, C. B. Eap and S. A. Shamsi, *J. Chromatogr. A*, 2015, **1420**, 119.
53. R. Iio, S. Chinaka, S. Tanaka, N. Takayama and K. Hayakawa, *Analyst*, 2003, **128**, 646.
54. S. Rudaz, S. Cherkaoui, P. Dayer, S. Fanali and J. L. Veuthey, *J. Chromatogr. A*, 2000, **868**, 295.
55. B. Yuan, H. Wu, T. Sanders, C. McCullum, Y. Zheng, P. B. Tchounwou and Y. M. Liu, *Anal. Biochem.*, 2011, **416**, 191.
56. X. Li, D. Xiao, X. M. Ou, C. McCullm and Y. M. Liu, *J. Chromatogr. A*, 2013, **1318**, 251.
57. X. Li, C. McCullum, S. Zhao, H. Hu and Y. M. Liu, *ACS Chem. Neurosci.*, 2015, **6**, 582.
58. M. Svidrnoch, A. Pribylka, V. Bekarek, J. Sevcik, V. Smolka and V. Maier, *J. Chromatogr. A*, 2016, **1467**, 383.
59. M. Moini, K. Klauenberg and M. Ballard, *Anal. Chem.*, 2011, **83**, 7577.
60. M. Moini, C. M. Rollman and C. A. France, *Anal. Chem.*, 2013, **85**, 11211.
61. A. Giuffrida, C. León, V. García-Cañas, V. Cuccinotta and A. Cifuentes, *Electrophoresis*, 2009, **30**, 1734.
62. K. Kalikova, T. Slechtova and E. Tesarova, *Separations*, 2016, **3**, 30.
63. K. Hamase, A. Morikawa and K. Zaitsu, *J. Chromatogr. A*, 2002, **781**, 73.
64. A. Prior, R. C. Moldovan, J. Crommen, A. C. Servais, M. Fillet, G. J. de Jong and G. W. Somsen, *Anal. Chim. Acta*, 2016, **940**, 150.
65. R. C. Moldovan, E. Bodoki, T. Kacso, A. C. Servais, J. Crommen, R. Oprean and M. Fillet, *J. Chromatogr. A*, 2016, **1467**, 400.
66. A. Prior, L. Sánchez-Hernández, J. Sastre-Toraño, M. L. Marina, G. J. de Jong and G. W. Somsen, *Electrophoresis*, 2016, **37**, 2410.
67. C. Simó, A. Rizzi, C. Barbas and A. Cifuentes, *Electrophoresis*, 2005, **26**, 1432.
68. L. Sánchez-Hernández, E. Domínguez-Vega, C. Montealegre, M. Castro-Puyana, M. L. Marina and A. L. Crego, *Electrophoresis*, 2014, **35**, 1244.
69. E. Sánchez-López, A. Marcos, E. Ambrosio, M. L. Marina and A. L. Crego, *J. Chromatogr. A*, 2016, **1467**, 372.

CHAPTER 10

Single-cell Metabolomics with Capillary Electrophoresis–Mass Spectrometry

ROSEMARY M. ONJIKO, ERIKA P. PORTERO AND P. NEMES*

Department of Chemistry & Biochemistry, University of Maryland, College Park, MD 20742, USA
*E-mail: nemes@umd.edu

10.1 Introduction and Overview

Single-cell analysis promises to be a new frontier in the study of how cells of identical genotype execute molecular programs to acquire important functional differences.[1–5] Molecular analysis with single-cell resolution provides biomolecular insights that would otherwise be masked due to signal averaging by conventional approaches that pool cells for analysis. The metabolome, the comprehensive suite of all metabolites produced in the cell, complements information on cell phenotype and physiological state. The metabolome amasses downstream products of intrinsic events, such as transcription, translation, and changing phases of the cell cycle, as well as extrinsic influences, including chemical, physical, and biological cues. Therefore, the single-cell metabolome provides a powerful descriptor of the cell's molecular state.

Characterization of the single-cell metabolome comes with analytical challenges. Metabolites have diverse physicochemical properties (*e.g.*, pK_a

New Developments in Mass Spectrometry No. 6
Capillary Electrophoresis–Mass Spectrometry for Metabolomics
Edited by Rawi Ramautar

Published by the Royal Society of Chemistry, www.rsc.org

and hydrophobicity), are highly dynamic, and cover several orders of magnitude in abundance.[1] For example, the Human Metabolome Database (HMDB) currently contains ~114 100 entries ranging from <1 nM to millimolar concentration. Compounding these challenges, single cells vary broadly in size across the phyla and even in the same organisms, offering minute amounts of material for detection.[6] Detection of the single-cell metabolome requires specialized analytical techniques, particularly those capable of low detection limits, high molecular specificity, quantification, and sufficient speed to probe metabolic changes with potentially high turn-over rates.

Single-cell mass spectrometry (MS) is one such technology that enables the characterization of metabolites in single cells. Technologies of single-cell MS employ diverse experimental conditions to provide small-molecule measurements with scalability in molecular coverage, spatial and temporal resolution, and/or operation under denaturing or native conditions. For an overview of single-cell MS, readers are referred to recent reviews in ref. 1 and 7–12. A subset of single-cell MS technologies employs some form of separation to advance the detectable coverage of the single-cell metabolome. Liquid-phase separation before electrospray ionization (ESI) MS aids molecular detection by obtaining compound-dependent information, *viz.* the time a compound spends in separation, and/or by removing interferences during ion generation or spectral detection. Additionally, gas-phase separation in an ion mobility analyzer helps reduce spectral interferences and quantify collision-cross sections as compound-dependent information. This strategy has detected ~400 ion signals and identified 23 metabolites and lipids from single *A. thaliana* cells.[13] The Metabolomics Standards Initiative recommends the use of separation time (*e.g.*, retention time in chromatography, migration time in electrophoresis) to advance the confidence of metabolite identifications.[14]

This chapter is focused on trace-sensitive CE-MS, which has demonstrated several advantages for single-cell metabolomics.[9] Perhaps most importantly, CE is compatible with the limited sample volumes that are contained by single cells. CE also obtains exquisite separation peak capacity to efficiently reduce the complexity of metabolite mixtures and help distinguish isobaric compounds prior to detection. We and others have coupled CE to electrospray ionization (ESI) MS and developed specialized workflows to detect hundreds of metabolite features in single cells.[2,4]

The general approach of single-cell studies is presented in Figure 10.1. The workflow begins with the identification and isolation of cells and is followed by sampling of their content before metabolites are extracted. In CE, extracted metabolites are separated based on differences in electrophoretic mobility, are ionized (usually by ESI), and metabolite ions are detected by MS–MS/MS. With a thoughtful experimental design, it is possible to identify metabolite signals and quantitatively compare their profiles between cells. By quantifying hundreds of metabolite features, CE-ESI-MS has enabled the differentiation of cells and cell types based on metabolic composition,

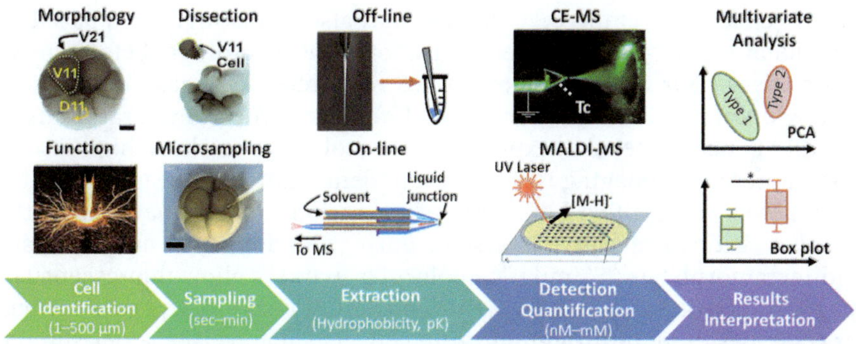

Figure 10.1 Workflow for single-cell metabolomics. Different technologies provide scalability across broad physical, temporal, and chemical dimensions to differentiate cells based on their metabolic composition. Representative technologies are depicted.[3,19,78,79] Adapted with permission from ref. 3 (http://pubs.acs.org/doi/abs/10.1021/ac500168d), Copyright (2014) American Chemical Society, further permissions related to the material excerpted should be directed to the ACS; ref. 19, Copyright (2017) American Chemical Society; ref. 78, Copyright (2014) American Chemical Society and ref. 79, Copyright (2010) American Chemical Society.

including single neurons in the central nervous system of the giant sea slug (*Aplysia californica*)[2,15,16] and embryonic cells of the 8-,[17,18] 16-,[4,19] and 32-cell[19] embryo of the South African clawed frog (*Xenopus laevis*). Furthermore, data from these studies have allowed us to design functional experiments that led to the discovery of small molecules capable of altering the normal tissue fate of embryonic cells.[4] In what follows, we give an overview of the major steps of the single-cell CE-ESI-MS analytical workflow (Figure 10.1). Where applicable, we make recommendations to advance the success of single-cell investigations.

10.2 Sampling of Single Cells

The goal of this step is to collect material from individual cells, providing the "sample" for further processing prior to CE-MS (Figure 10.1). Depending on the purpose of the study, cell sampling may involve an initial step to define cells of a particular phenotype before sampling commences. Morphology and established cell maps have allowed the distinction of cell types in powerful biological models, such as individual neurons in the central nervous system of *Aplysia californica*,[15] crabs,[20,21] and *Drosophila melanogaster* (fruit fly).[22] Likewise, based on cell morphology and cell fate maps,[23–25] we have identified cells occupying the left–right, dorsal–ventral, and animal–vegetal axes of the *Xenopus laevis* embryo and used single-cell CE-ESI-MS to compare their metabolomic composition. These studies revealed that the neural-tissue fated D11 cell, the epidermal-fated V11 cell, and the hindgut-fated V21 cell produced distinct metabolomes in the 16-cell *X. laevis* embryo.[4] Cell types may also be distinguished based on functional studies, such as electrophysiological recordings, before chemical analysis. For example, metabolites were recently measured in excitatory glutamatergic and inhibitory GABAergic neurons in the rat brain by CE-MS.[3] For cell types that express known molecular markers, high-resolution fluorescence imaging facilitates cell identification. The success of this approach is exemplified by *in situ* hybridization marking neurons in the *A. californica* central nervous system during development, neurogenesis, and metamorphosis.[26] Similarly, molecular markers allow fluorescence or magnetic activated cell sorting (FACS or MACS, resp.) to separate particular cell types in high-throughput from cell suspensions (see reviews in ref. 27–33). In developing biological systems, optical imaging may be necessary to identify cells in a precursor–progenitor relationship. For example, we have used morphology (*e.g.*, pigmentation and size) and fluorescent dyes to track cell clones in the 8-, 16-, 32-, 64-, and 128-cell *X. laevis* embryo before characterizing their small molecules[4,17,19,34] and proteins[35–37] by CE-ESI-MS. Alternatively, high-throughput imaging by single-cell MALDI MS enables the recognition of cell types based on chemical composition without information of phenotype, morphology, or location, as recently demonstrated for cells in rat pancreatic islets.[38]

To collect a portion of whole cells, different technologies have been developed with scalability in size and collection speed (Figure 10.2). Manual dissection[39,40] is readily amenable to larger cells. As shown in Figure 10.2A, sharp tungsten needles, fine forceps, or finer tools have been used to isolate 50–500 μm cells from *A. californica*,[15] a mouse ganglion,[3] or *X. laevis* embryos.[4,17,41] With a modest throughput, microdissection offers the benefit of preserving information on the spatial origin of isolated cells, which can add important information to address biological questions, such as signaling molecules expressed in neuronal circuits[3,15,16] or metabolites implicated in cell-fate decisions.[4] For smaller cells, which challenge dissection accuracy, approaches capable of finer resolution were introduced. Sharpened or pulled capillaries were found to be able to reproducibly target and aspirate picoliter-to-nanoliter volumes from rat thalamic neurons,[3] tomato trichome cells,[42] and frog embryonic cells,[19] as shown in Figure 10.2A. Capillaries containing dual barrels helped enhance the throughput of these measurements. Figure 10.2B exemplifies how this strategy was able to collect metabolites from multiple *Allium cepa* cells.[43] Fluid force microscopy takes an alternative approach, in which a cantilever is driven by an atomic force microscope to probe subpicoliter volumes from human cervical cancer (HeLa) cells (Figure 10.2C).[44]

Capillary sampling has also been coupled to CE-MS. Representative examples include the metabolic study of single neurons in the rat thalamus[3] and single *X. laevis* cells.[3,19] Microsampling allowed the same cell to be sampled multiple times, and surprisingly, *in situ* microsampled cells were found to continue cell division as the embryo developed to the next stage.[19] Alternatively, fluid force microscopy[44] or laser capture microdissection[45] may be

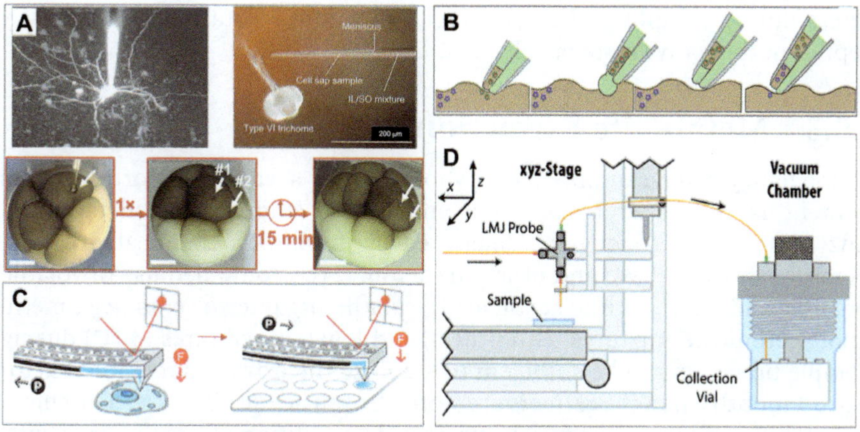

Figure 10.2 Examples of approaches enabling single-cell sampling. (A) Fine-pulled capillaries were used to withdraw the cytoplasm of (top, left) a single thalamic neuron in the rat brain, (top, right) a single type VI stalk cell from intact tomato trichomes, and (bottom) the right ventral-animal (V1) cell in the 8-cell *Xenopus laevis* embryo. The embryo continued to divide to the 16-cell stage with spots marking the locations of subsequent samplings on the cell. (B) A dual-barrel pipette for sampling individual *A. cepa* cells. (C) A fluid force microscopy probe was used to aspirate the cytoplasm from HeLa cells (left), which was then dispensed on a microarray spot for MS analysis (right). (D) Using a liquid junction interface, metabolites were extracted from identified cells in a rat pancreatic islet for MALDI.[3,19,38,42–44] Adapted with permission from ref. 3 (http://pubs.acs.org/doi/abs/10.1021/ac500168d), Copyright (2014) American Chemical Society, further permissions related to the material excerpted should be directed to the ACS; ref. 19, Copyright (2017) American Chemical Society; ref. 38 (http://pubs.acs.org/doi/abs/10.1021%2Facs.analchem.7b01782%20), Copyright (2017) American Chemical Society, further permissions related to the material excerpted should be directed to the ACS; ref. 42, Copyright (2016) American Chemical Society; ref. 43 with permission from The Royal Society of Chemistry; and ref. 44, Copyright (2017) American Chemical Society.

used to sample cells. Recently, a combination of MALDI-MS and CE-ESI-MS was used to differentiate single α and β cells from a rat pancreatic islet based on peptide content (Figure 10.2D).[38]

High-fidelity sampling helps results interpretation in single-cell studies. With sufficiently fine spatial resolution, cross-contamination can be eliminated or, at least, assigned to neighboring cells. Fast sampling can minimize metabolic changes arising from endogenous metabolism, transitions in cell cycle, or metabolic degradation, thus providing a credible snapshot of the cell's metabolic activity state. Using capillary microsampling, we recently aspirated a portion of identified embryonic cells with a 5 s per cell throughput, compared to 5 min per cell by manual dissection. With minimal damage to the cell, capillary microsampling CE-MS significantly lowered oxidative stress during sampling, thus yielding metabolomic data that more closely represent the native state of cells in *X. laevis*.[19]

10.3 Metabolite Extraction

Immediately after sampling, the whole cell or a collected portion of its content is processed to preserve and extract endogenous metabolites. Enzymes may be deactivated/denatured by inhibitors, chemicals, or temperature-shock.[46–48] We and others have found the combined use of volatile organic acids (*e.g.*, formic/acetic acid), volatile organic solvents (*e.g.*, methanol/acetonitrile mixtures with water), and low temperatures (4 °C) during sample processing to be sufficient to stabilize the extracted metabolites for up to 1 month when the extracts are stored frozen at −80 °C.[4,17,19,49] The chemical stability of the extract may be improved by addition of stabilizing agents. Glycerol, a known cryogenic preservative, has improved the reproducibility of metabolite quantification by 4-fold in single *A. californica* neurons.[16]

Next, metabolites are extracted from the collected specimen using solvents. To broaden detection coverage, extraction typically exploits a combination of solvents that are tailored to the physicochemical properties of the compound classes of interest (*e.g.*, amino acids, lipids, or sugars). In addition, the pH of the solvent is adjusted to facilitate ionization. The extraction solution may consist of miscible or immiscible solvents with the latter allowing for subsequent fractionation. Because cells can exhibit varying physical properties and chemical compositions, even in the same tissue or organisms, we recommend optimizing and validating the method of extraction using the cells of interest. We found modeling the physicochemical properties to be helpful in guiding the initial design of single-cell experiments. Figure 10.3A shows an example in which the octanol/water distribution coefficients (D) were calculated at different pHs, which helped select a series of polar–apolar solvent mixtures to enhance the extraction of select metabolite classes from single *X. laevis* cells (Figure 10.3B).[17] In addition, we also found the use of chemical standards to be ideal to establish the optimal pH and solvent composition for enhanced quantitative recovery and ionization efficiency, as well as to tune the ESI mass spectrometer for sensitivity.

To enhance the analytical performance of single-cell CE-ESI-MS, extraction may be supplemented with additional steps. Internal standards added to the extraction solvent system or the resulting extract can help with qualitative and quantitative testing of the robustness of sampling, quantitative recovery, and the reproducibility of CE separation, as well as correcting for fluctuations in ESI-MS performance (*e.g.*, by normalization). An isotopologue of the analyte serves as the internal standard of choice, because it has similar physicochemical properties to the metabolite of interest without interference in detection. Depending on endogenous concentration and detection sensitivity, pre-concentration steps (*e.g.*, off-column or on-column) may be necessary to enrich metabolites with low abundances. Desalting or fractionation steps may be performed to reduce sample complexity and matrix interferences,

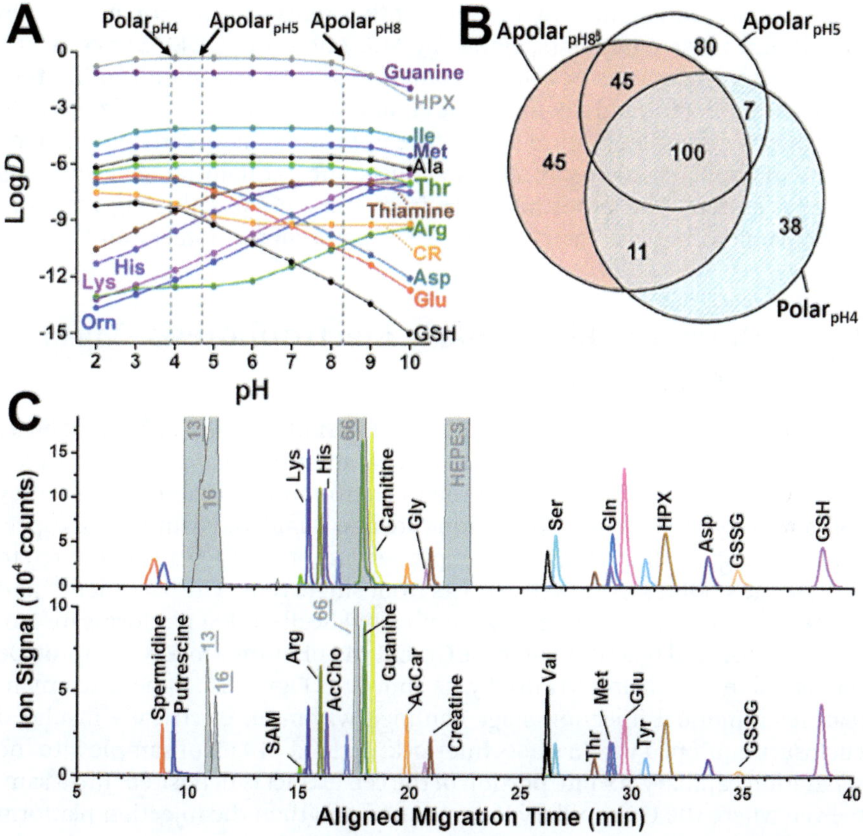

Figure 10.3 Sample processing strategies for improved detection sensitivity in single-cell CE-MS. (A) Modeling of physicochemical properties, such as the distribution coefficient (D), helps guide the selection of extraction solvents. (B) Extraction under different pH and hydrophobic conditions enhanced metabolic coverage in single *X. laevis* cells. Key: Polar$_{pH4}$, 50% methanol with 0.5% acetic acid; Apolar$_{pH4}$, 40% acetonitrile with 40% methanol; Apolar$_{pH8}$, 40% acetonitrile with 40% methanol at pH 8. (C) Representative electropherograms comparing the separation of select metabolites between cell extracts prepared by whole-cell dissection (top panel) and capillary microsampling (bottom panel). Minimal contamination by salts (see grey peaks) improved separation and detection performance by microprobe single-cell CE-ESI-MS.[17,19] Adapted from ref. 17 with permission from The Royal Society of Chemistry and with permission from ref. 19, Copyright (2017) American Chemical Society.

albeit at the risk of analyte losses and/or modifications to the metabolome. We have found simplicity and speed to be efficient guides during the design of our single-cell metabolomic workflows.[4,17]

Another purpose of extraction is to improve separation by CE and detection by ESI-MS. Extraction has the benefit of simplifying the molecular complexity of the extract, thus removing potential interferences during ion generation by ESI and mass analysis (spectral) by MS. Microprobe CE-ESI-MS, which extracts small amounts of cellular content (~10 nL), was found to improve the separation and yield a comparable signal-to-noise ratio to whole-cell (~90 nL) dissected samples (Figure 10.3C). The improvement was attributed to the ability of capillary microsampling to collect substantially smaller amounts of salts, buffers, and other additives from the embryo culturing medium, which would otherwise interfere with CE separation and ionization.[17,19]

10.4 Detection by Capillary Electrophoresis–Mass Spectrometry

Next, the cellular extract is analyzed by CE-ESI-MS, which entails a subset of tasks: the sample is injected into a separation capillary, metabolites are separated based on differences in their electrophoretic mobility, ionized by ESI, and the generated metabolite ions are mass-analyzed using a mass spectrometer (Figure 10.1). With ~nL sample consumption, CE is ideally suited to analyze metabolites in limited extracts from single cells. The Sweedler[2,50] and Nemes[4,35] laboratories have custom-built single-cell CE-ESI platforms for MS. Construction and operation of these CE-ESI-MS platforms follows a prototype published in ref. 50 and 51. Briefly, as shown in Figure 10.4, the instrument features a manual injection stage confined within an electrically insulated enclosure equipped with a safety interlock. To load ~10 nL of sample into the separation capillary, a <1 μL portion of the cell extract is deposited into a sample vial where the CE capillary is positioned and then the injection platform is elevated. CE separation is performed by translating the capillary inlet end into the background electrolyte (BGE) and applying ~20 kV across the ~1 m long capillary.

ESI is the most commonly used technology to ionize CE-separated metabolites for MS detection. CE-ESI interfaces vary greatly in design and performance (see reviews in ref. 52–54). We apply a modified form of the original co-axial sheath-flow design,[55] which has been found to provide robust operation since inception. In our design, the nebulizer gas is removed to help direct the electrospray plume to the mass spectrometer, the dimensions of the emitter are scaled to the separation capillary to minimize sample dilution in the ion source, and the geometry of the electrospray emitter and the operation parameters are carefully chosen to maintain the electrospray in the stable cone-jet regime, which provides the most efficient ionization.[56] This CE-ESI interface provides a 60 amol lower limit of detection and 4–5 log-order dynamic range for quantification.[4] Furthermore, we have most recently developed a second-generation version of this interface, which uses a tapered-tip emitter to achieve 260 zmol (156 000 copies) sensitivity for peptide standards.[57] These CE-ESI-MS platforms have allowed the detection of hundreds of metabolite signals from single cells that were isolated from *A. californica* neurons[51] and *X. laevis* embryonic cells,[4,16,17] including the identification of

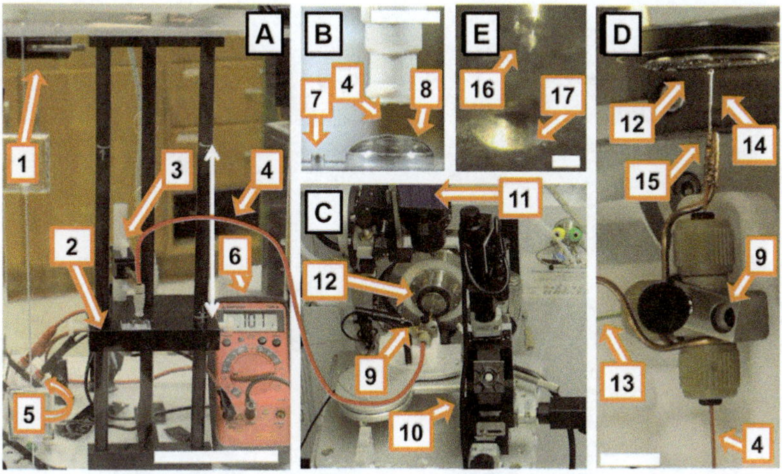

Figure 10.4 A custom-built CE-ESI-MS platform for characterizing metabolites in single-cells. (A) Major components include: (1) a safety-interlock-enabled enclosure for electrical isolation, (2) a sample loading platform, (3) a holder allowing translation of (4) a fused silica separation capillary, (5) a resistor, and (6) a digital multimeter for measurement of current. See the text for operation and details. (B) A close-up view of the sample-loading platform consisting of: (7) a sampling vial and (8) a background electrolyte vial. (C) The CE-ESI-MS ionization source consisting of: (9) the CE-ESI interface on a (10) three-axis translation stage, (11) a camera to monitor performance of the ESI interface, and (12) a mass spectrometer featuring a sampling plate. (D) A magnified view of the T-union that is part of the interface, and connects (4) the separation capillary, (13) the secondary capillary supplying sheath solvent, and (14) the metal-emitter for generating the electrospray in the (15) cone-jet regime (see Taylor cone). (E) A close-up view of the (16) Taylor cone in front of the (17) mass spectrometer orifice. Adapted with permission from ref. 50. Copyright © 2013 Nature Publishing Group.

~100 metabolites with significant metabolic differences between cell types. Development and recent commercialization of alternative designs, such as low-flow sheath[58] and sheathless[59] interfaces, raise opportunities to further the sensitivity of CE-ESI for single-cell metabolomics.

Various strategies can be applied to enhance the throughput of single-cell measurements using CE-ESI-MS. Capillary batch injection enables multisample measurements and up to 80–100% injection efficiency.[60] To improve sensitivity for low abundance metabolites, online preconcentration methods, such as sample stacking,[61] can be incorporated into the workflow. Figure 10.5A shows large-volume sample stacking allowing nucleotides to be detected from a 60 µm *A. californica* neuron.[62] For higher throughput, multi-segmented injection offers significant benefits. Figure 10.5B presents one such example, in which human plasma extracts were measured 10-times faster, with ~3 min per sample throughput without compromising identification or quantification.[63]

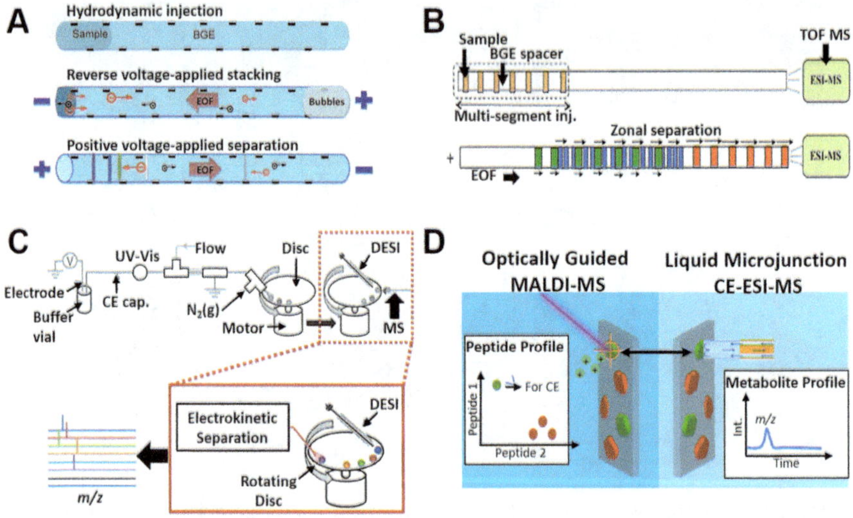

Figure 10.5 Recent advances to enhance CE-MS measurement throughput and sensitivity. (A) Large-volume sample stacking improved detection sensitivity. (B) Multi-segmented injection enhanced the throughput to 3 min per sample. (C) CE with desorption electrospray ionization (DESI) minimized ion suppression effects due to salts and detergents. (D) Optically guided MALDI-CE-MS screened large numbers of cells based on peptide and metabolite composition.[38,62–64] Adapted with permission from ref. 38 (http://pubs.acs.org/doi/abs/10.1021%2Facs.analchem.7b01782%20), Copyright (2017) American Chemical Society; ref. 62 with permission from The Royal Society of Chemistry; ref. 63, Copyright (2013) American Chemical Society and ref. 64, Copyright (2011) American Chemical Society.

Other ionization sources provide complementary benefits for single-cell CE-MS. Figure 10.5C shows the hyphenation of CE to desorption electrospray ionization (DESI): the separated compounds are dried onto a rotating Teflon disk, whence they are desorbed and ionized by focused charged droplets with improved tolerance for salts and detergents.[64] Figure 10.5D presents another setup in which MALDI was off-line coupled to CE-MS to profile cells in rat pancreatic islets.[38] After α- and β-cell types were distinguished based on their peptide content,[65,66] the dried samples were microextracted for metabolic analysis by CE-ESI-MS. To broaden the coverage of the metabolome, both negative[62] and positive ionization modes may be used and/or multiple analytical platforms combined.

10.5 Data Processing

The last step of the workflow is aimed at identifying and/or quantifying metabolite signals between single cells (Figure 10.1). CE-ESI-MS studies utilize multiple dimensions of information to this end, including accurate

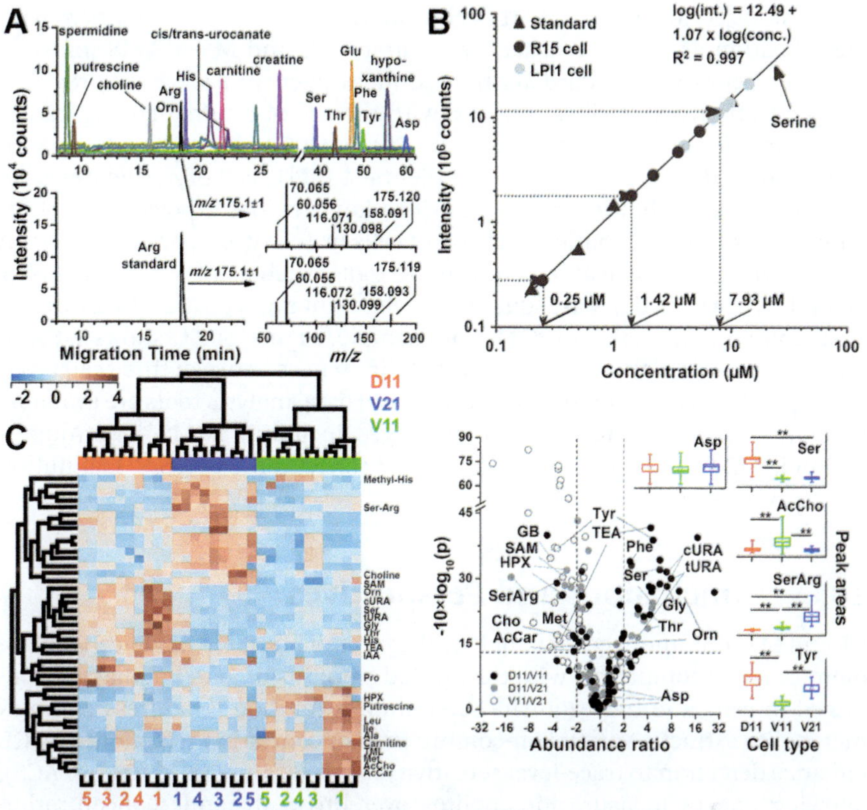

Figure 10.6 Identification and quantification of small molecules in single cells. (A) An electropherogram exemplifying the separation of select identified small molecules measured from an individual cell derived from a 16-cell *Xenopus laevis* embryo (top panel). Representative identification of a metabolite signal as arginine in a *X. laevis* cell based on the comparison of accurate mass, fragmentation behavior, and migration time against those measured for a chemical standard (bottom panel). (B) Quantification of a targeted metabolite, serine, in different R15 and LPl1 neurons of *A. californica.* (C) Hierarchical cluster analysis metabolically differentiating the D11, V11, and V21 cells in 16-cell *Xenopus laevis* embryos (left panel). Volcano plot validating significant metabolite differences between the cell types (right panel). Adapted with permission from R. M. Onjiko, S. A. Moody and P. Nemes, *Proc. Natl. Acad. Sci. U. S. A.*, 2015, 112, 6545–6550;[4] and ref. 50. Copyright © 2013 Nature Publishing Group.

m/*z*, migration time, fragmentation, and peak intensity/area. Representative separations are demonstrated for select metabolite signals in Figure 10.6A. Comparison of these orthogonal pieces of information makes confident identification possible for any metabolite that can be obtained as a chemical standard or for which tandem MS information is available in a tandem MS database, such as mzCloud and Metlin.[67] For example, Figure 10.6A shows a signal at *m*/*z* 175.1188 in the D11 cell from the 16-cell *X. laevis* embryo with identification as arginine based on accurate *m*/*z* and MS–MS/MS information acquired from standard arginine. Under-the-curve peak areas provide a proxy for relative quantification, and with the use of external concentration calibration, for absolute quantification. Figure 10.6B exemplifies the quantification of serine in the R15 and left pleural 1 (LPl1) cell of *A. californica via* concentration calibration.[4,15,50] Quantification provides a powerful tool for targeted or discovery characterization of cells or cell types, particularly when combined with statistical or multivariate tools, such as hierarchical cluster analysis (HCA), partial least squares discriminant analysis (PLS-DA), or principal component analysis (PCA). For example, Figure 10.6C shows HCA of metabolites with significant enrichment differences between three cell types in the same 16-cell embryo of *X. laevis*. These data analysis tools are available in many publicly available software packages, including XCMS,[68] MetAlign,[69] MSFind, MS-DIAL,[70] MetaboAnalyst,[71] MZmine,[72] and GNPS,[73] facilitating the analysis of single-cell MS data.

10.6 Conclusions and Perspectives

CE and MS combine excellent analytical figures of merit for single-cell metabolomics. CE is compatible with the limited amounts of material that are contained in cells and has exquisite separation resolution to address complex metabolite extracts. Various on-column pre-concentration techniques by CE enhance detection to trace-level sensitivity (*e.g.*, ~60 amol or 10 nM by FACS), allowing one to measure metabolites over broad dynamic concentration ranges exhibited in biological systems. MS detection complements CE with high molecular specificity, label-free operation, and capability for quantification to characterize broad types of metabolites in cells. Data resulting from single-cell CE-MS has shed new light on small-molecule differences between cells in various tissues and model organisms, ranging from metabolites in early stages of developing vertebrate embryos[4] to signaling molecules in the nervous system.[74]

Continuing technological advances promise exciting opportunities to advance our understanding of cell biology.[77] Sampling approaches with faster operation and scalability in space and time allow a larger number of cells to be probed, thus enhancing statistical power to uncover differences between cells and to characterize cells at the level of the whole population. These developments in fast sampling can be supported by analyses capable of higher throughput, including sample processing on microfluidic devices,[75] separation by microchip electrophoresis,[76] and ionization by MALDI[9,38] or ESI[2,4,50] mass spectrometers. To analyze the large amounts of complex data from these studies, software tools capable of statistical and multivariate analyses are needed, particularly those that can address the narrow and often low-level signals that are expected from single-cell studies. Parallel advances in the steps of the single-cell CE-MS workflow can help extend this powerful bioanalytical technology to systems biology studies on cells and limited cell populations, promoting the understanding of molecular mechanisms that affect cell and developmental processes.

Acknowledgements

This work was supported by the National Science Foundation Grant DBI-1832968 (previously DBI-1651388) (to P. N.) and the COSMOS Club Foundation Award (to E. P. P.). The opinions and conclusions expressed in this publication are solely those of the authors and do not necessarily represent the official views of the funding sources.

References

1. R. Zenobi, *Science*, 2013, **342**, 1243259.
2. T. Lapainis, S. S. Rubakhin and J. V. Sweedler, *Anal. Chem.*, 2009, **81**, 5858–5864.
3. J. T. Aerts, K. R. Louis and S. R. Crandall, *et al.*, *Anal. Chem.*, 2014, **86**, 3203–3208.
4. R. M. Onjiko, S. A. Moody and P. Nemes, *Proc. Natl. Acad. Sci. U. S. A.*, 2015, **112**, 6545–6550.
5. M. Fessenden, *Nature*, 2016, **540**, 153–155.
6. D. Cohen, J. A. Dickerson and C. D. Whitmore, *et al.*, *Annu. Rev. Anal. Chem.*, 2008, **1**, 165–190.
7. A. Svatos, *Anal. Chem.*, 2011, **83**, 5037–5044.
8. M. Heinemann and R. Zenobi, *Curr. Opin. Biotechnol.*, 2011, **22**, 26–31.
9. T. J. Comi, T. D. Do, S. S. Rubakhin and J. V. Sweedler, *J. Am. Chem. Soc.*, 2017, **139**, 3920–3929.
10. S. S. Rubakhin, E. J. Lanni and J. V. Sweedler, *Curr. Opin. Biotechnol.*, 2013, **24**, 95–104.
11. T. H. Ong, E. G. Tillmaand and M. Makurath, *et al.*, *Biochim. Biophys. Acta, Proteins Proteomics*, 2015, **1854**, 732–740.
12. D. G. Spiller, C. D. Wood, D. A. Rand and M. R. H. White, *Nature*, 2010, **465**, 736–745.
13. L. W. Zhang, D. P. Foreman and P. A. Grant, *et al.*, *Analyst*, 2014, **139**, 5079–5085.
14. L. W. Sumner, A. Amberg and D. Barrett, *et al.*, *Metabolomics*, 2007, **3**, 211–221.
15. P. Nemes, A. M. Knolhoff, S. S. Rubakhin and J. V. Sweedler, *Anal. Chem.*, 2011, **83**, 6810–6817.
16. P. Nemes, A. M. Knolhoff, S. S. Rubakhin and J. V. Sweedler, *ACS Chem. Neurosci.*, 2012, **3**, 782–792.

17. R. M. Onjiko, S. E. Morris, S. A. Moody and P. Nemes, *Analyst*, 2016, **141**, 3648–3656.
18. R. M. Onjiko, D. O. Plotnick, S. A. Moody and P. Nemes, *Anal. Methods*, 2017, **9**, 4964–4970.
19. R. M. Onjiko, E. P. Portero, S. A. Moody and P. Nemes, *Anal. Chem.*, 2017, **89**, 7069–7076.
20. K. Jeanne Dit Fouque, A. Garabedian and J. Porter, *et al.*, *Anal. Chem.*, 2017, **89**, 11787–11794.
21. A. Buchberger, Q. Yu and L. J. Li, *Annu. Rev. Anal. Chem.*, 2015, **8**, 485–509.
22. S. Neupert, S. S. Rubakhin and J. V. Sweedler, *Chem. Biol.*, 2012, **19**, 1010–1019.
23. S. A. Moody, *Dev. Biol.*, 1987, **122**, 300–319.
24. S. A. Moody, *Dev. Biol.*, 1987, **119**, 560–578.
25. L. Dale and J. M. W. Slack, *Development*, 1987, **99**, 527–551.
26. A. Heyland, Z. Vue and C. R. Voolstra, *et al.*, *J. Exp. Zool., Part B*, 2011, 113–134.
27. J. P. McCoy, *Hematology*, 2002, **16**, 229–243.
28. S. F. Ibrahim and G. van den Engh, *Curr. Opin. Biotechnol.*, 2003, **14**, 5–12.
29. K. Makker, A. Agarwal and R. K. Sharma, *Indian J. Exp. Biol.*, 2008, **46**, 491–497.
30. C. W. Shields, C. D. Reyes and G. P. Lopez, *Lab Chip*, 2015, **15**, 1230–1249.
31. S. J. Lo and D. J. Yao, *Int. J. Mol. Sci.*, 2015, **16**, 16763–16777.
32. N. Wen, Z. Zhao and B. Y. Fan, *et al.*, *Molecules*, 2016, **21**, 881.
33. C. Cecala and J. V. Sweedler, *Analyst*, 2012, **137**, 2922–2929.
34. E. P. Portero, R. M. Onjiko, S. A. Moody and P. Nemes, *65th American Society for Mas Spectrometry Conference on Mass Spectrometry and Allied Topics*, Indianapolis, IN, 2017.
35. C. Lombard-Banek, S. A. Moody and P. Nemes, *Angew. Chem., Int. Ed.*, 2016, **55**, 2454–2458.
36. C. Lombard-Banek, S. Reddy, S. A. Moody and P. Nemes, *Mol. Cell. Proteomics*, 2016, **15**, 2756–2768.
37. C. Lombard-Banek, A. B. Baxi, S. A. Moody and P. Nemes, *65th American Society for Mass Spectrometry Conference on Mass Spectrometry and Allied Topics*, Indianapolis, IN, 2017.
38. T. J. Comi, M. A. Makurath and M. C. Philip, *et al.*, *Anal. Chem.*, 2017, **89**, 7765–7772.
39. P. Hu, W. Zhang, H. Xin and G. Deng, *Front. Cell Dev. Biol.*, 2016, **4**, 116.
40. M. Qi, M. C. Philip, N. Yang and J. V. Sweedler, *ACS Chem. Neurosci.*, 2018, **9**, 40–50.
41. P. A. Grant, M. B. Herold and S. A. Moody, *J. Visualized Exp.*, 2013, **71**, e4458.
42. T. Nakashima, H. Wada and S. Morita, *et al.*, *Anal. Chem.*, 2016, **88**, 3049–3057.
43. A. Saha-Shah, C. M. Green, D. H. Abraham and L. A. Baker, *Analyst*, 2016, **141**, 1958–1965.

44. O. Guillaume-Gentil, T. Rey and P. Kiefer, *et al.*, *Anal. Chem.*, 2017, **89**, 5017–5023.
45. E. Kummari, S. X. Guo-Ross and J. B. Eells, *J. Visualized Exp.*, 2015, **96**, e52336.
46. M. M. Chen, A. L. Li and M. C. Sun, *et al.*, *J. Zhejiang Univ., Sci., B*, 2014, **15**, 333–342.
47. M. Faijes, A. E. Mars and E. J. Smid, *Microb. Cell Fact.*, 2007, **6**, 27.
48. R. V. Kapoore, R. Coyle and C. A. Staton, *et al.*, *Analyst*, 2017, **142**, 2038–2049.
49. A. B. Canelas, C. Ras and A. ten Pierick, *et al.*, *Metabolomics*, 2008, **4**, 226–239.
50. P. Nemes, S. S. Rubakhin, J. T. Aerts and J. V. Sweedler, *Nat. Protoc.*, 2013, **8**, 783–799.
51. R. M. Onjiko, E. P. Portero, S. A. Moody and P. Nemes, *J. Visualized Exp.*, 2017, **130**, e56956.
52. J. Krenkova and F. Foret, *Proteomics*, 2012, **12**, 2978–2990.
53. W. Zhang, T. Hankemeier and R. Ramautar, *Curr. Opin. Biotechnol.*, 2017, **43**, 1–7.
54. X. F. Zhong, Z. C. Zhang, S. Jiang and L. J. Li, *Electrophoresis*, 2014, **35**, 1214–1225.
55. R. D. Smith, J. A. Olivares, N. T. Nguyen and H. R. Udseth, *Anal. Chem.*, 1988, **60**, 436–441.
56. P. Nemes, I. Marginean and A. Vertes, *Anal. Chem.*, 2007, **79**, 3105–3116.
57. S. B. Choi, M. Zamarbide, M. C. Manzini and P. Nemes, *J. Am. Soc. Mass Spectrom.*, 2017, **28**, 597–607.
58. L. L. Sun, G. J. Zhu and Y. M. Zhao, *et al.*, *Angew. Chem., Int. Ed.*, 2013, **52**, 13661–13664.
59. J. T. Whitt and M. Moini, *Anal. Chem.*, 2003, **75**, 2188–2191.
60. M. Grundmann and F. M. Matysik, *Anal. Bioanal. Chem.*, 2012, **404**, 1713–1721.
61. A. Slampova, Z. Mala, P. Gebauer and P. Bocek, *Electrophoresis*, 2017, **38**, 20–32.
62. J. X. Liu, J. T. Aerts and S. S. Rubakhin, *et al.*, *Analyst*, 2014, **139**, 5835–5842.
63. N. L. Kuehnbaum, A. Kormendi and P. Britz-McKibbin, *Anal. Chem.*, 2013, **85**, 10664–10669.
64. G. K. Barbula, S. Safi and K. Chingin, *et al.*, *Anal. Chem.*, 2011, **83**, 1955–1959.
65. T. H. Ong, D. J. Kissick and E. T. Jansson, *et al.*, *Anal. Chem.*, 2015, **87**, 7036–7042.
66. E. T. Janson, T. J. Comi, S. S. Rubakhin and J. V. Sweedler, *ACS Chem. Biol.*, 2016, **11**, 2588–2595.
67. C. A. Smith, G. O'Maille and E. J. Want, *et al.*, *Ther. Drug Monit.*, 2005, **27**, 747–751.
68. C. A. Smith, E. J. Want and G. O'Maille, *et al.*, *Anal. Chem.*, 2006, **78**, 779–787.

69. A. Lommen, *Anal. Chem.*, 2009, **81**, 3079–3086.
70. H. Tsugawa, T. Cajka and T. Kind, *et al.*, *Nat. Methods*, 2015, **12**, 523–526.
71. J. G. Xia, N. Psychogios, N. Young and D. S. Wishart, *Nucleic Acids Res.*, 2009, **37**, W652–W660.
72. M. Katajamaa, J. Miettinen and M. Oresic, *Bioinformatics*, 2006, **22**, 634–636.
73. M. X. Wang, J. J. Carver and V. V. Phelan, *et al.*, *Nat. Biotechnol.*, 2016, **34**, 828–837.
74. E. V. Romanova, J. T. Aerts, C. A. Croushore and J. V. Sweedler, *Neuropsychopharmacology*, 2014, **39**, 50–64.
75. J. S. Mellors, K. Jorabchi, L. M. Smith and J. M. Ramsey, *Anal. Chem.*, 2010, **82**, 967–973.
76. X. T. Li, S. L. Zhao, H. K. Hu and Y. M. Liu, *J. Chromatogr., A*, 2016, **1451**, 156–163.
77. C. Lombard-Banek, E. P. Portero, R. M. Onjiko and P. Nemes, *Genesis*, 2017, **55**, e23012.
78. N. Pan, W. Rao and N. R. Kothapalli, *et al.*, *Anal. Chem.*, 2014, **86**, 9376–9380.
79. A. Amantonico, P. L. Urban and S. R. Fagerer, *et al.*, *Anal. Chem.*, 2010, **82**, 7394–7400.

CHAPTER 11

Capillary Electrophoresis–Mass Spectrometry for Lipid Analysis

PETER LINDENBURG[a,b], KASPAR VULLA[c], ISABELLE KOHLER[b] AND RUTH SHIMMO*[c]

[a]Professorship Metabolomics, University of Applied Sciences Leiden, Leiden, The Netherlands; [b]Division of Systems Biomedicine and Pharmacology, Leiden Academic Centre for Drug Research, Leiden University, Leiden, The Netherlands; [c]School of Natural Sciences and Health, Tallinn University, Narva mnt. 29, 10120 Tallinn, Estonia
*E-mail: ruth.shimmo@tlu.ee

11.1 Introduction

11.1.1 Lipids

The International Union for Pure and Applied Chemistry (IUPAC) defines lipids as 'a loosely defined term for substances of biological origin that are soluble in nonpolar solvents'.[1] The diversity of lipids is enormous. For example, almost 600 molecular species can be quantified in human plasma, covering the six main mammalian lipid categories: fatty acids, glycerolipids, glycerophospholipids, sphingolipids, sterol lipids, and prenol lipids.[2] The diversity of lipid species reflects their wide variety of biological functions. Amongst others, lipids regulate the ion traffic through cell membranes, arachidonic acid affects the activity of mechanosensitive ion pores, and sphingomyelin is important in the assembly of lipid rafts influencing the stabilization of the

New Developments in Mass Spectrometry No. 6
Capillary Electrophoresis–Mass Spectrometry for Metabolomics
Edited by Rawi Ramautar

Published by the Royal Society of Chemistry, www.rsc.org

clustering of neurotransmitter receptors. Lipids regulate receptors and other proteins involved in signal transduction by direct binding. They also play a crucial role in inflammatory processes.

Fluctuations in lipid concentrations have been associated with a wide variety of diseases, including neurodegenerative and infectious diseases, cancer, diabetes, *etc.*[3,4] For example, Alzheimer's disease has been shown to be linked to an abnormal metabolism of cholesterol.[5] Gangliosides, *i.e.*, complex phospholipids with three or more sugars on a ceramide backbone, promote the aggregation of β-amyloid protein on membrane surfaces.[6] Exposure to polyunsaturated fatty acids has been demonstrated to increase α-synuclein levels in cases of early onset Parkinson's disease, while this is decreased by saturated fatty acids.[7] Many of the neurodegenerative disorders are linked to imbalanced protein–lipid (or protein–protein) interactions, which increase the possibility of intracellular formation of protein deposits.[8] These few examples underline the need for reliable analytical technologies that are capable of covering the wide existing variety of lipid species, allowing for a better understanding of their multiple roles in a large variety of diseases and (patho)physiological conditions.

11.1.2 Lipid Analysis

Mass spectrometry (MS) is nowadays considered the gold standard detection technique for analyzing complex biological samples due to its excellent selectivity and sensitivity. The wide use of MS in lipidomics has also fostered the emergence of several lipid-based databases using experimental MS data, such as LipidMaps, LipidBank, LipidBlast, *etc.*, which all play an essential role in lipid identification using cross-comparisons.[9]

The ongoing improvements in MS technology achieved over the last two decades have also significantly increased its sensitivity, resolution, accuracy, speed and dynamic range, making it the detector of choice in lipid analysis.[9] One of the most widely used approaches in lipidomics is direct infusion MS (DI-MS), also referred to as shotgun-lipidomics, which has gained popularity due to the significantly increased resolution and linear dynamic range of modern MS instrumentation. In shotgun-lipidomics, samples are directly infused into the MS without prior separation, allowing high-throughput analysis. However, DI-MS does not allow for discrimination between isomeric and isobaric lipids, and low-abundance lipids might not be detected due to ion suppression caused by matrix interferences.[9] The former issue can be resolved by using ion mobility mass spectrometry (IMS), which allows the separation of isobaric and even isomeric analytes. However, this technique is often not sufficient to overcome the potential matrix effects and ion suppression. The hyphenation of MS to a separation technique, including liquid chromatography (LC), thin layer chromatography (TLC), gas chromatography (GC), capillary electrophoresis (CE) and supercritical fluid chromatography (SFC), is therefore considered the ideal strategy for lipid analysis.

LC-MS offers robust, reproducible, and reliable analysis of lipids. Although normal phase LC and hydrophilic interaction liquid chromatography (HILIC) have been reported for lipid analysis, reversed phase LC (RPLC) is the most popular LC mode.[10] Compared to HILIC, which allows for discrimination between lipid classes, RPLC enables separation within classes, based on chain length and double bonds of the acyl chains. The power of RPLC-MS in lipidomics has been clearly demonstrated in the work of Slatter *et al.*, who detected *circa* 5600 lipid species in human platelets using LC-MS.[11] RPLC-MS is thus often the method of choice for lipid analysis due its ease of use and high reproducibility.

Thin layer chromatography (TLC) is, from a historical perspective, the classical standard in lipid analysis. It is cost-effective and fast, but its separation power is limited and it only allows for identification of lipid classes, and not individual lipid species.[12] A promising development is the combination of TLC with matrix-assisted laser desorption/ionization (MALDI) MS,[13,14] where TLC plates are used as MALDI target plates, thus enabling sensitive and more comprehensive lipid analysis.

GC-MS has mainly been applied for the analysis of fatty acids, typically short chain fatty acids.[12] It generally offers higher separation resolution than LC and at a higher speed (approximately 15 min for one analysis).[15] On the other hand, lipid analysis by GC requires sample derivatization, which is time-consuming and can, in the case of fatty acids, lead to the loss and/or degradation of compounds, as well as poor reproducibility.[16]

The use of CE-MS is highly underrepresented in lipid analysis. Dozens of promising CE-based methods, including capillary electrochromatography (CEC) and micellar electrokinetic chromatography (MEKC), have already been reported for lipid analysis, however mostly using background electrolytes (BGEs) that are not compatible with MS and (thus) coupled to spectrophotometric detection. Nonetheless, CE-MS offers various advantages in lipidomics-based research. First of all, CE is famous for its unmatched separation efficiency. Furthermore, its separation mechanism is based on a completely orthogonal principle (*i.e.*, based on differences in electrophoretic mobility of the analytes) compared to chromatography (*i.e.*, based on differences in the distribution of analytes between mobile and stationary phases).[17] Recently, the number of individual lipid species present in mammalian cells has been estimated to be 10 000.[9] An arsenal of complimentary analytical techniques is therefore essential to be able to analyze this large number of lipids, which also present a wide range of physico-chemical properties. CE demonstrates several beneficial features for lipid analysis. Firstly, it is particularly well-suited for the analysis of charged compounds – a notable advantage since many lipid species are charged or can be charged depending on the pH of the sample. Secondly, the typical flow rate of CE, *i.e.*, in the nL min^{-1} range, is compatible with electrospray ionization (ESI), while recent developments in the hyphenation of CE to MS have resulted in significant improvements in sensitivity.[18] Furthermore, various in-capillary concentration enhancement techniques, such as field-enhanced sample stacking, isotachophoresis, pH-mediated sample stacking, *etc.*, are at hand to further improve detection limits.[19] CE separation takes place in narrow-bore capillaries (typically 30–100 µm i.d.), which makes this technique highly suitable for the analysis of volume-limited samples. Finally, CE method development can be straightforward and uncomplicated, requiring nothing more than regular replacement of the BGE or capillary. CE is also considered cost-effective and environmentally friendly, as the consumption of toxic and expensive organic solvents is very low.

11.2 CE-MS of Lipids

This section reports and discusses the CE-MS-based studies carried out in the field of lipid analysis over the past 20 years, and is structured according to distinct lipid classes, *i.e.*, phospholipids, glycolipids, acylcarnitines, surfactants, and fatty acids.

11.2.1 Phospholipids

Phospholipids are the main constituents of biological membranes and have already been used for more than 30 years as a model to study the behavior of membranes and membrane-bound macromolecules.[20] Phospholipids consist of two fatty acid residues, representing the hydrophobic part of the molecule, and a hydrophilic phosphate group, which is linked *via* a glycerol molecule. Figure 11.1 shows examples of phospholipids from four phospholipid classes, *i.e.*, phosphatidylcholines, phosphatidylethanolamines, phosphatidylglycerols and phosphatidylserines.

Raith *et al.*[21] demonstrated the utility of CE-MS for the analysis of some phospholipid classes. The separation was based on their head groups, including 1-palmitoyl-2-oleyl-phosphatidylcholine (POPC), 1-palmitoyl-2-oleyl-phosphatidylethanolamine (POPE), 1-palmitoyl-2-oleyl-phosphatidylglycerol (POPG) and 1-palmitoyl-2-oleyl-phosphatidylserine (POPS), as illustrated in Figure 11.1. The separation was achieved using Methyl-Sil deactivated capillaries (75 µm i.d., 60 cm length), a BGE consisting of acetonitrile/2-propanol/*n*-hexane (57 : 38 : 5, v/v) containing 20 mM ammonium acetate, and a separation voltage of 30 kV. This rather apolar BGE was necessary to meet the solubility requirements of phospholipids. Since this non-aqueous CE approach (NACE) does not create any electro-osmotic flow (EOF), a small pressure difference was therefore applied during the experiments to allow for MS detection. The concentration of the phospholipids was *circa* 0.65 µM. The CE method was coupled to an ion trap MS operated in negative mode, using a sheath liquid consisting of a mixture of methanol/water (80 : 20, v/v) at a flow rate of 0.5 µL min^{-1}. Figure 11.2 illustrates the obtained separation between the different phospholipid classes using this approach.

Interestingly, Gao *et al.*[22] managed to separate phospholipids in bare fused silica capillaries (50 µm i.d., 60 cm length) using a BGE composed of methanol/acetonitrile (40:60, v/v) containing 20 mM ammonium acetate and 0.5%

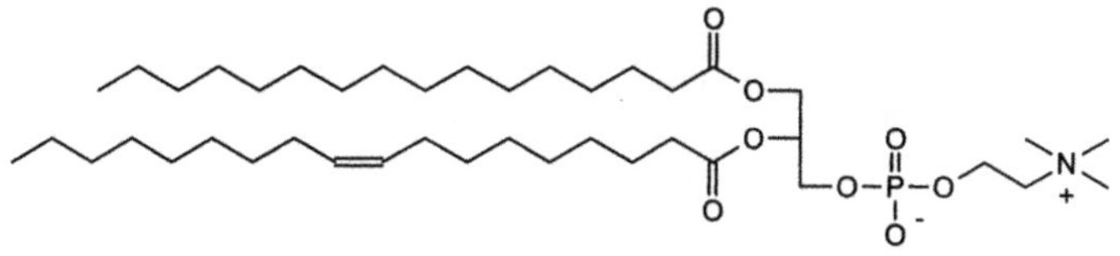

1-Palmitoyl-2-oleoylphosphatidylcholine (POPC) (M_r = 760.1)

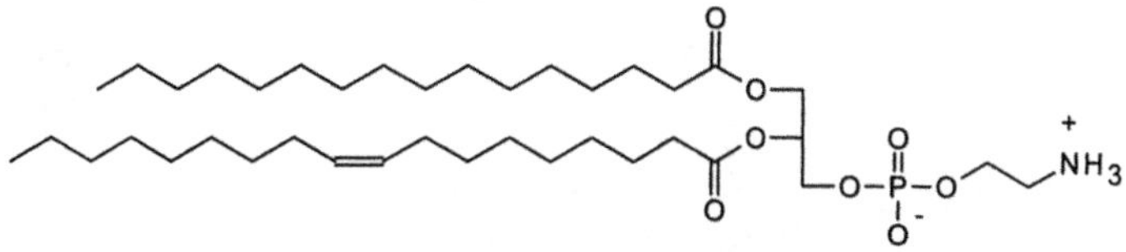

1-Palmitoyl-2-oleoylphosphatidylethanolamine (POPE) (M_r = 718.01)

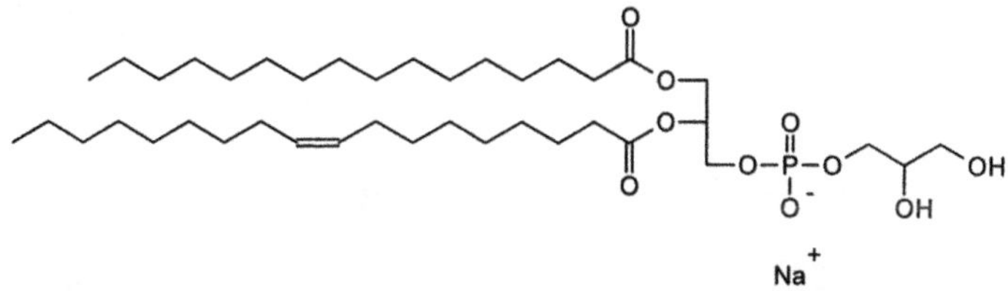

1-Palmitoyl-2-oleoylphosphatidylglycerol (Sodium salt) (POPG) (M_r = 771.0)

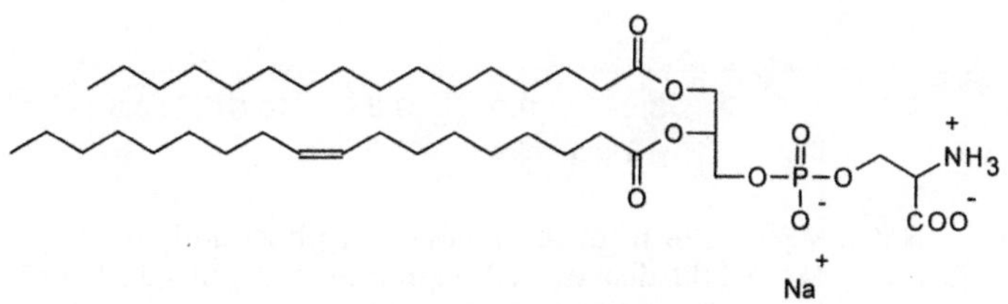

1-Palmitoyl-2-oleoylphosphatidylserine (Sodium salt) (POPS) (M_r = 784.0)

Figure 11.1 An example of phospholipid species from four phospholipid classes. Reprinted from *J. Chromatogr. A*, **802**, K. Raith, R. Wolf, J. Wagner and R. H. H. Neubert, Separation of phospholipids by nonaqueous capillary electrophoresis with electrospray ionisation mass spectrometry, 185–188.[21] Copyright 1998, with permission from Elsevier.

acetic acid, and a separation voltage of 30 kV. This NACE-MS method was coupled to an ion trap MS operated in negative mode and allowing for tandem MS, using a sheath liquid composed of methanol/acetonitrile, 40 : 60, v/v, containing 50 mM ammonium acetate at a flow rate of 2 μL min^{-1}. The tandem MS mode enabled discrimination between co-migrating individual phospholipid species within the phospholipid classes based on their fatty acid tail.

Figure 11.3 shows the separation of phospholipids (phosphatidylcholine (PtdCho), glycerophosphoethanolamine (PE), phosphatidylserine (PtdSer), phosphatidylinositol (PtdIns), and 1-palmitoyl-2-oleoyl-phosphatidylglycerol (PtdGro)) obtained with this method.

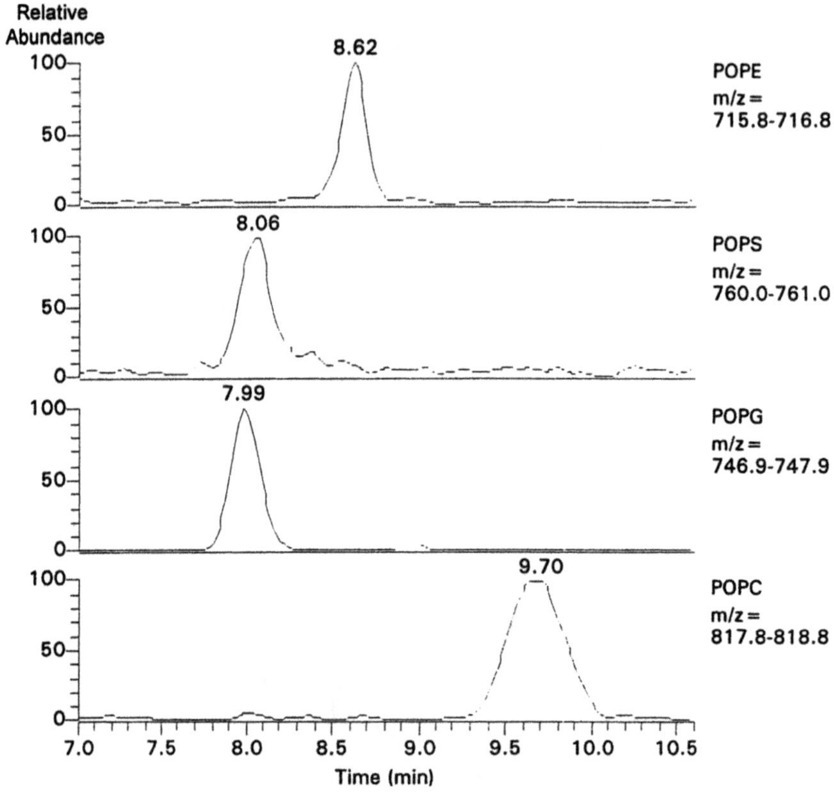

Figure 11.2 NACE-MS results from the analysis of phospholipids. CE-MS conditions: Methyl-Sil deactivated capillaries (75 μm i.d., length 60 cm), separation voltage 30 kV, BGE 20 mM ammonium acetate in acetonitrile/2-propanol/*n*-hexane (57:38:5, v/v). MS is in negative ion mode. CE coupled to ion trap MS (negative mode) with a sheath liquid (methanol/water, 80:20, 5.0 μL min^{-1}). Reprinted from *J. Chromatogr. A*, **802**, K. Raith, R. Wolf, J. Wagner and R. H. H. Neubert, Separation of phospholipids by nonaqueous capillary electrophoresis with electrospray ionisation mass spectrometry, 185–188.[21] Copyright 1998, with permission from Elsevier.

Montealegre *et al.*[23] used a very similar NACE-MS approach for the analysis of phospholipids in olive fruit. The BGE consisted of 100 mM ammonium acetate in methanol/acetonitrile (60:40, v/v) with 0.5% acetic acid. The separation took place in bare fused silica capillaries (50 μm i.d., 60 cm length) with an applied voltage of 25 kV. The system was coupled to an ion trap MS using a sheath liquid composed of methanol/0.5% acetic acid (80:20, v/v) at a flow rate of 6 μL min^{-1}. Contrary to Raith *et al.*[21] and Gao *et al.*,[22] the MS was operated in the positive mode. Both negative and positive modes were investigated but the latter gave better detection limits as well as more stable CE currents. The NACE-MS system allowed the separation and detection of the following phospholipid classes: lysophosphatidic

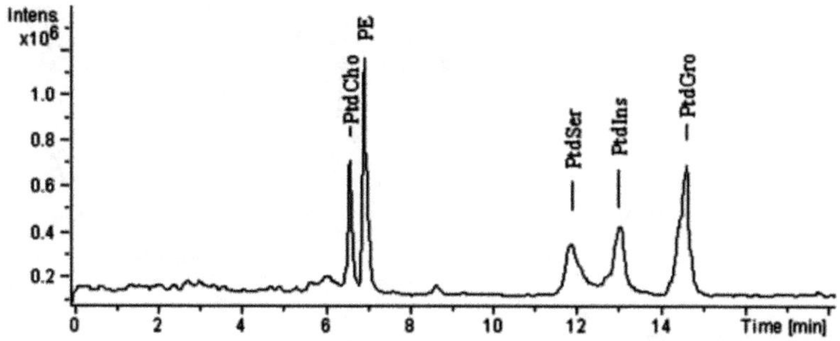

Figure 11.3 NACE-MS results of phospholipid analysis. Capillary: uncoated fused silica, 60 cm length, 50 µm i.d. BGE: methanol/acetonitrile (40:60, v/v) containing 20 mM ammonium acetate and 0.2% HOAc. MS in negative ion mode. Separation voltage 30 kV. Coupled to ion trap MS (negative mode) using a sheath liquid (methanol/acetonitrile (40:60 v/v) containing 50 mM ammonium acetate, 2 µL min^{-1}). Experimental conditions: see main text. PtdCho = phosphatidylcholine, PE = glycerophosphoethanolamine, PtdSer = phosphatidylserine, PtdIns = phosphatidylinositol, PtdGro = 1-palmitoyl-2-oleoyl-phosphatidylglycerol. Reprinted with permission from ref. 22. Copyright © 2007 WILEY-VCH Verlag GmbH & Co. KGaA, Weinheim.

acid, lysophosphatidylcholine, phosphatidylcholine, lysophosphatidylethanolamine, phosphatidylethanolamine, phosphatidylserine, phosphatidylinositol, phosphatidic acid and phosphatidylglycerol. The detection limits were between 2.2 µg mL^{-1} (phosphatidylcholine) and 40.9 µg mL^{-1} (phosphatidylethanolamine). Using MS/MS, two to five molecular species per phospholipid class could be identified in olive fruit, mainly stearic (18:0), oleic (18:1), and palmitic (16:0) acids. Figure 11.4 shows the obtained NACE-MS results for the analysis of phospholipids in olives.

Nalesso *et al.*[24] developed a NACE-MS method for the analysis of phosphatidylethanol, aiming for a fast method for the analysis of this alcohol biomarker. This study focused on two phosphatidylethanol species, *i.e.*, 16:0/16:0 (analyte of interest) and 9:0/9:0 (internal standard). The BGE was carefully optimized by studying variable percentages of acetonitrile, methanol, 2-propanol, dichloromethane, and water along with variable concentrations of ammonium acetate, which resulted in a BGE consisting of acetonitrile (45%, v/v), 2-propanol (20%, v/v), dichloromethane (20%, v/v), methanol (10%, v/v), water (5%, v/v) and 25 mM ammonium acetate. Uncoated fused-silica capillaries (75 µm i.d., 90 cm length) as well as a separation voltage of 30 kV were used. A sheath liquid interface was used for hyphenation to an ion trap MS, using a sheath liquid consisting of 5 mM ammonium acetate (pH 9) in methanol (80:20, v/v) at a flow rate of 1.5 µL min^{-1}. The method was capable of detecting phosphatidylethanol 16:0/16:0 in blood with a limit of detection (LOD) of 0.1 µM, but individual phosphatidylethanol species could not be separated. However, using MS/MS, several phosphatidylethanol species were identified.

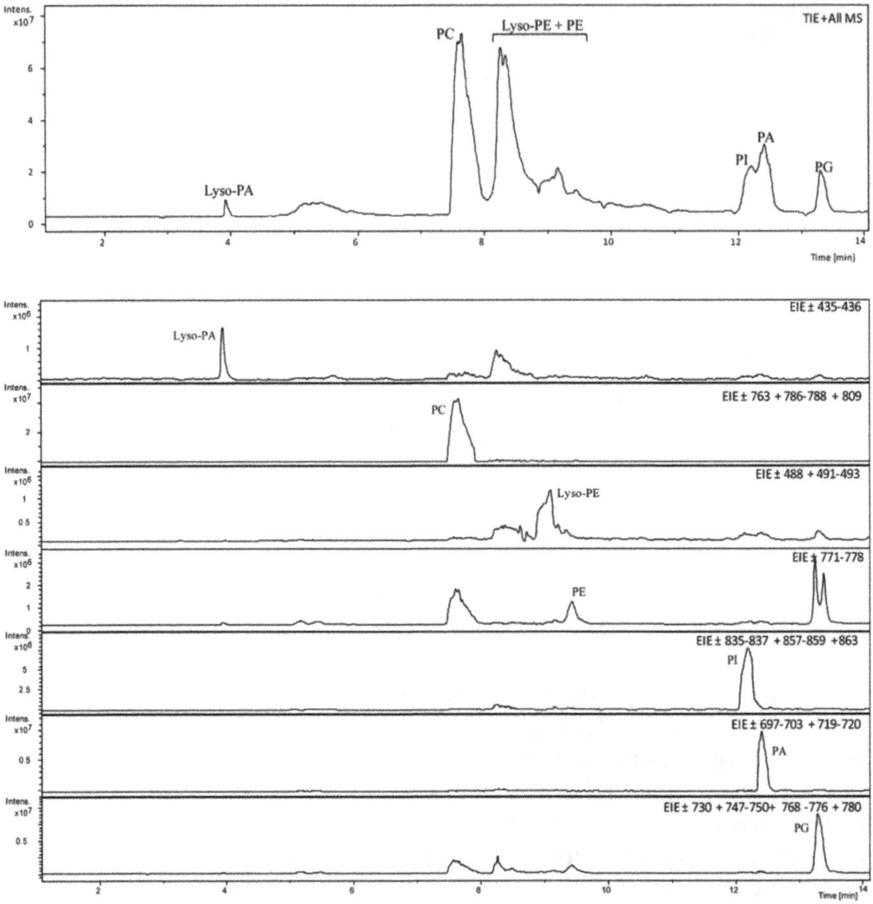

Figure 11.4 NACE-MS results of phospholipid analysis of olives. Compounds detected: lyso-PA, lysophosphatidic acid; lyso-PC, lysophosphatidylcholine; PC, phosphatidylcholine; lyso-PE, lysophosphatidylethanolamine; PE, phosphatidylethanolamine; PS, phosphatidylserine; PI, phosphatidylinositol; PA, phosphatidic acid; PG, phosphatidylglycerol. Capillary: bare fused silica, 50 μm i.d.; total length 60 cm. BGE: 100 mM ammonium acetate in methanol/acetonitrile (60 : 40, v/v) with 0.5% acetic acid. Separation voltage 25 kV. MS in positive mode. Coupled to ion trap MS with a sheath liquid (80 : 20 (v/v) methanol/water with 0.5% (v/v) acetic acid, flow rate 6 μL min^{-1}). Reprinted with permission from ref. 23. Copyright (2013) American Chemical Society.

Jang *et al.*[25] added extra separation power to a CE-MS system by using capillaries (50 μm i.d., 30 cm length) with a monolithic layer of molecular imprinted polymer, thus developing an open tubular capillary electrochromatography (OT-CEC) method. The monolithic layer was based on methacrylic acid, ethylene glycol dimethacrylate, and 4-styrenesulfonic acid. The separation took place in a BGE consisting of isopropanol/acetonitrile/50 mM

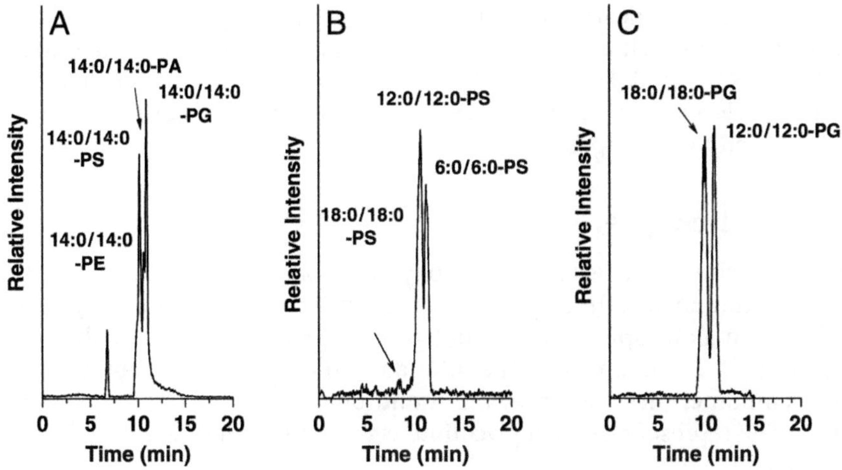

Figure 11.5 OT-CEC-ESI-MS electropherograms of (A) four phospholipids with different head groups having identical acyl chains (14:0 for both) (PE = phosphatidylethanolamine, PS = phosphatidylserine, PA = phosphatic acid, PG = phosphatidylglycerol); (B) PSs; and (C) PGs with different chain lengths. An OT-CEC column with a polymeric monolithic layer was used in a solution of 60:30:10 isopropanol/acetonitrile/50 mM ammonium acetate at 18 kV. The CE system was coupled to ion trap MS (negative mode) using a sheath liquid (80:20 methanol/water and 0.05% ammonium hydroxide 500 nL min^{-1}). Reprinted with permission from ref. 25. Copyright © 2011 WILEY-VCH Verlag GmbH & Co. KGaA, Weinheim.

ammonium acetate (60:30:10, v/v) under a separation voltage of 18 kV. The OT-CEC system was coupled to an ion trap MS using a laboratory-built sheath liquid-assisted sprayer delivering a sheath liquid composed of methanol/water 80:20 (v/v) with 0.05% ammonium hydroxide, at a flow rate of 500 nL min^{-1}. The system was able to separate phospholipid classes (*i.e.*, based on the phospholipid head group) as well as individual phospholipids (*i.e.*, based on the acyl chain).

Figure 11.5 shows the separation of four phospholipid classes (Figure 11.5A) with the same acyl groups (all 14:0/14:0), *i.e.*, phosphatidylethanolamine, phosphatidylserine, phosphatic acid and phosphatidylglycerol, the separation of phosphatidylserine species (Figure 11.5B), *i.e.*, 18:0/18:0, 12:0/12:0 and 6:0/6:0, and the separation of phosphatidylglycerol species (Figure 11.5C), *i.e.*, 18:0/18:0 and 12:0/12:0. Application of the method to a lipid extract from human urine using tandem MS resulted in the separation and identification of 18 phospholipid species.

So far, this is the only CE-MS method reported in the literature that is capable of separating phospholipids by their polar head group as well as their acyl chain length, therefore underlining that choosing an OT-CEC-MS for analyzing complex phospholipid samples is very promising. The selected phospholipid standards differed from each other significantly (*i.e.*, by 6 carbon atoms), thereby indicating that the separation of phospholipid species differing less from each other will be a challenging endeavor by the developed OT-CEC-MS method. However, there is room for improvement, for instance by using a longer capillary or by further optimization of both the column material and BGE composition.

11.2.2 Glycolipids

Glycolipids are molecules with a carbohydrate bound to the lipid moiety. The lipid part contains, in the majority of cases, either a sphingosine or a glycerol backbone, making sphingolipids and glyceroglycolipids the most abundant glycolipids.[26] The main role of glycolipids is to maintain the stability of the cell membrane; therefore, they are found on the surface of all eukaryotic cells.[27] The representatives of glycolipid classes are lipopolysaccharides and gangliosides.

11.2.2.1 Lipopolysaccharides

Lipopolysaccharides are large molecules found in gram-negative bacteria that consist of a lipid and a large polysaccharide part. Lipopolysaccharides bind to the CD14/TLR4/MD2 receptor complex in many cell types, promoting the secretion of toxic compounds like reactive oxygen species. Analysis of lipopolysaccharides often focuses on the analysis of the carbohydrates after the removal of the lipid moiety, described for example in ref. 28–30. As the lipopolysaccharide-related CE-MS methodology is not dedicated to the analysis of lipids, it remains outside the scope of this book chapter. We refer readers with a special interest in this to ref. 31.

Tuffal *et al.*[32] used CE-MS to study molecular complexing of mycobacterial methylglucose lipopolysaccharides with palmitoyl-CoA, using uncoated fused silica capillaries (75 µm i.d., 106 cm length) with a BGE composed of 20 mM ammonium acetate at pH 9. The separation voltage was 30 kV. CE-MS coupling was achieved with a sheath liquid-assisted sprayer and a sheath liquid composition of 2-propanol/BGE (80 : 20, v/v). The MS, a two-sector instrument, was operated in the negative mode. Using this approach, strong indications of the formation of a noncovalent complex between lipopolysaccharides and a lipidic ligand were demonstrated.

11.2.2.2 Gangliosides

Gangliosides are a family of glycosphingolipids containing a hydrophilic oligosaccharide chain of variable constitution linked to a hydrophobic ceramide moiety that may have a diverse composition with respect to the sphingosine type and fatty acid residues (Figure 11.6).

Zamfir *et al.*[34] carried out offline CE-MS experiments, where they collected CE fractions and analyzed them with a nanoESI-QTOF-MS/MS system. This offline approach allowed the benefits of optimized CE separation and

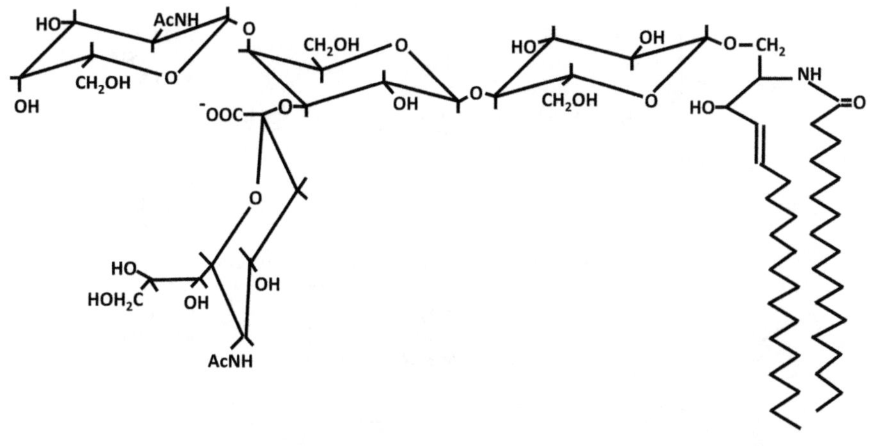

Figure 11.6 Example of a ganglioside containing one sialic acid (GM2). Adapted from *Neurochemical Research*, GM2 ganglioside and pyramidal neuron dendritogenesis, **20**, 1998, 1287–1299, S. U. Walkley, D. A. Siegel and K. Dobrenis,[33] (© Plenum Publishing Corporation 1995), with permission of Springer.

nanoESI conditions to be combined. Since an MS-compatible BGE was used (50 mM ammonium acetate/ammonium hydroxide at pH 12), it would be interesting to see the online CE-MS results, especially because the experimental conditions led to at least 11 individual ganglioside peaks using UV detection.

The group of Her *et al.* focused on developing a robust online CE-MS method for the analysis of gangliosides.[35–38] CE separation of gangliosides was achieved in uncoated fused silica capillaries (50 μm i.d., 100 cm length) using complex BGEs, for example 50 mM borate and 50 mM phosphate buffer containing 20 mM α-cyclodextrin at pH 9.9. With this BGE, four major ganglioside forms (GM1, GDla, GDlb, GTlb) were separated, with two peaks observed for each ganglioside based on the difference in chain length of the ceramide moiety. However, coupling this CE method to MS is cumbersome, as several components (*i.e.*, α-cyclodextrin and sodium borate) of the BGE are not volatile and would clog the spray needle. With more volatile BGEs, such as ammonium acetate or 2-[*N*-cyclohexylamino]-ethanesulfonic acid, CE-MS analysis of gangliosides could be achieved, using a sheath liquid composed of water/isopropanol (80:20, v/v) at 10 μL min^{-1}, but two major ganglioside forms (GDla and GDlb) could not be separated anymore.[35] However, a low-flow CE-MS interface with a beveled tip enabled the analysis of gangliosides using a BGE consisting of 40 mM borate and 20 mM α-cyclodextrin.[36,37] The geometry of the spray tip and the low-flow regime make it more resistant to non-volatile BGE components.

With this approach, the four major ganglioside forms, including the two disialoganglioside isomers (GD 1a, GD 1b) and compounds with different chain lengths of the ceramide moiety for GM1, GD1a, and GD1b, could be

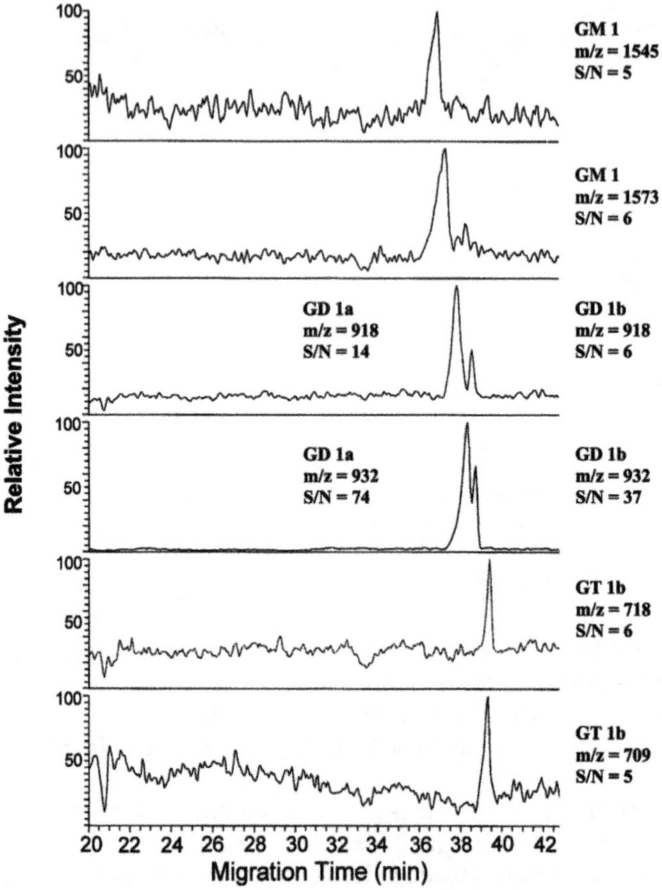

Figure 11.7 Electropherograms of ganglioside analysis. Capillary: bare fused silica, length 100 cm, 50 μm i.d. BGE: 40 mM borate and 20 mM α-cyclodextrin, pH adjusted to 10.6 with ammonia. Separation voltage 20 kV. MS in the negative mode. Coupled to time-of-flight MS (positive mode) with a low-flow laboratory-built sheath liquid-assisted interface (sheath liquid isopropanol/water (80:20, v/v), 400 nL min^{-1}). Reprinted with permission from ref. 36. Copyright (2004) American Chemical Society.

separated and detected, as shown in Figure 11.7. However, the detection limits were not favorable, showing the need for further improvement. This was achieved by adjusting the BGE composition, *i.e.*, by replacing α-cyclodextrin with isopropanol, and sodium borate with ammonium borate to lower ion suppression.[38] Furthermore, a liquid junction was included in the CE-MS interface to remove borate ions before they could enter the ESI process. With this approach, the ganglioside detection limits could be enhanced up to 11 times.

11.2.3 Acylcarnitines

Free carnitine and acylcarnitines play an important role in fatty acid metabolism. They act as a mitochondrial carrier, by electroneutral exchange between acylcarnitine (entering mitochondria), and free carnitine (exiting mitochondria). Carnitine/acylcarnitine carrier deficiency is an innate error of metabolism that is, among others, characterized by life-threatening comas upon fasting. This innate error can be detected in neonatal blood and treated with dietary interventions.[39]

The publications covered in the current chapter concentrate on lipophilic acetylcarnitines and do not include CE-MS analysis of free carnitine and short chain acylcarnitines – unless long chain acylcarnitines are also involved.

Heinig *et al.* developed a CE-MS method for the analysis of acylcarnitines and applied it to plasma, urine, and dried blood spot samples.[40] Bare fused-silica capillaries (75 µm i.d., 80 cm length) were used with a separation voltage of 30 kV. The CE method was coupled to a triple quadrupole MS using a laboratory-built coaxial sheath-flow interface with a sheath liquid consisting of 2-propanol/water (50 : 50, v/v) containing 0.3% acetic acid at a flow rate of 6 µL min^{-1}.

Figure 11.8 shows the results of various BGEs that were studied, using a set of 12 model compounds ranging from free carnitine (C_0) to stearoylcarnitine (C_{18}). Figure 11.8A shows the results obtained using an aqueous BGE consisting of 0.8% formic acid where the most apolar acylcarnitines, *i.e.*, palmitoylcarnitine (C_{16}) and stearoylcarnitine (C_{18}), could not be detected, while tetradecanoylcarnitine (C_{14}) showed a poorly detectable and tailing peak. The addition of 50% organic modifier, *i.e.*, acetonitrile (Figure 11.8B) or methanol (Figure 11.8C), and 3.2% acetic acid resulted in detection of these most apolar acylcarnitines, and, moreover, isomeric separation of butyrylcarnitine. The optimal results were obtained using a fully non-aqueous BGE, which resulted in optimal separation of all acylcarnitines and the highest sensitivity.

Since the carnitine/acylcarnitine compound class encompasses very polar to very apolar molecules, it was not possible to develop a method that was capable of analyzing all acylcarnitines in one experiment using common sample preparation and separation conditions. For the analysis of the most apolar acylcarnitines, the sample preparation consisted of liquid–liquid extraction using ethyl acetate/acetonitrile for urine or ethyl acetate only for plasma. CE-MS analysis was carried out with a BGE composed of acetonitrile/methanol (50 : 50, v/v), 10 mM ammonium acetate, and 6.4% acetic acid. With this method, C_8 to C_{18} acylcarnitines could be determined in urine (LOD 31–330 ng mL^{-1}) and plasma (LOD 29–640 ng mL^{-1}).

In a subsequent study from the same group,[41] a CE microchip was developed and tested on urine spiked with short-chain acylcarnitines (*i.e.*, free carnitine, acetylcarnitine, butyrylcarnitine and octanoylcarnitine), using an aqueous buffer consisting of 10 mM ammonium acetate and 5% acetic acid in deionized water (pH 3).

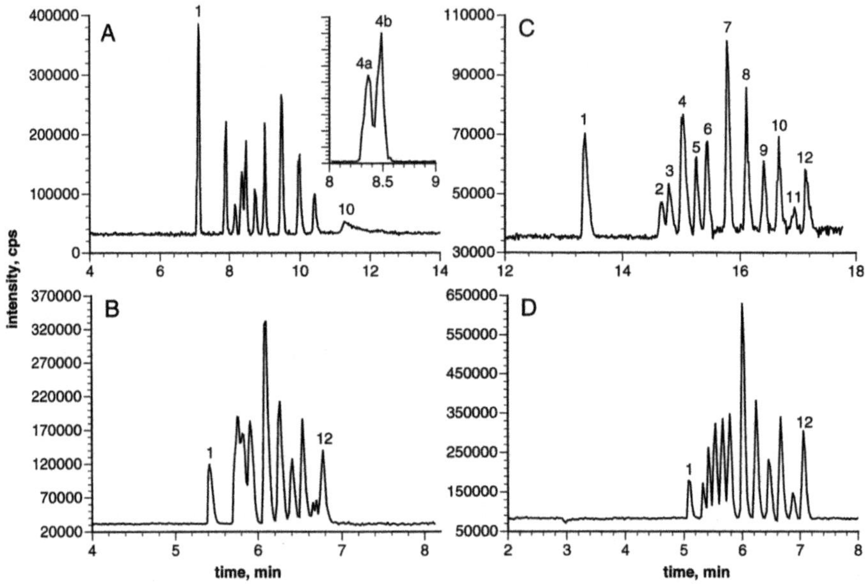

Figure 11.8 CE-MS analysis of carnitine/acylcarnitines using different BGE compositions, combined with a triple quadrupole MS with a sheath liquid composed of 2-propanol/water (50:50, v/v) with 0.3% acetic acid at a flow rate of 6 μL min^{-1}. (A) An aqueous BGE composed of 10 mM ammonium acetate with 0.8% formic acid; (B) BGE composed of 10 mM ammonium acetate, 50/50 water/acetonitrile and 3.2% acetic acid; (C) BGE composed of 10 mM ammonium acetate, 50/50 water/methanol and 3.2% acetic acid, and (D) non-aqueous BGE composed of acetonitrile/methanol (50:50, v/v), 10 mM ammonium acetate and 6.4% acetic acid. Peak numbers: (1) carnitine (C_0), (2) acetylcarnitine (C_2), (3) propionylcarnitine (C_3), (4) butyrylcarnitine (C_4), (5) valeroylcarnitine (C_5), (6) hexanoylcarnitine (C_6), (7) octanoylcarnitine (C_8), (8) decanoylcarnitine (C_{10}), (9) dodecanoylcarnitine (C_{12}), (10) tetradecanoylcarnitine (C_{14}), (11) palmitoylcarnitine (C_{16}), (12) stearoylcarnitine (C_{18}). (a) isobutyrylcarnitine, (b) *n*-butyrylcarnitine. Reprinted from *J. Chromatogr. B*, **735**, K. Heinig and J. Henion, Determination of carnitine and acylcarnitines in biological samples by capillary electrophoresis-mass spectrometry, 171–188.[40] Copyright 1999, with permission from Elsevier.

11.2.4 Lipidic Surfactants

Surfactants are amphiphilic molecules that have a nonpolar tail and a polar head, forming aggregates called micelles, which are usually used as emulsifiers, or wetting and foaming agents. Typically, only the longer C_6–C_{18} chains, which aggregate to form micelles at a concentration greater than the critical micelle concentration, are classified as surfactants.[42]

A surfactant molecule contains one or two hydrocarbon chains, which may be linear, branched or aromatic. The head group can be negatively or positively charged, zwitterionic or without a charge. Surfactants lower the surface tension and are widely used in soaps, *e.g.*, sodium stearate, which represents 50% of the commercial surfactant production. Another important application of surfactants is the preparation of organs prior to transplantation. Surfactants are also produced *in vivo*, *e.g.*, pulmonary surfactants such as phospholipoproteins, which are secreted by type II alveolar cells.[43] Norton *et al.* have developed different capillary electrochromatography (CEC)-MS approaches for zwitterionic,[44] cationic[45] and nonionic[46] surfactant analysis. Zwitterionic surfactants frequently serve as ingredients for personal care products such as shampoos, conditioners and cosmetics. The ionic charge balance of zwitterionic surfactants makes them well suited as detergents to solubilize cell membranes.[44]

The analysis of zwitterionic species (*i.e.*, C_8–C_{18} Zwittergent surfactants) has been performed using a packed CEC column with an internally tapered tip, coupled to mass spectrometry. The optimal separation conditions were achieved with a CEC column (75 μm i.d., 65 cm length) packed with 3 μm C_{18} particles and a BGE consisting of 75% acetonitrile (v/v) and 5 mM Tris at pH 8.0. The separation voltage was 22 kV. The CEC column lasted for *circa* 150 injections. Hyphenation to a single quadrupole MS was achieved using a sheath liquid-assisted sprayer. The MS was operated in the positive mode. Different ionization sources, namely ESI and atmospheric pressure chemical ionization (APCI), were also compared. The sheath liquid was composed of 1 mM ammonium formate in methanol/water (80 : 20, v/v) delivered at a flow rate of 3 μL min^{-1} in the case of ESI and 50 μL min^{-1} for APCI. ESI resulted in less fragmentation of the analytes than APCI and has at least three orders of magnitude lower detection limits. Finally, the CEC-MS method was successfully applied to the analysis of a real industrial Zwittergent sample containing lipid species with C_8 to C_{16} acyl chains (*i.e.*, Zwittergent Z8-Z16). Figure 11.9 shows the obtained electropherograms for the analysis of Zwittergent Z8-Z16, as well as the influence of the percentage of acetonitrile on the separation.

Cationic surfactants contain a hydrophobic alkyl chain and a hydrophilic group, the latter often being a quaternary amine, rendering these compounds permanently positively charged in solution. Alkyltrimethylammonium ($ATMA^+$) compounds are widely used in industry as well as for cosmetics and pharmaceutical applications.[45]

A CEC-MS method for the analysis of $ATMA^+$ ions with chain lengths of C_1–C_{18} was described by Norton *et al.*[45] An internally tapered capillary was packed with mixed mode column material (particle size 3 μm) containing C_6 alkyl chains and sulfonate groups (*i.e.*, reversed phase mode combined with strong cation exchange). With a BGE consisting of 90% acetonitrile (v/v) and 0.04% trimethylamine, C_6–C_{18} $ATMA^+$ ions could be separated within 15 min, using a separation voltage of 18 kV. Coupling with single quadrupole MS (ESI positive mode) was achieved using a sheath liquid composed of methanol/water (70 : 30, v/v) with 10 mM NH_4OAc at pH 6.7 and at a flow rate of 5 μL min^{-1}. Decreasing the acetonitrile percentage to 70% (v/v) in the BGE resolved C_1 and C_2 $ATMA^+$, but at the cost of a prolonged analysis time (40 min). The CEC column could be operated for 165 runs, corresponding to

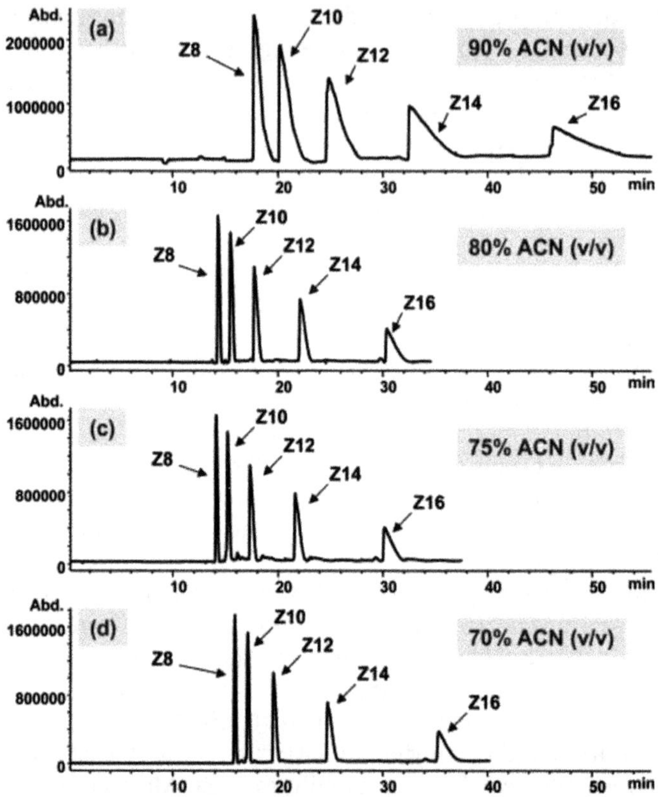

Figure 11.9 Electropherograms of CEC-MS analysis of Zwittergents Z8-Z16. Column: 64 cm length, 75 μm i.d., of which 20 cm is packed with 3 μm C_{18} particles, and with an internally tapered tip. BGE composition: varying percentage of acetonitrile, 5 mM Tris, pH 8.0. Separation voltage of 22 kV. CE coupled to single quadrupole MS (positive mode) with a sheath liquid (1 mM ammonium formate in methanol/water, 80 : 20, 7.5 μL min^{-1}). (a) 90% acetonitrile in the BGE, (b) 80% acetonitrile, (c) 75% acetonitrile, and (d) 70% acetonitrile. Reprinted with permission from ref. 54. Copyright (2005) American Chemical Society.

111 hours. The LODs ranged between 5 ng mL^{-1} (C_{18}-TMA^+) down to 500 pg mL^{-1} (C_{12}- and C_{14}-TMA^+). The optimized CEC-MS method was successfully applied to the analysis of a commercial Arquad S-50 $ATMA^+$ detergent sample containing soybean fatty acid derivatives, leading to the results shown in Figure 11.10. In this sample, C_{12}–C_{18} $ATMA^+$ species could be separated. Interestingly, for C_{18}-$ATMA^+$ the *cis–trans* isomers of double and single unsaturated species could also be detected.

Using a column packed with a 3 μm CEC Reliasil ODS-1 stationary phase (100 Å pore size), the same CEC-MS strategy developed in ref. 44 and 45 was followed for the analysis of nonionic surfactants.[46] The so-called Triton X solutions are mixtures of alkylphenol polyethoxylate nonionic surfactants of

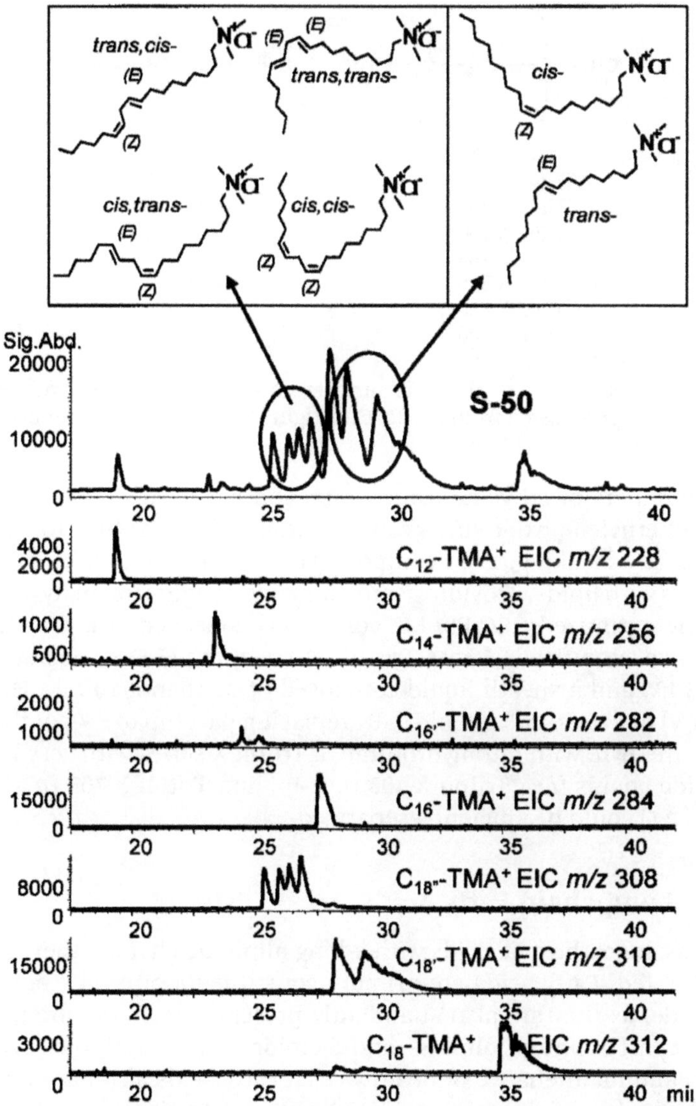

Figure 11.10 CEC-MS results obtained for the analysis of an Arquad S-50 industrial sample (15 μg mL^{-1}). Column: 64 cm length, 75 μm i.d., of which 27 cm is packed with 3 μm mixed mode particles, and with an internally tapered tip. BGE composition: 70% acetonitrile (v/v) with 0.04% TEA (v/v) at pH 3.0 and 15 mM ammonium acetate. Separation voltage 15 kV. CE coupled to single quadrupole MS (positive mode) with a sheath liquid composed of 10 mM ammonium acetate in methanol/water (50 : 50, v/v), at a flow rate of 7 μL min^{-1}. Top trace: monitoring of C_{12}–C_{18}-TMA^+; left circle: separation of four *cis–trans* isomers of C_{18}-TMA^+ with two degrees of unsaturation; right circle: separation of two *cis–trans* isomers of C_{18}-TMA^+ with one degree of unsaturation. Other traces represent individual TMA^+ ions. Reprinted with permission from ref. 45. Copyright © 2006 WILEY-VCH Verlag GmbH & Co. KGaA, Weinheim.

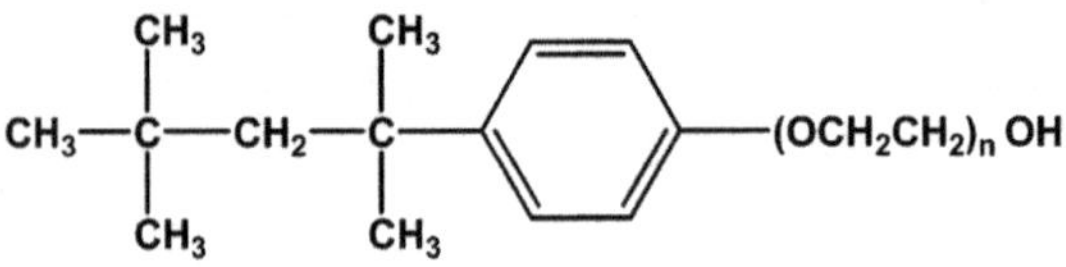

Triton X-15	$n = 1$
Triton X-45	$n = 4–5$
Triton X-100	$n = 9–10$
Triton X-165	$n = 16$
Triton X-305	$n = 30$
Triton X-405	$n = 40$
Triton X-705	$n = 70$

Figure 11.11 Structure of nonionic surfactants known as Triton X. Adapted with permission from ref. 46. Copyright (2007) American Chemical Society.

varying numbers of ethylene oxide units, and are used as detergents. The number of ethylene oxide units can vary from 1 (Triton X-15) to 70 (Triton X-705). Strictly speaking, these compounds are not lipids but they do present similarities with lipids, providing some valuable insights for analysis of 'true' lipids. The optimized CEC-ESI-MS conditions consisted of a mobile phase containing acetonitrile/2.5 mM Tris (90 : 10, v/v) at pH 8, a separation voltage of 18 kV, and a sheath liquid composed of methanol/10 mM HCO_2NH_4 (50 : 50, v/v) delivered at 5 µL min^{-1}. By replacing part (up to 5%) of the acetonitrile in the BGE with tetrahydrofuran, a Triton X series with very long ethylene oxide chains (*i.e.*, Triton X-405 (n = 40) and Triton X-705 (n = 70), see Figure 11.11) could be efficiently separated.

11.2.5 Long-chain Fatty Acids

Fatty acids are carboxylic acids with a long aliphatic chain, either saturated or unsaturated. Fatty acids are primary constituents of cells – both inside and outside, as they are also abundantly present in cell membranes as an important part of phospholipids – and therefore play crucial roles in the construction and maintenance of healthy cells. Fatty acids also have numerous biological functions including oxidative stress reduction, as well as neuro- and cardiovascular protection. Many of the fatty acids are endogenously produced, with some exceptions like linolenic acid and α-linolenic acid, referred to as essential fatty acids. Fatty acids are often categorized into the following groups: (1) short-chain fatty acids (C_1–C_5), (2) medium-chain fatty acids (C_6–C_{12}) (3), long-chain fatty acids (C_{13}–C_{21}) and very long-chain fatty acids (C_{22} and larger). The analysis of fatty acids is classically performed by GC, but also often with LC.[47] In this chapter, the analysis of long-chain fatty acids is solely discussed, due to their highly apolar nature. It is worth mentioning that no CE-MS method for the analysis of very long chain-fatty acids has been reported so far.

A CE-MS method has been developed by Petersson *et al.* for the analysis of some fatty acids and prostaglandins using a special sheathless CE-MS interface.[48] Prostaglandins are a group of physiologically active lipid mediators derived enzymatically from fatty acids, playing an essential role in inflammation and having diverse hormone-like effects. The authors used an in-house built CE system and 19 μm i.d. × 150 μm o.d. fused silica capillaries, with lengths varying between 29 and 50 cm, and a specially designed stainless steel liner with a smooth sharpened tip. The compounds analyzed were lauric acid (C_{12}), myristic acid (C_{14}), palmitic acid (C_{16}) and stearic acid (C_{18}), separated using a voltage of 32 kV. The BGE consisted of 2 mM acetic acid in ammonium acetate : acetonitrile (50 : 50, v/v) at pH 5.2. The CE system was coupled to a triple quadrupole MS using a laboratory-built sheathless CE-MS sprayer. The separation voltage was applied at the interface while the electric circuit necessary for electrophoretic separation was closed *via* liquid that built up during CE between the liner and the capillary tip. Figure 11.12 shows a CE-MS electropherogram obtained for the separation of four fatty acid standards. Since fatty acids carry a negative charge, while a positive separation voltage was used, the

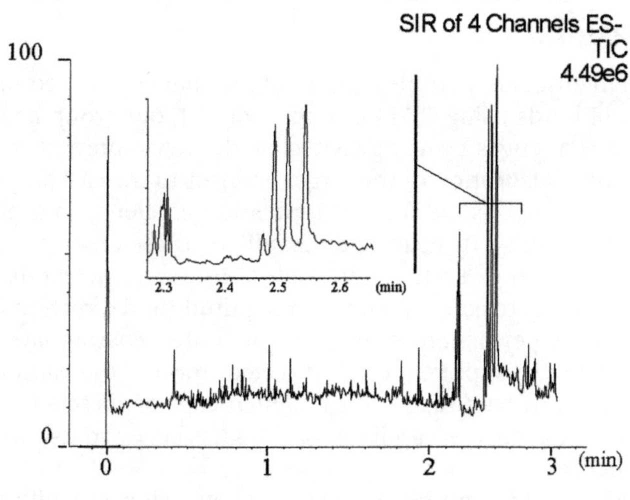

Figure 11.12 CE-MS results of the analysis of lauric acid (C_{12}), myristic acid (C_{14}), palmitic acid (C16) and stearic acid (C18). Capillary: uncoated fused silica, 50 cm, 19 μm i.d. with the tip sharpened. BGE composed of 2 mM acetic acid–ammonium acetate/acetonitrile (50 : 50, v/v) at pH 5.2. Separation voltage 32 kV. CE coupled to triple quadrupole MS (negative mode) *via* a home-built sheathless CE-MS sprayer. Reprinted from *J. Chromatogr. A*, **854**, M. A. Petersson, G. Hulthe and E. Fogelqvist, New sheathless interface for coupling capillary electrophoresis to electrospray mass spectrometry evaluated by the analysis of fatty acids and prostaglandins, 141–154.[48] Copyright 1999, with permission from Elsevier.

first fatty acid detected was the largest one, stearic acid, showing the lowest electrophoretic mobility. In addition, this method also led to the separation of 4 prostaglandin stereoisomers, of a total of 6 injected.

Lee *et al.*[49] reported an interesting approach for the analysis of 21 fatty acids, based on addition of the dicationic ion pair *N,N'*-dibutyl 1,1'-pentylenedipyrrolidium *via* the sheath liquid, allowing for enhanced sensitivity. Separations were carried out in bare fused-silica capillaries (50 μm i.d., 85 cm length), using a 30 kV separation voltage. The CE system was coupled to a time-of-flight MS using a coaxial sheath liquid-assisted sprayer, delivering a sheath liquid consisting of methanol/water (50 : 50, v/v) at a flow rate of 10 μL min^{-1}. The optimal BGE was composed of 30 mM ammonium formate in 40% acetonitrile at pH 10. The high acetonitrile content and high pH were necessary to enhance the solubility of the largest fatty acids. Figure 11.13 illustrates the benefits of adding *N,N'*-dibutyl 1,1'-pentylenedipyrrolidium to the system, in particular to the sheath liquid, with separation and satisfactory peak shape for all 21 fatty acids ranging from C_2 to C_{22}, even though the separation resolution decreased with increased acyl length.

11.3 CE-MS Analysis of Lipids Using Coated Capillaries

All the aforementioned examples highlight the challenges encountered with the analysis of lipids using CE-MS. In this context, our group has chosen to address such challenges by using coated capillaries to prevent hydrophobic interactions and adsorption of the target analytes to the capillary wall. Such interactions are responsible for low separation efficiency and poor migration time repeatability. By coating the capillary, these analyte–wall interactions can be reduced or even suppressed.[50] Moreover, the capillary coating offers the possibility to control both the magnitude and direction of the EOF, depending on the permanent charges of the (outer) coating layer. The EOF can be, for instance, suppressed,[51,52] directed towards the cathode, similar to what happens in bare-fused silica capillaries,[53] or directed towards the anode.[54–58] The relevance of capillary coatings has been extensively described by Huhn *et al.*[59]

Lipid analysis adds another challenge when using a capillary coating, *i.e.*, that it should be stable when using BGEs with high contents of organic solvents, or purely organic BGEs (NACE), which are often necessary in lipid analysis to ensure complete dissolution of the compounds of interest. The coating material must also be compatible with MS detection, ensuring that no ion suppression or fouling of the MS caused by leaking components of the coating occurs. A common opinion on coated capillaries is that coating procedures can be very time-consuming, or that commercially available coated capillaries are expensive. Recently, however, simple-to-prepare and yet well-performing capillary coatings have been introduced, as underlined with the following examples.[60–62] Raith *et al.* were able to separate and detect

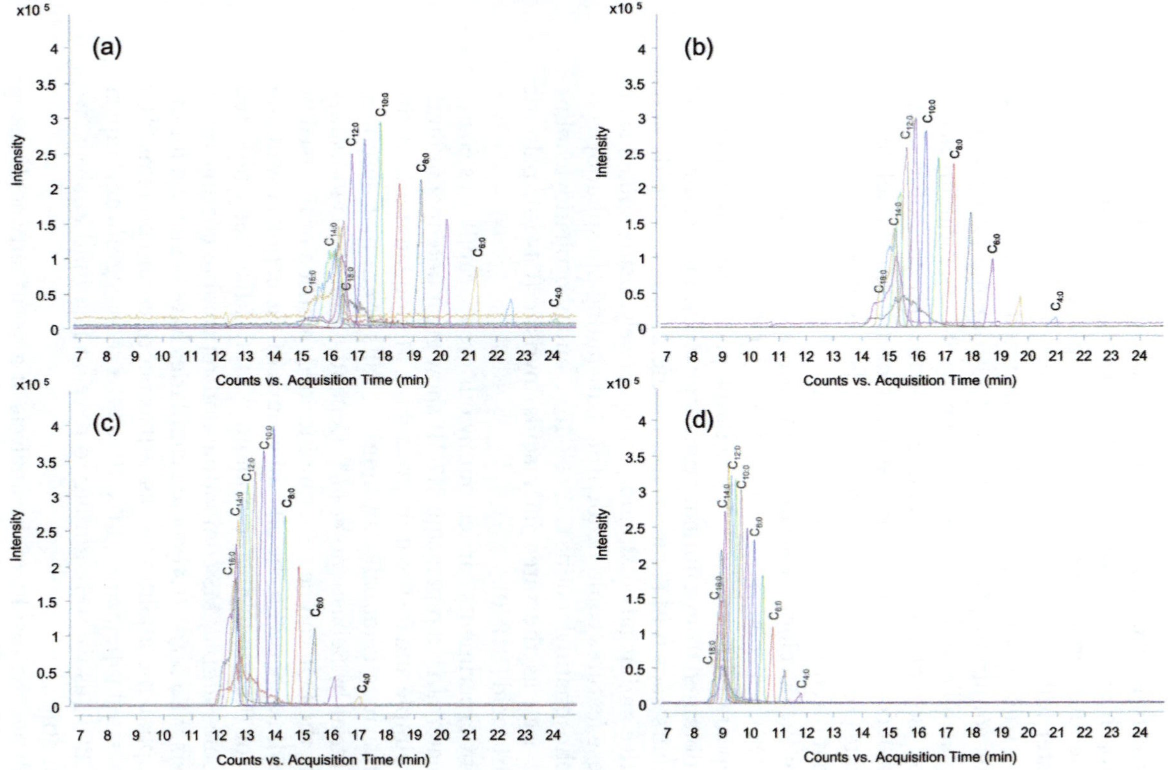

Figure 11.13 CE-MS results obtained for the analysis 21 fatty acids (C_2–C_{22}) using a BGE composed of 40% acetonitrile and 30 mM ammonium formate at pH 10. Capillary: uncoated fused silica (length 85 cm, 50 μm i.d.). Separation voltage 30 kV. CE coupled to time-of-flight MS using a sheath liquid composed of methanol/water, 50 : 50, v/v, and at a flow rate of 10 μL min^{-1}. (a) Detection using negative ESI mode; (b) positive ESI mode with addition of 250 mM *N*,*N*′-dibutyl 1,1′-pentylenedipyrrolidium to the sample prior to injection; (c) positive ESI mode, with addition of 250 mM *N*,*N*′-dibutyl 1,1′-pentylenedipyrrolidium to the BGE; and (d) positive ESI mode, with addition of 250 mM *N*,*N*′-dibutyl 1,1′-pentylenedipyrrolidium to the sheath liquid. Reprinted from *Anal. Chim. Acta*, **984**, J. H. Lee, S. J Kim, S. Lee, J. K. Rhee, S. Y. Lee and Y. C. Na, Saturated fatty acid determination method using paired ion electrospray ionization mass spectrometry coupled with capillary electrophoresis, 223–231.[49] Copyright 2017, with permission from Elsevier.

phospholipids with a CE-MS system using Methyl-Sil deactivated capillaries that were originally used for GC and can thus be easily purchased. However, no data with regards to the coating stability and performance have been reported.[21]

Martma *et al.* demonstrated the feasibility of polyE-323 as a coating material for CE analysis of lipids.[60] PolyE-323 is a positively charged polyamine, containing amino, chloride and hydroxyl groups, resulting in an anodic EOF. The coating procedure consisted of flushing the capillary with 1 M NaOH, 1 M HCl and water followed by rinsing it for 45 min with 7.5% polyE-323 coating solution. This coating has been demonstrated to be stable under near non-aqueous conditions, *i.e.*, with up to 94.5% organic solvents in the BGE. The coating stability towards different organic solvents, including acetonitrile, methanol, isopropanol, propylene carbonate and chloroform, was investigated. A BGE consisting of a mixture of acetonitrile/methanol/20 mM ammonium acetate (56.5 : 38 : 5.5, v/v) led to no background MS signal, which would suggest that bleeding of the polyE-323 coating was observed.

The stability of the developed CE-MS method was evaluated using sphingomyelin as an EOF marker and showed excellent migration time (RSD 1.3%) and peak area repeatability (RSD 6.2%). The developed method was applied to a lipid extract obtained from a human primary glioblastoma cell line (U-87 MG) with the results shown in Figure 11.14. 67 and 23 putative lipids were detected in positive and negative MS mode, respectively. Using spiked samples, some of these features could be identified, including phosphatidylcholine 18 : 1/16 : 0, phosphatidylcholine C18 : 1/C16 : 1, and phosphatidylcholine C18 : 1/C18 : 1, as well as the three fatty acids myristic (C14 : 0), palmitic (C16 : 0) and stearic acid (C18 : 0).

Based on recent experiments carried out by our group, capillaries coated with the monoquaternarized piperazine M7C4I show great potential in lipids analysis, using a simple and fast coating procedure. Monoquaternarized piperazines (also known as ω-iodoalkylammonium salts) were first described as a coating material by Sebastiano *et al.*[61] Depending on the ω-iodoalkylammonium used, the EOF can be either maintained, reduced or reversed by forming covalent links between the silanol groups of the capillary wall and the iodine atom of the modifying compound.[61] M7C4I has been reported as a highly suitable, simple, MS-compatible and repeatable coating material for CE analysis.[62] So far, M7C4I-coated capillaries have been used for the analysis of tryptophan metabolites,[61,62] as well as peptides and proteins.[63] Of all monoquaternarized piperazines, M7C4I has the most significant impact in reversing the EOF; as we were aiming to develop fast lipid analysis, we selected it for our study.

Figure 11.15 shows the results of the analysis of a set of fatty acids using M7C4I-coated capillaries (50 and 75 μm i.d., 80 cm length), and also compared to a polyE-323-coated capillary (75 μm i.d., 80 cm length). The BGE consisted of a mixture of acetonitrile/methanol/20 mM ammonium acetate buffer (56.5 : 38 : 5.5, v/v) at pH 7.1.[60] The CE system was coupled to a time-of-flight MS using a sheath-flow CE-MS and a sheath liquid composed of

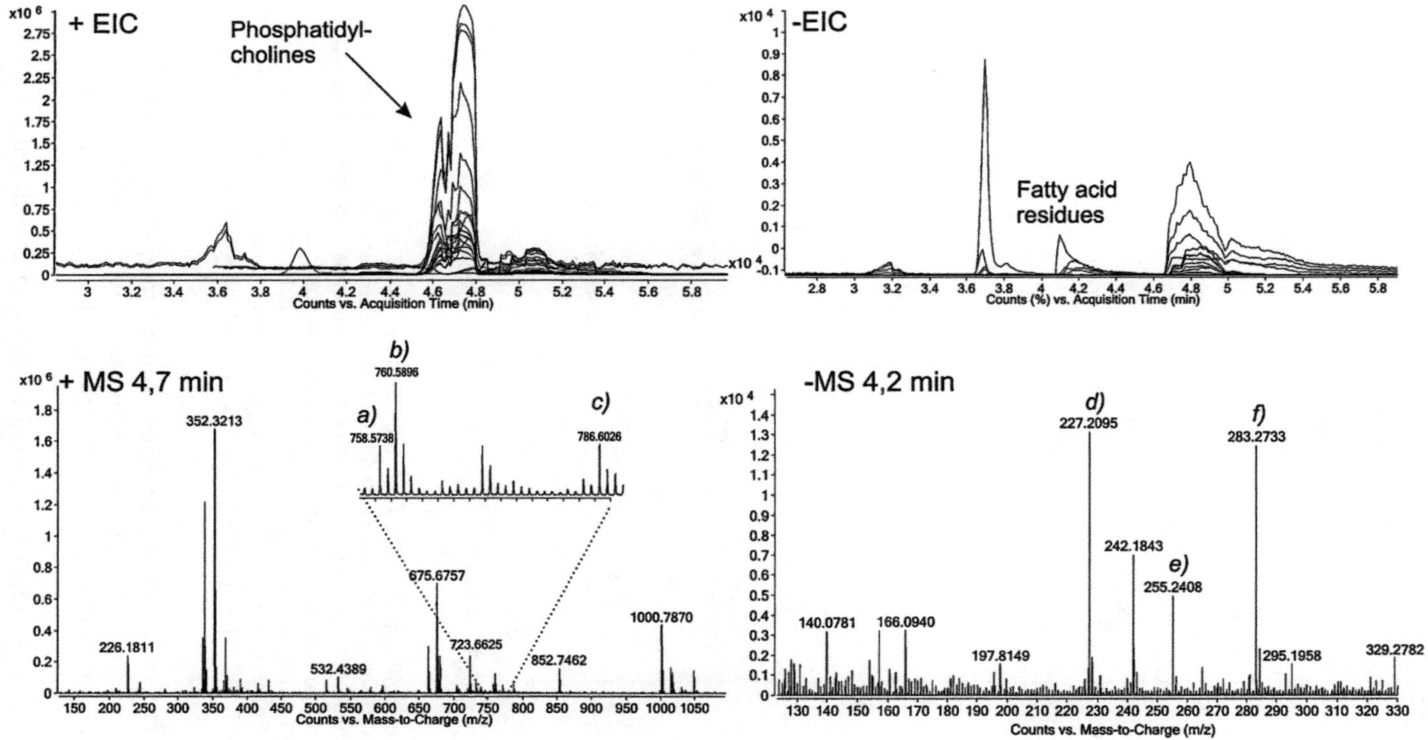

Figure 11.14 CE-MS results for the analysis of a lipid extract obtained from a human primary glioblastoma U-87 MG cell line, showing extracted ion electropherograms (upper part) and examples of mass spectra in positive and negative mode (lower part). (a) Putative *m/z* for phosphatidylcholine 18:2/16:0; (b) putative mass for phosphatidylcholine 18:1/16:0; (c) putative mass for phosphatidylcholine 18:2/18:0; (d) putative mass for myristic acid residues; (e) putative mass for palmitic acid residues; (f) putative mass for stearic acid residues. Capillary: length 90 cm, 75 µm *i.e.*, coated with polyE-323. BGE composition: acetonitrile/methanol/20 mM ammonium acetate (56.5:38:5.5, v/v). Separation voltage −30 kV. CE coupled to time-of-flight MS using a sheath liquid composed of water/methanol (50:50, v/v) with 10 mM ammonium acetate at a flow rate of 1 µL min^{-1}. Reprinted from *J. Chromatogr. A*, **1317**, K. Martma, P. W. Lindenburg, K. L. Habicht, K. Vulla, K. Resik, G. Kuut and R. Shimmo, Potential of polyE-323 coated capillaries for capillary electrophoresis of lipids, 193–198.[60] Copyright 2013, with permission from Elsevier.

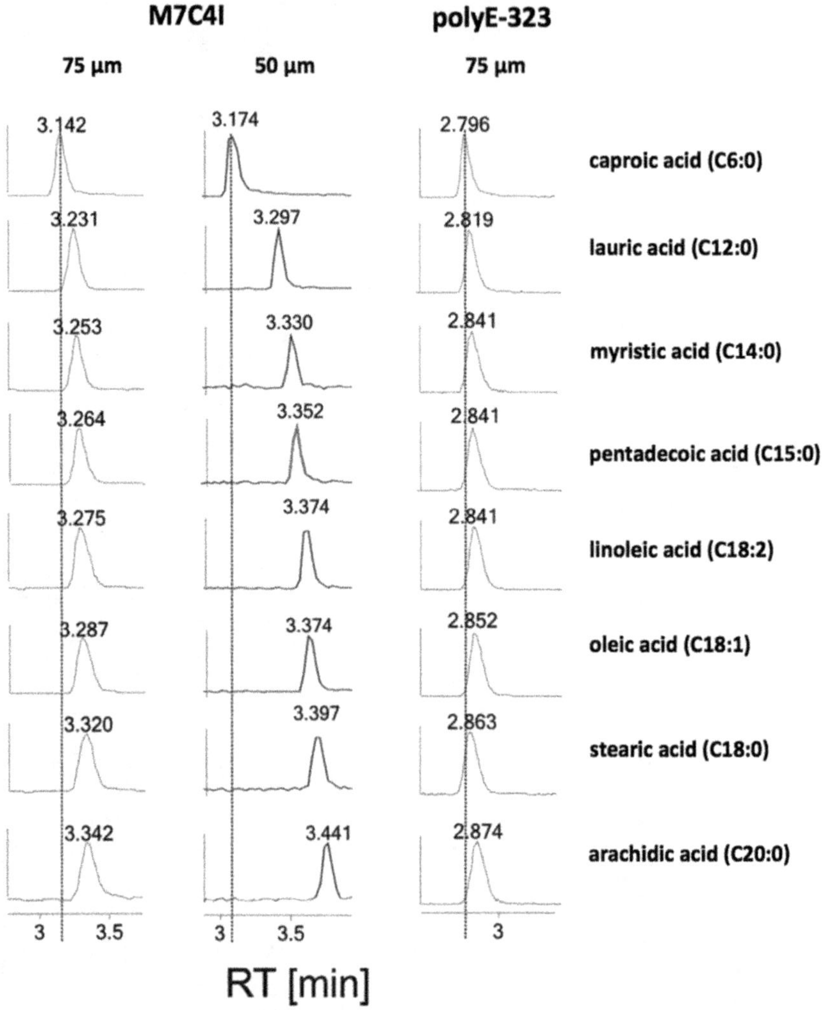

Figure 11.15 Electropherograms obtained for the analysis of fatty acids using M7C4I- and polyE-323-coated capillaries. Capillary: 80 cm length, 50 or 75 μm i.d., coated with M7C41. BGE composition: acetonitrile, methanol and 20 mM ammonium acetate buffer (56.5 : 38 : 5.5, v/v) at pH 7.1. Separation voltage −30 kV. CE coupled to time-of-flight MS (negative mode) using a sheath liquid composed of methanol/water (98 : 2, v/v), 20 mM ammonium acetate, at a flow rate of 0.5 μL min^{-1}.

methanol/water (98 : 2, v/v) with 20 mM ammonium acetate, at a flow rate of 0.5 μL min^{-1}. The coating procedure for M7C4I was adapted from the work of Sebastiano *et al.*[61] and consisted of rinsing the capillary for 2 min with a M7C4I solution.

As illustrated in Figure 11.15, M7C4I-coated capillaries markedly improved the electrophoretic separation compared to the polyE-323 coating. Using a 50 µm i.d. capillary further improved the separation, however at the cost of lower signal-to-noise ratios. The peak shapes of the individual fatty acid traces were symmetrical, indicating no significant interaction between the analytes and the capillary wall, nor with the coating material. No MS contamination was observed when using the M7C4I coating. The linear dynamic range was estimated to be from 0.1 to 100 µM for all the fatty acids, with R^2-values above 0.99. The LOD values of the fatty acids, with a 75 µm i.d. capillary, were all approximately 0.1 µM, corresponding to a 1 fmol injected amount. These LODs are similar to the ones reported using GC-MS (*e.g.*, 0.250 fmol for lauric acid[15]). The typical LOD values for LC-MS analysis of fatty acids are close to 1 nM concentration, corresponding to 4–30 fmol.[64,65] The migration times were repeatable, with RSDs ranging between 1.9 and 2.3%, while the peak area RSDs ranged between 3.2 and 4.5%. These results, comparable to those obtained with polyE-323 capillaries,[60] indicate that no analyte loss due to adsorption to the capillary wall took place in the M7C44I-coated capillaries.

The developed method was subsequently applied to the analysis of a complex biological sample. Cell membrane suspensions were obtained from U-87 cells by a modified procedure based on that of Karelson *et al.*[66] and lipids were extracted using Bligh and Dyer liquid–liquid extraction.[67] Figure 11.16 shows the results obtained for the lipid extract analysis, including the extracted ion electropherograms of the fatty acids, which migrated between 4.0 and 4.2 min (Figure 11.16A) and the averaged mass spectrum of the migration time span (Figure 11.16B), revealing 17 putatively identified fatty acids. The putative identity of these fatty acids is supported by both their exact mass and migration order. The number of putatively identified fatty acids found with our method is comparable to other chromatographic methods. For instance, Zhu *et al.* detected 20 fatty acids in rape bee pollen[64] using LC with fluorescence detection and Wang *et al.* detected 12 fatty acids in mouse liver tissue.[65] Quehenberger *et al.* detected 21 fatty acids in blood plasma with their GC-MS method.[15]

In order to evaluate the potential of the developed method in untargeted lipidomics, all molecular features with a signal-to-noise rate above 10 were extracted, leading to 53 molecular features detected in negative ionization mode, including the 17 already putatively identified fatty acids. In positive MS mode, 143 molecular features were detected, which represents a major improvement in comparison with the results obtained with polyE-323-coated capillaries, which led to 67 and 23 molecular features detected in positive and negative mode, respectively. This reveals the better coating performance of M7C4I, with less loss of analyte by adsorption and/or less ion suppression due to bleeding of the coating. Based on their mass, further lipids could be putatively identified, including phosphatidylglycerol 32 : 1, as well as phosphatidylcholines C36 : 2 and C34 : 1. The latter co-migrated due to their zwitterionic properties, therefore having no electrophoretic mobility. The fact that we are able to analyze compounds with very long acyl chains shows the suitability of the developed CE-MS method for very apolar compounds.

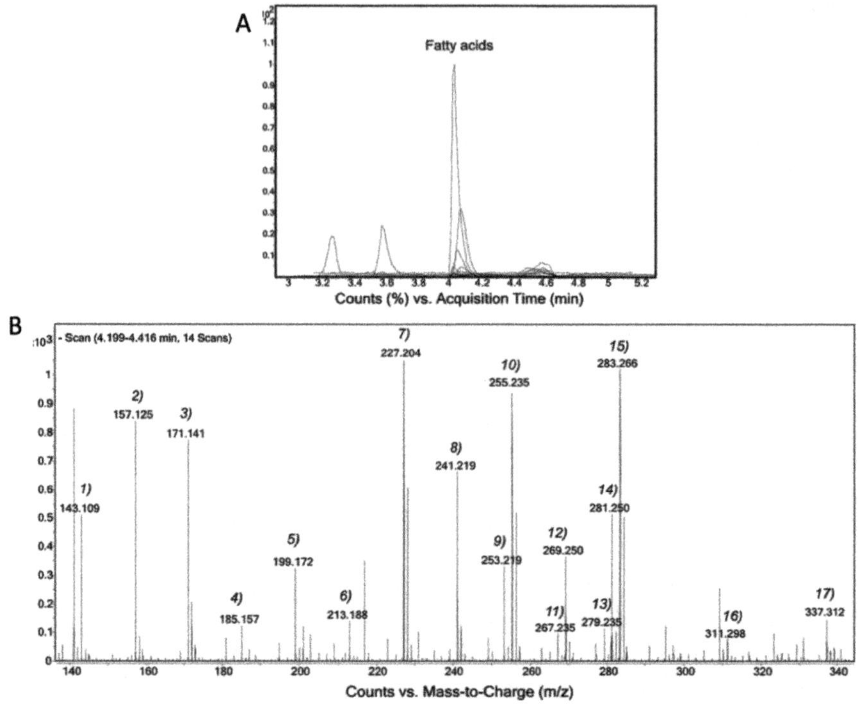

Figure 11.16 CE-MS results obtained for the analysis of lipid extracts from U-87 cell membranes using M7C41-coated capillaries. (A) Extracted ion electropherograms of a U-87 cell extract using M7C4I. Capillary: 80 cm length, 75 μm i.d., coated with M7C41. BGE composition: acetonitrile/methanol/20 mM ammonium acetate buffer pH 7.1 (56.5 : 38 : 5.5, v/v). Separation voltage −30 kV. CE coupled to time-of-flight MS using a sheath liquid composed of methanol/water (98 : 2, v/v) and 20 mM ammonium acetate at a flow rate of 0.5 μL min^{-1}. (B) ESI-MS mass spectrum extracted from the fatty acid migration region, with the list of putatively identified fatty acids and their corresponding carbon numbers. (1) caprylic acid (8 : 0); (2) pelargonic acid (9 : 0); (3) capric acid (10 : 0); (4) undecyclic acid (11 : 0); (5) lauric acid (12 : 0); (6) tridecylic acid (13 : 0); (7) myristic acid (14 : 0); (8) pentadecyclic acid (15 : 0); (9) palmitoleic (*cis*) or palmitoelaidic (*trans*) acid (16 : 1); (10) palmitic acid (16 : 0); (11) heptadecanoic acid (17 : 1); (12) margaric acid (17 : 0); (13) linoleic acid (18 : 2); (14) oleic (*cis*) or elaidic (*trans*) acid (18 : 1); (15) stearic acid (18 : 0); (16) arachidic acid (20 : 0); (17) erucic acid (22 : 1).

11.4 Concluding Remarks

CE-MS has been used for a variety of lipid classes, *i.e.* phospholipids, glycolipids, acylcarnitines, surfactants, and fatty acids. Several CE-modes have been employed, *i.e.* conventional CZE, NACE and CEC and combined with several MS techniques, the most prevailing being ESI ionization with a single or triple quadrupole analyzer. The highest separation power, allowing for separation of lipids within their class (*i.e.* based on differences in their acyl chains), has been obtained with CEC. However, this work was carried out using laboratory-built CEC columns, according to time-consuming and rather complex procedures.

The BGEs of all the work discussed in this chapter share one characteristic, *i.e.* that they are apolar (conventional CE using a highly organic BGE (<50% aqueous) to NACE with a very apolar BGE), both because of the solubility issues and to allow for the suppression of analyte–wall interactions. As a consequence of this, generally high separation voltages are used in CE-MS of lipids (sometimes in combination with short capillaries), since only low currents are generated in such low conductivity BGEs and decrease of the separation efficiency by Joule heating is not an issue. The use of high separation voltages can be an advantage to decrease the analysis time. Indeed, in several cases short analysis times (3–10 min) for lipids are reported.

Two capillary coating materials (M7C4I and PolyE-323) were demonstrated to be suitable for CE-MS analysis of lipids in highly organic media, using fatty acids as an example. The most promising results were obtained with M7C4I-coated capillaries. With this coating, lower detection limits, better separation efficiencies, and faster analysis could be achieved. This CE-MS platform can, with respect to detection limits, compete with other analytical chromatography-based platforms, with LODs in the low nM (0.5 fmol injected) range. Also, a good number of fatty acids in cell extracts could be (putatively) identified.

The separation power of CE-MS discussed in this chapter is in general outmatched by other hyphenated separation technologies such as GC-MS and LC-MS. CE-MS is a complementary lipid analysis technique that allows for adding another separation mechanism to the wide-spread chromatography-based methods. One of the charming features of CE as a separation method is the ease of modifying the separation system, either by changing the BGE, or by modifying the capillary wall, or a combination of both. MS detection adds sensitivity, identification power and selectivity. An interesting option for future research would be to hyphenate CE with ion mobility mass spectrometry (IMS) for the analysis of fatty acids. Such a system would allow for separation in three dimensions: CE (electrophoretic mobility), IMS (structure and size) and MS (mass-to-charge ratio). Within this combination, the limited separation that is achieved in the CE dimension might be enough to significantly contribute to a high throughput CE-IMS-MS lipidomics platform that is capable of measuring a multitude of fatty acids and distinguishing their isomers.

The authors have declared no conflict of interest.

Acknowledgements

Financial support was provided by the European Union through the European Regional Development Fund (Centre of Excellence "Mesosystems: Theory and Applications", TK114) (K. V. and R. S.), and Centre of Excellence in

Behavioural and Neural Sciences (financed by project TU TEE) (R. S.). P. W. L. was supported by Internationalization Programme DoRa, which is carried out by Foundation Archimedes. Prof. Roberto Sebastiano from the Polytechnic of Milan kindly donated solid M7C4I.

References

1. http://goldbook.iupac.org/html/L/L03571.html, accessed 06-09-2017.
2. E. Fahy, S. Subramaniam, R. C. Murphy, M. Nishijima, C. R. Raetz, T. Shimizu, F. Spener, G. van Meer, M. J. Wakelam and E. A. Dennis, *J. Lipid Res.*, 2009, **50**, S9.
3. M. R. Wenk, *Nat. Rev. Drug Discovery*, 2005, **4**, 594.
4. M. Orešič, V. A. Hänninen and A. Vidal-Puig, *Trends Biotechnol.*, 2008, **26**, 647.
5. R. G. Cutler, J. Kelly, K. Storie, W. A. Pedersen, A. Tammara, K. Hatanpaa, J. C. Troncoso and M. P. Mattson, *Proc. Natl. Acad. Sci. U. S. A.*, 2004, **101**, 2070.
6. K. Yanagisawa, A. Odaka, N. Suzuki and Y. Ihara, *Nat. Med.*, 1995, **1**, 1062.
7. R. Sharon, I. Bar-Joseph, M. P. Frosch, D. M. Walsch, J. A. Hamilton and D. J. Selkoe, *Neuron*, 2003, **37**, 583.
8. D. J. Selkoe, *Nat. Cell Biol.*, 2004, **6**, 1054.
9. S. Tumanov and J. J. Kamphorst, *Curr. Opin. Biotechnol.*, 2017, **43**, 127.
10. K. Jurowski, K. Kochan, J. Walczak, M. Baranska, W. Piekoszewski and B. Buszewski, *Crit. Rev. Anal. Chem.*, 2017, **17**, 418.
11. D. A. Slatter, M. Aldrovand, A. O'Connor, S. M. Allen, C. J. Brasher, R. C. Murphy, S. Mecklemann, S. Ravi, V. Darley-Usmar and V. B. O'Donnell, *Cell Metab.*, 2016, **23**, 930.
12. H. C. Köfeler, A. Fauland, G. N. Rechberger and M. Trötzmüller, *Metabolites*, 2012, **2**, 19.
13. H. Luftmann, M. Aranda and G. E. Morlock, *Rapid Commun. Mass Spectrom.*, 2007, **21**, 3772.
14. K. Dreisewerd, J. Müthing, A. Rohlfing, I. Meisen, Z. Vukelić, J. Peter-Katalinić, F. Hillenkamp and S. Berkenkamp, *Anal. Chem.*, 2005, **77**, 4098.
15. O. Quehenberger, A. Armando and E. A. Denni, *Biochim. Biophys. Acta*, 2011, **1811**, 648.
16. G. L. Wei and E. Zeng, *Trends Anal. Chem.*, 2011, **30**, 1429.
17. C. R. Evans and J. W. Jorgenson, *Anal. Bioanal. Chem.*, 2004, **378**, 1952.
18. P. W. Lindenburg, R. Haselberg, G. Rozing and R. Ramautar, *Chromatographia*, 2015, **78**, 367.
19. A. Šlampová, Z. Malá, P. Gebauer and P. Boček, *Electrophoresis*, 2017, **38**, 20–32.
20. T. H. Watts, A. A. Brian, J. W. Kappler, P. Marrack and H. M. McConnell, *Proc. Natl. Acad. Sci. U. S. A.*, 1984, **81**, 7564.
21. K. Raith, R. Wolf, J. Wagner and R. H. H. Neubert, *J. Chromatogr. A*, 1998, **802**, 185.

22. F. Gao, Z. Zhang, X. Fu, W. Li, T. Wang and H. Liu, *Electrophoresis*, 2007, **28**, 1418.
23. C. Montealegre, L. Sánchez-Hernández, A. L. Crego and M. L. Marina, *J. Agric. Food Chem.*, 2013, **61**, 1823.
24. A. Nalesso, G. Viel, G. Cecchetto, G. Frison and S. D. Ferrara, *Electrophoresis*, 2010, **31**, 1227.
25. R. Jang, K. H. Kim, S. A. Zaidi, W. J. Cheong and M. H. Moon, *Electrophoresis*, 2011, **32**, 2167.
26. W. W. Christie and X. Han, *Lipid Analysis*, The Oily Press, 4th edn, 2010.
27. https://www.nature.com/subjects/glycolipids#research-and-reviews, accessed on 06-10-2017.
28. S. Auriola, P. Thibault, I. Sadovskaya and E. Altman, *Electrophoresis*, 1998, **19**, 2665.
29. J. Li, P. Thibault, A. Martin, J. C. Richards, W. W. Wakarchuk and W. van der Wilp, *J. Chromatogr. A*, 1998, **817**, 325.
30. X. Zhu, K. Hatanaka, T. Yamagata and T. Sato, *Electrophoresis*, 2009, **30**, 3519.
31. J. Li and J. C. Richards, *Mass Spectrom. Rev.*, 2007, **26**, 35.
32. G. Tuffal, A. Tuong, C. Dhers, F. Uzabiaga, M. Riviere, C. Picard and G. Puzo, *Anal. Chem.*, 1998, **70**, 1853.
33. S. U. Walkley, D. A. Siegel and K. Dobrenis, *Neurochem. Res.*, 1995, **20**, 1287.
34. A. Zamfir, Z. Vukelić and J. Peter-Katalinić, *Electrophoresis*, 2002, **23**, 2894.
35. D. D. Ju, C. C. Lai and G. R. Her, *J. Chromatogr. A*, 1997, **779**, 195.
36. M. C. Tseng, Y. R. Chen and G. R. Her, *Anal. Chem.*, 2004, **76**, 6306.
37. Y. R. Chen, M. C. Tseng and G. R. Her, *Electrophoresis*, 2005, **26**, 1376.
38. Y. H. Hsueh, J. L. Huang, M. C. Tseng and G. R. Her, *Electrophoresis*, 2010, **31**, 1138.
39. F. Palmieri, *Biochim. Biophys. Acta*, 2008, **1777**, 564.
40. K. Heinig and J. Henion, *J. Chromatogr. B: Biomed. Sci. Appl.*, 1999, **735**, 171.
41. Y. Deng, J. Henion, J. Li, P. Thibault, C. Wang and D. J. Harrison, *Anal. Chem.*, 2001, **73**, 639.
42. K. Kosswig, Surfactants, in *Ullmann's Encyclopedia of Industrial Chemistry*, Wiley-VCH, Weinheim, 2005.
43. J. B. West, *Respiratory Physiology– the Essentials*, Williams & Wilkins, Baltimore, 1994.
44. D. Norton, J. Zheng, N. D. Danielson and S. A. Shamsi, *Anal. Chem.*, 2005, **77**, 6874.
45. D. Norton, S. A. Rizvi and S. A. Shamsi, *Electrophoresis*, 2006, **27**, 4273.
46. D. Norton and S. A. Shamsi, *Anal. Chem.*, 2007, **79**, 9459.
47. T. Hajri and N. A. Abumrad, *Annu. Rev. Nutr.*, 2002, **22**, 383.
48. M. A. Petersson, G. Hulthe and E. Fogelqvist, *J. Chromatogr. A*, 1999, **854**, 141.
49. J. H. Lee, S. J. Kim, S. Lee, J. K. Rhee, S. Y. Lee and Y. C. Na, *Anal. Chim. Acta*, 2017, **984**, 223.

50. I. Rodriguez and S. F. Y. Li, *Anal. Chim. Acta*, 1999, **383**, 1.
51. S. Hjerten, *J. Chromatogr.*, 1985, **347**, 191.
52. T. J. Thompson, F. Foret, P. Vouros and B. L. Karger, *Anal. Chem.*, 1993, **65**, 900.
53. Z. Wang and A. Prange, *Anal. Chem.*, 2002, **74**, 626.
54. M. A. Moseley, L. J. Deterding, K. B. Tomer and J. W. Jorgenson, *Anal. Chem.*, 1991, **63**, 109.
55. D. R. Goodlet, J. H. Wahl, H. R. Udseth and R. D. Smith, *J. Microcolumn Sep.*, 1993, **5**, 57.
56. E. Hardenborg, A. Zuberovic, S. Ullsten, S. L. Söderberg, E. Heldin and K. E. Markides, *J. Chromatogr. A*, 2003, **1003**, 217.
57. S. Garza, S. Chang and M. Moini, *J. Chromatogr. A*, 2007, **1159**, 14.
58. K. Harada, E. Fukusaki and A. Kobayashi, *J. Biosci. Bioeng.*, 2006, **101**, 403.
59. C. Huhn, R. Ramautar, M. Wuhrer and G. W. Somsen, *Anal. Bioanal. Chem.*, 2010, **396**, 297.
60. K. Martma, P. W. Lindenburg, K. L. Habicht, K. Vulla, K. Resik, G. Kuut and R. Shimmo, *J. Chromatogr. A*, 2013, **1317**, 193.
61. R. Sebastiano, C. Gelfi, P. G. Righetti and A. Citterio, *J. Chromatogr. A*, 2001, **924**, 71.
62. B. Arvidsson, N. Johannessen, A. Citterio, P. G. Righetti and J. Berquist, *J. Chromatogr. A*, 2007, **1159**, 154.
63. A. Elhamili, M. Wetterhall, B. Arvidsson, R. Sebastiano, P. G. Righetti and J. Berquist, *Electrophoresis*, 2008, **29**, 1619.
64. F. Zhu, W. C. Zhao, X. E. Zhao, Y. R. Suo, S. J. Liu and J. M. You, *Chin. J. Anal. Chem.*, 2007, **35**, 489.
65. F. H. Wang, X. J. Xiong, X. F. Guo, H. Wang and H. S. Zhang, *J. Chromatogr. A*, 2013, **1291**, 84.
66. E. Karelson, S. Fernaeus, K. Reis, N. Bogdanovic and T. Land, *J. Neurosci. Res.*, 2005, **79**, 368.
67. E. G. Bligh and W. J. Dyer, *Can. J. Biochem. Physiol.*, 1959, **37**, 911.

CHAPTER 12

Multisegment Injection–Capillary Electrophoresis–Mass Spectrometry: A Robust Platform for High Throughput Metabolite Profiling with Quality Assurance

PHILIP BRITZ-MCKIBBIN

Department of Chemistry and Chemical Biology, McMaster University, Hamilton, ON, L8S 4M1, Canada
*E-mail: britz@mcmaster.ca

12.1 Introduction

Recent advances in nontargeted metabolite profiling (*i.e.*, metabolomics) offer a revolutionary approach for phenotyping individuals at a molecular level, which is needed for new breakthroughs in precision medicine.[1] Metabolomics not only serves to identify and validate clinically relevant biomarkers of disease at early stages of latency,[2] but can also provide new insights into the mechanisms of disease pathophysiology,[3] predict disease risk outcomes in critically ill infants,[4] and reveal the phenotype of "silent" genetic

New Developments in Mass Spectrometry No. 6
Capillary Electrophoresis–Mass Spectrometry for Metabolomics
Edited by Rawi Ramautar

Published by the Royal Society of Chemistry, www.rsc.org

mutations.[5] The "metabolome" consists of all small molecules (<1.5 kDa) that define the functional activity of a cell or organism.[6] As a result, metabolites represent endogenous by-products of cellular metabolism, as well as xenobiotics and bioactive compounds derived from the diet, environment and gut microflora. For instance, the human urine metabolome consists of over 3000 chemically diverse compounds over a wide dynamic range with less than 15% comprising known metabolites annotated in public databases with reported concentrations.[7] Since metabolites are "real-world" end-products of gene expression and lifelong environmental exposures, metabolomics offers a promising approach for biomarker discovery.[8] Thus, metabolomics represents a powerful "hypothesis-generating" paradigm for scientific discovery that can evaluate the dynamics of gene–environment interactions when using NMR and increasingly high resolution and accurate mass spectrometry (MS) technology. Separation science plays an essential role in enhancing the performance of electrospray ionization (ESI)-MS, which is required for the resolution, identification and accurate quantification of metabolites in complex biological samples.[9]

The major goal of metabolomics is to measure statistically different metabolite responses between two or more sample cohorts based on their measured signal intensities. Complementary multivariate and univariate statistical methods are then used to reduce data dimensionality while ranking a sub-set of molecular features that are associated with group classification.[10] Unknown metabolites of clinical significance can then be subsequently identified by acquiring high resolution MS/MS with confirmation by comparison with authentic standards, thus allowing for their biochemical interpretation.[11] However, high efficiency separations based on liquid chromatography (LC), gas chromatography (GC), capillary electrophoresis (CE) and supercritical fluid chromatography (SFC) constitute a major bottleneck to high throughput metabolomic studies as compared to direct infusion/flow injection analysis-MS[12,13] or ultrafast ion mobility-MS strategies.[14] Indeed, sample throughput for global metabolite profiling is fundamentally limited by the time required for sample elution in order to achieve adequate peak capacity and metabolome coverage while using one or more column types. Separations also necessitate additional time to ensure adequate column conditioning/equilibration to reduce sample carry-over and elution time variations, including specific sample workup/sample pretreatment requirements (*e.g.*, chemical derivatization, liquid extraction) and data post-processing using algorithms for time alignment and peak picking.[15] Moreover, intermittent analysis of quality control (QC) samples is critical to reduce false discoveries and system drift when performing large-scale metabolomic studies.[16] Collectively, these issues represent a major throughput and cost barrier to widespread application of MS-based metabolomics in large-scale clinical or epidemiological studies,[17] especially when using multiple separation platforms to expand the metabolome coverage.[18] Indeed, fast analysis times (<5 min per sample) in LC-MS are typically reserved for targeted classes of metabolites, such as rapid characterization of polyphenols in red wine using data independent acquisition.[19] In all cases, conventional MS-based separations rely on a "single" sample injection prior to ionization per run, whereas optimization of gradient elution when using small analytical column dimensions at high flow rates and/or column switching programs in LC-MS offer only modest gains to overall sample throughput for selected metabolites.[20]

CE-MS offers excellent selectivity for resolving a wide range of highly polar and ionic metabolites[21] as compared to reversed-phase ion pairing LC[22] or hydrophilic interaction chromatography (HILIC),[23] which suffer from poor retention time reproducibility, complex separation mechanisms, skewed peak shapes and/or ion suppression effects. Additionally, CE-MS relies on isocratic elution using a homogeneous background electrolyte (BGE) together with steady-state ionization conditions when using a coaxial sheath liquid interface, including recent nanoflow interface designs with improved sensitivity.[24–26] These attributes allow for the design of customized serial sample injection formats to greatly boost sample throughput and data fidelity when using multi-segment injection-capillary electrophoresis-mass spectrometry (MSI-CE-MS) without complicated hardware modifications or added infrastructure costs.[27] Importantly, MSI-CE-MS offers a higher throughput platform (<2–3 min per sample) and accelerated data workflow for biomarker discovery with quality assurance that is critical to reduce the bias and false discoveries that have plagued earlier "-omics" research initiatives.[28] In this chapter, we outline the development and validation of MSI-CE-MS as a novel multiplexed separation format for metabolomics that enables unambiguous identification and reliable quantification of metabolites in volume-limited bio-banked samples (<3–5 μL), including dried blood spots (DBSs)[29] and sweat specimens from infants.[30] Moreover, recent developments in expanding metabolomic coverage in MSI-CE-MS to include ionic yet non-polar lipids will also be described when using non-aqueous buffer systems.[31] In all cases, MSI-CE-MS methods will be supported by rigorous method validation, preventative maintenance protocols, integrated QC/QA strategies, intermethod comparisons of quantitative performance and batch-correction algorithms for achieving high technical precision over long periods of operation. These technical advances are urgently needed to enable high throughput, and low cost, yet robust metabolomic studies when using CE-MS, which is amenable to population health studies,[32] as well as comprehensive drug surveillance testing as an alternative to two-tiered screening by immunoassays and LC-MS/MS methods.[33]

12.2 CE-MS in Metabolomics: Overcoming Poor Robustness

CE is a mature technology that has played key roles in genetic sequencing for early completion of the Human Genome,[34] automated clinical screening of hemoglobinopathies,[35] and tryptic peptide mapping for bottom-up proteomics,[36] with growing applications in the biopharmaceutical industry, including glycan analysis and intact therapeutic protein characterization.[37] The translation of robust CE assays for routine analysis has been contingent on inter-method comparison studies with previously validated methods (*e.g.*, MS/MS, GC/MS, ICP-MS),[29,38,39] as well as inter-laboratory or round robin studies[40,41] that are supported by proficiency testing with accredited laboratories/agencies (*e.g.*, the CDC).[29,39] Regrettably, rigorous method validation and long-term quality control/quality assurance (QC/QA) practices have not been routinely implemented in CE-MS metabolomic studies published to date, including evidence for accurate metabolite quantification. Pioneering work in metabolomics using CE-MS was first demonstrated by Soga and colleagues in the early 2000s for analysis of a wide range of cationic[42] and anionic[43,44] metabolites under positive and negative ion mode conditions, respectively. Although analysis of cationic metabolites in bare fused-silica capillaries using a strongly acidic aqueous background electrolyte (BGE) (*e.g.*, formic acid or acetic acid) is frequently used by several other research groups,[22] routine analysis of acidic metabolites by CE-MS has proven far more elusive and controversial. For instance, Büscher *et al.*[45] conducted a cross-platform comparison of the quantitative performance of GC/MS, LC-MS and CE-MS for analysis of primary metabolites from cell extracts, which reported that CE-MS methods using commercial kits with coated capillaries for acidic metabolites in negative ion mode suffered from poor robustness with frequent current errors. This report unfortunately perpetuated an impression among metabolomic research groups that CE-MS is not reliable for routine analysis by non-specialists. Soga *et al.*[46] later reported an improved method for anionic metabolite profiling by CE-MS when using a platinum needle sprayer to prevent the incidental corrosion of stainless steel emitters and capillary blockages when performing separations under reversed polarity; however, this technical advancement emerged seven years after their initial publications[43,44] highlighting the need for careful method validation and inter-laboratory studies at the earliest stages of method development. Inadequate method validation and the lack of preventative maintenance and QC/QA protocols represent common sources of bias contributing to false discoveries in published biomedical research, including selective reporting, inadequate study power and lack of replication.[47]

12.3 Capillary Preparation and Conditioning: Buffer Compatibility Issues in CE-MS

Migration time variations are widely claimed to be one of the major limitations of robust CE-MS methods in metabolomics due to changes in inner capillary surface properties that impact the electro-osmotic flow (EOF) between runs and between capillaries. Although separations under EOF-suppressed buffer systems ($pH < 2$) or use of dynamic coating methods to stabilize the

EOF have been reported (see also Chapter 3),[48] proper conditioning and rinsing of the fused-silica capillary is essential while ensuring temperature control of the CE-MS instrumentation, including the capillary inlet (above the sample tray) and emitter sections (prior to the ion source) that are often not actively thermostated. Time warping algorithms can also be applied to correct for non-linear changes in apparent migration times in CE-MS during data post-processing of large-scale metabolomic studies.[49] Alternatively, the electrophoretic mobility of a metabolite (a fundamental analyte parameter in CE as opposed to its apparent migration time) can be directly measured for greater reliability of CE-based analyses.[50] Unfortunately, these two procedures are not offered in vendor software used for data processing in commercial CE-MS instrumentation, which is largely derived from LC-MS platforms. Figure 12.1 provides an overview of a series of standardized procedures established in our laboratory for the preparation of polyimide-coated fused-silica capillaries prior to CE-MS metabolomic studies. This begins with careful capillary cutting *via* a diamond blade device followed by polyimide removal (7 mm length) at both ends of the capillary *via* a burner, and subsequent capillary polishing to ensure a smooth and reproducible surface. These steps are

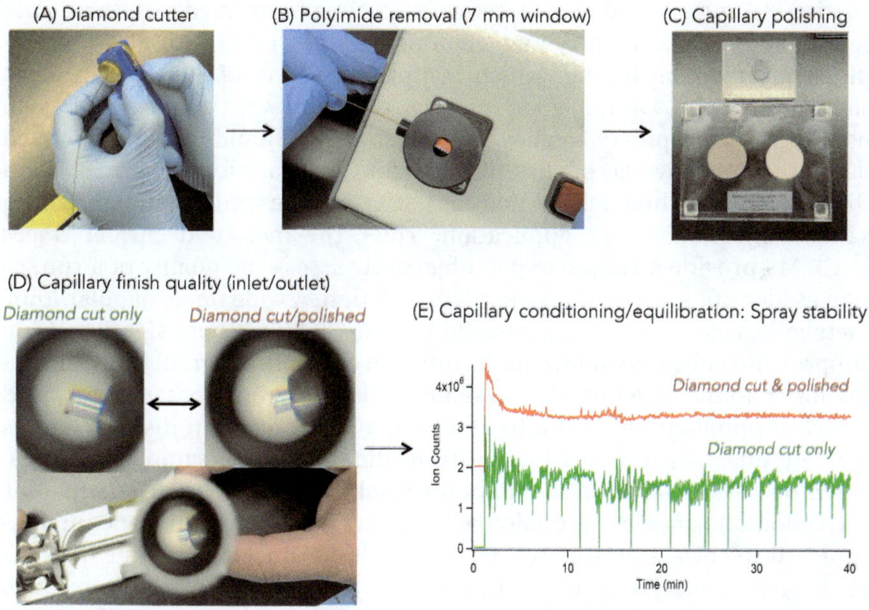

Figure 12.1 Standardized method for the preparation of polyimide-coated fused-silica capillaries for CE-MS, including (A) capillary cutting, (B) polyimide removal at sections of inlet/outlet, (C) capillary tip polishing and (D) visual inspection prior to (E) capillary conditioning and voltage application in background electrolyte. This approach is critical to robust performance in CE-MS by reducing sample carry-over effects, polyimide swelling or hydrolysis while producing a stable spray with low background noise.

critical to reduce sample carry-over effects following repeated hydrodynamic injections at the capillary inlet, as well as allowing for stable electrospray formation with low background noise when using a conventional coaxial sheath liquid interface in CE-MS. Similarly, commercial porous sprayer emitter tips developed for sheathless CE-MS applications with higher sensitivity[51] require careful handling and long-term testing given their inherent fragility with potential batch variations to ensure robust performance. In all cases, adequate capillary conditioning is needed for equilibration of the capillary inner surface, which is achieved by repeated rinsing and separation of representative samples (*e.g.*, quality controls) under an applied voltage prior to analysis of randomized batches of samples, including blanks and QC samples as controls.

Although polyimide swelling of capillaries has long been reported as a major constraint to the robustness of non-aqueous CE (NACE) when using organic solvent-rich electrolyte solutions (*e.g.*, acetonitrile) to better solubilize hydrophobic drugs,[52] aqueous alkaline buffers containing ammonia (ammonium acetate or ammonium bicarbonate, pH > 9.0) widely used as volatile electrolytes in CE-MS have recently been shown to contribute to incidental capillary fractures *via* irreversible aminolysis of the outer polyimide coating.[53] Figure 12.2 depicts a series of 42 runs performed over three days for anionic metabolite profiling when using MSI-CE-MS with a seven sample serial injection format, which comprised a total of 238 human urine samples analyzed. Overall, good inter-day precision was demonstrated when using different capillary batches and buffer/sheath liquid preparations each day based on an overlay of current traces for CE separation (CV = 2.5%) with the exception of three failed runs caused by capillary misalignment during sample injection/voltage application. Thus, the measured current traces in CE-MS provide a simple way to objectively assess the quality of a run, as well as identify sources of bias for data rejection. A wide range of polar/ionic metabolites can be directly measured by MSI-CE-MS without significant ion suppression effects, as shown in the total ion electropherogram (TIE), including lower abundance urinary metabolites shown in representative extracted ion electropherograms (EIE). In all cases, a series of seven distinct signals that are offset in time originating from different urine samples are measured by ESI-MS, which also allows for resolution of isobaric/isomeric ions with good intermediate precision while using a single non-isotope internal standard to correct for differences in on-column injection volume (*i.e.*, relative ion response) as reflected by their corresponding control charts, such as the uremic toxin, indoxyl sulfate.[53] Although targeted analysis of metabolites by CE-MS with matching stable-isotope internal standards can further improve precision and accuracy, this approach is not feasible when performing discovery-based metabolomics unless using chemical isotope-labeling strategies for relative quantification.[54] The images of exposed capillaries demonstrate distortion/slipping of the polyimide coating and its eventual dissolution when in contact with an aqueous ammonia buffer for separation

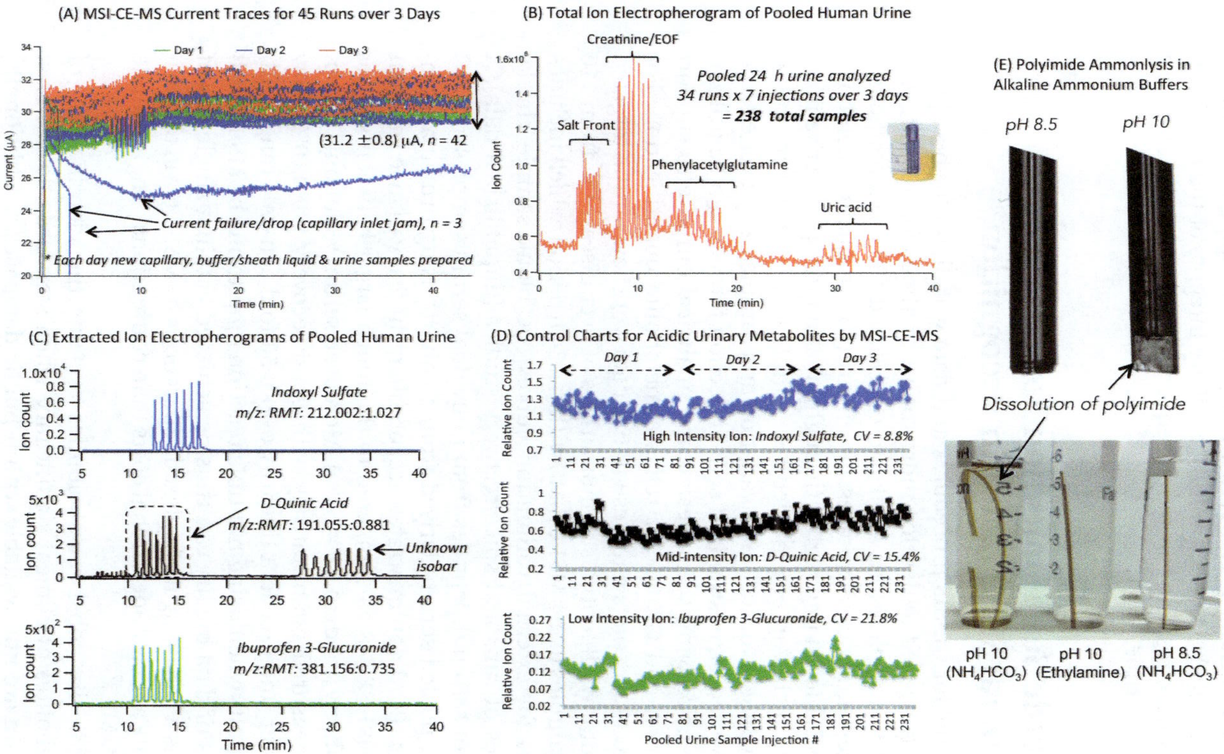

Figure 12.2 The long-term performance of MSI-CE-MS for routine metabolite profiling of human urine with higher sample throughput and quality assurance, including (A) overlay of current traces highlighting consistent inter-day reproducibility, (B) representative total ion electropherogram of the urine metabolome under negative ion mode conditions, (C) extracted ion electropherograms for three acidic urinary metabolites together with their (D) corresponding control charts over three days of operation, and (E) irreversible polyimide degradation by ammonia under aqueous alkaline conditions (pH > 9) contributing to capillary fractures, which is pH and electrolyte dependent. Reproduced with permission from ref. 53. Copyright (2016) American Chemical Society.

(pH 10), unlike non-reactive secondary amines (ethylamine, pH 10) and less alkaline ammonia buffers (pH = 8.5), where ammonium (NH_4^+) is the predominate species in solution. Caution is also needed when using moderately caustic ammonia solutions together with organic solvents as the sheath liquid in CE-MS, which can compromise capillary integrity at the distal end near sprayer interface resulting in more frequent capillary breakages with a reduced average lifetime.

12.4 Versatile Serial Injection Formats with Temporal Signal Pattern Recognition

Detailed protocols for CE-MS procedures and data workflows for best practices are urgently needed to develop more robust methods for non-targeted metabolite profiling applications similar to established UPLC-MS methods.[55] MSI-CE-MS offers unique advantages for ensuring high quality metabolomic data that is achieved in a higher throughput and more productive manner, such as the inclusion of a pooled QC sample within each run and/or blank control to assess for technical variation and sample carry-over effects, respectively.[27] Moreover, customized serial injection formats can be utilized that reflect the underlying study design and analytical purpose to support method validation and QC/QA for reliable metabolite quantification. For instance, Figure 12.3 depicts a representative data workflow for the targeted analysis of γ-glutamyl dipeptides as biomarkers of liver disease[56,57] from 116 serum extract samples using MSI-CE-MS within a 24 h period, which includes additional runs for external calibration curves, QC/blank sample checks between batches of runs and spike/recovery studies (unpublished data). A 2D scores plot of serum dipeptides from individual serum samples and QC specimens (*i.e.*, pooled serum derived from a cohort) using principal component analysis (PCA) demonstrates acceptable technical variance (median CV = 13%) of MSI-CE-MS as compared to the much larger between-subject biological variability (median CV = 60%). Moreover, serum γ-glutamyl dipeptides can be measured quantitatively with good recovery (90–110%), low sample carry-over, and excellent linearity over a 50-fold dynamic range with low detection limits when using multiple reaction monitoring by MS/MS. All these attributes of data quality can thus be achieved within a single day of continuous runs when using multiplexed separations by MSI-CE-MS to maximize productivity at lower operational costs than conventional single sample injections in CE-MS.[57]

A major hurdle in global metabolomics remains the identification of unknown metabolites of clinical significance when standards do not exist or MS/MS spectra are not available within public databases.[58] Additionally, a major fraction of detected molecular features when using ESI-MS (>90%) represent spurious peaks, background signals and redundant signals (*e.g.*, in-source fragments, isotopic signals, salt adducts) when performing

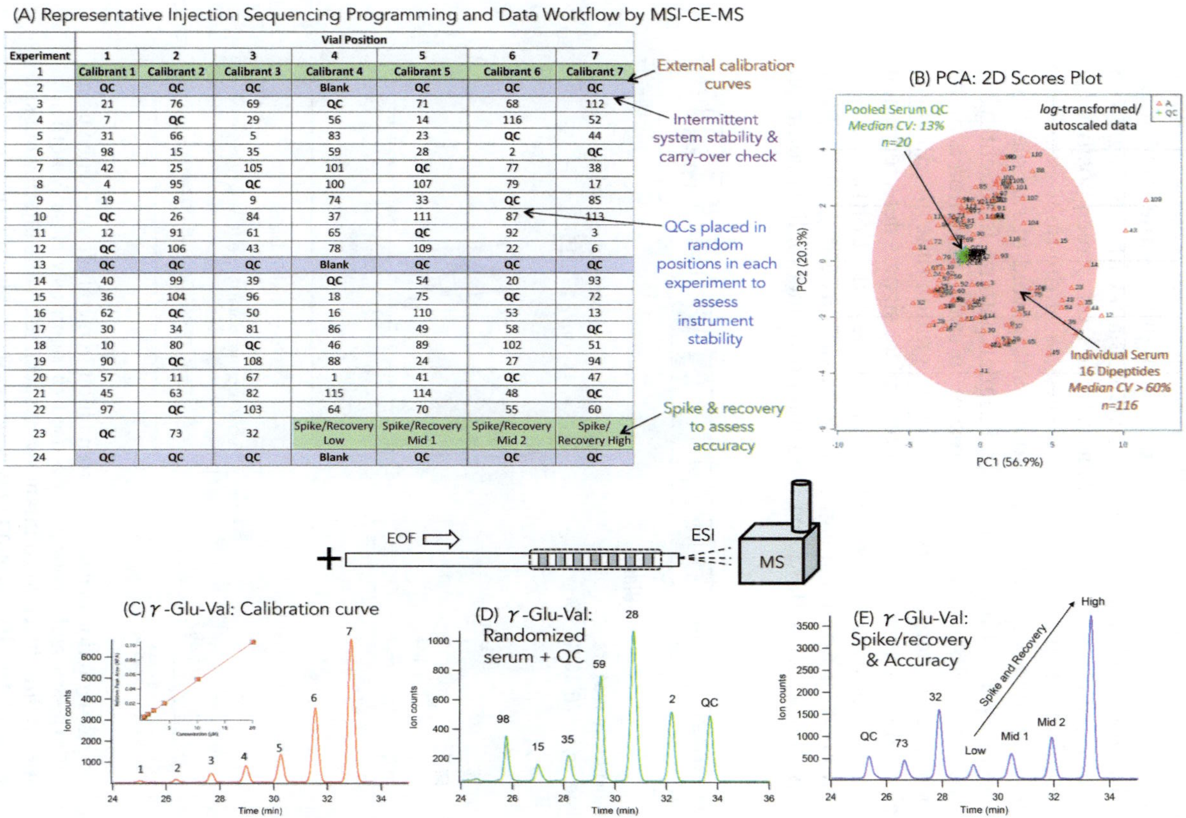

Experiment	Vial Position 1	2	3	4	5	6	7
1	Calibrant 1	Calibrant 2	Calibrant 3	Calibrant 4	Calibrant 5	Calibrant 6	Calibrant 7
2	QC	QC	QC	Blank	QC	QC	QC
3	21	76	69	QC	71	68	112
4	7	QC	29	56	14	116	52
5	31	66	5	83	23	QC	44
6	98	15	35	59	28	2	QC
7	42	25	105	101	QC	77	38
8	4	95	QC	100	107	79	17
9	19	8	9	74	33	QC	85
10	QC	26	84	37	111	87	113
11	12	91	61	65	QC	92	3
12	QC	106	43	78	109	22	6
13	QC	QC	QC	Blank	QC	QC	QC
14	40	99	39	QC	54	20	93
15	36	104	96	18	75	QC	72
16	62	QC	50	16	110	53	13
17	30	34	81	86	49	58	QC
18	10	80	QC	46	89	102	51
19	90	QC	108	88	24	27	94
20	57	11	67	1	41	QC	47
21	45	63	82	115	114	48	QC
22	97	QC	103	64	70	55	60
23	QC	73	32	Spike/Recovery Low	Spike/Recovery Mid 1	Spike/Recovery Mid 2	Spike/ Recovery High
24	QC	QC	QC	Blank	QC	QC	QC

Figure 12.3 Overview of the data workflow in MSI-CE-MS for efficient analysis of large batches of serum extracts ($n = 116$) with quality assurance over a 24 h period, including (A) a randomized serial injection program comprising six serum extracts and one quality control (QC) sample in each run, (B) a data summary using principal component analysis (PCA) comparing the magnitude of biological to technical variance as determined by repeated analysis of a randomized QC for each run, as well as extracted ion electropherograms for the γ-glutamyl dipeptide, γ-Glu-Val, derived from runs based on (C) an external calibration curve, (D) randomized serum extracts with QC and (E) spike-recovery studies at four concentration levels.

non-targeted metabolomics of complex biological samples, such as bacterial extracts. Rigorous data annotation with accurate mass, retention time and MS/MS reference spectra matching may be applied for known metabolites;[59] alternatively, a dilution trend filter may be applied to identify reproducible yet authentic/non-background metabolites in samples not in blank samples, including unknown metabolites.[60] Figure 12.4 highlights the implementation of a dilution trend filter when using MSI-CE-MS within a single run based on serial injection of a pooled sweat sample from screen-positive CF infants,[30] including a triplicate injection at the lowest dilution, as well as several samples at increasing dilution and a blank devoid of sample. This approach generates a distinctive temporal signal pattern for identifying authentic yet reproducible metabolites present in sweat as denoted by its characteristic mass-to-charge ratio and relative migration time (*m*/*z*:RMT) as shown for citrulline. In contrast, spurious or artifact signals detected when using peak picking algorithms (*e.g.*, molecular feature extraction) are readily rejected based on their non-ideal signal pattern, such as a large fraction of unknown ions (*m*/*z* 178.1588). Known and unknown metabolites common

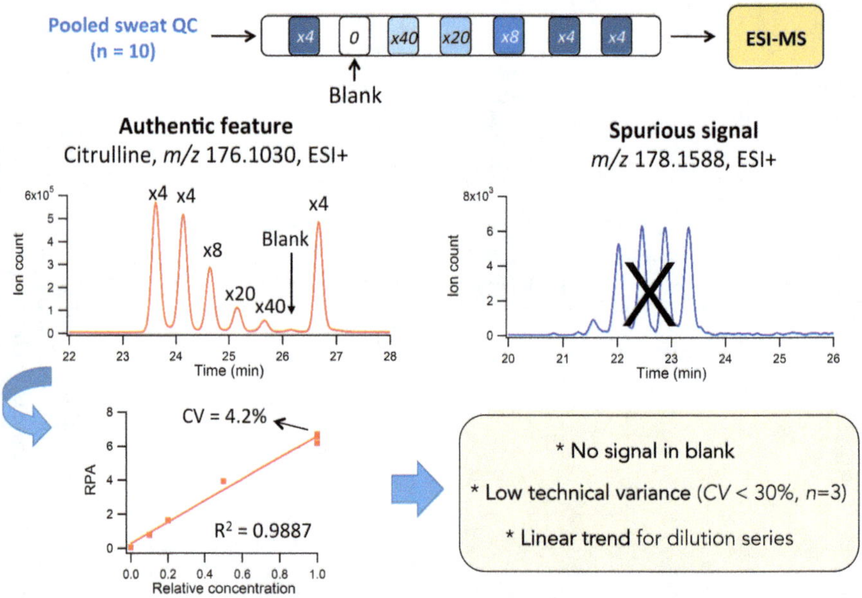

Figure 12.4 Implementation of a dilution trend filter in metabolomics when using MSI-CE-MS and major criteria used to rigorously classify authentic metabolites in a pooled infant sweat specimen while filtering out spurious signals, background ions and irreproducible molecular features. Reproduced with permission from ref. 30 (http://pubs.acs.org/doi/abs/10.1021/acscentsci.7b00299). Copyright (2017) American Chemical Society. Further permissions related to the material excerpted should be directed to the ACS.

to the study cohort are thus rigorously selected as they lack a significant signal in the blank while demonstrating acceptable precision for screening (CV < 30–40%) and a linear signal reduction upon dilution. This method offers a rapid, convenient and efficient process for filtering out artifact and irreproducible signals at the earliest stages of the metabolomics data workflow prior to multivariate data analysis. Furthermore, data post-processing is further streamlined since a targeted extraction of pre-filtered metabolites of unknown clinical significance can be performed as part of an accelerated data workflow for biomarker discovery.[61] This is a critical element in metabolomics as a way to go beyond differentiating signals of validated biomarkers with known chemical structures and established concentrations/reference ranges that demonstrate clinical utility for the screening, diagnosis or prognosis of treatable human diseases.[62]

Sample pretreatment represents an important pre-analytical step and potential source of bias in nearly all metabolomic studies, notably for the analysis of heterogeneous specimens, such as solvent extraction of skeletal muscle tissue biopsies.[63] Although various organic solvent combinations and pH conditions have been explored to maximize metabolite recovery in various sample types, the Bligh–Dyer extraction method[64] remains one of the most widely used liquid extraction procedures that allows for fractionation of polar metabolites (top aqueous methanol layer) from lipids (bottom chloroform layer) following centrifugation. However, different extraction procedures display differences in metabolome coverage, extraction efficiency, reproducibility and suitability for high throughput screening, which is also dependent on the specific specimen type. Figure 12.5 depicts a five-segment serial sample injection format by MSI-CE-MS that was applied to maximize the recovery of polar/ionic metabolites from lyophilized muscle tissue, which were exhaustively extracted using four repeated methanol : water portions as compared to a blank extract (unpublished data). Overall, excellent recovery exceeding 90% was achieved for a wide range of intra-cellular compounds when pooling together the first two extracts, including cationic (*e.g.*, acetylcarnitine) and anionic (*e.g.*, phosphocreatine) metabolites associated with energy metabolism. In all cases, muscle-derived metabolites were quantified and identified by MS/MS together with spiking with authentic standards with measured concentrations normalized to dried mass. Additionally, this approach allowed for efficient extraction of polar/ionic metabolites with no residual sample carry-over (*i.e.*, blank extract), including labile metabolites prone to oxidation (*e.g.*, reduced glutathione) and hydrolysis (*e.g.*, ATP) when performing the Bligh–Dyer extraction using pre-chilled solvents with sample workup (*i.e.*, ultracentrifugation) at +4 °C.

Although various organic solvent combinations, temperatures and pH conditions have been explored to maximize metabolite recovery (*e.g.*, Bligh–Dyer extraction), methyl *tert*-butyl ether (MTBE) has been proposed as an alternative extraction solvent for lipids from biological samples since

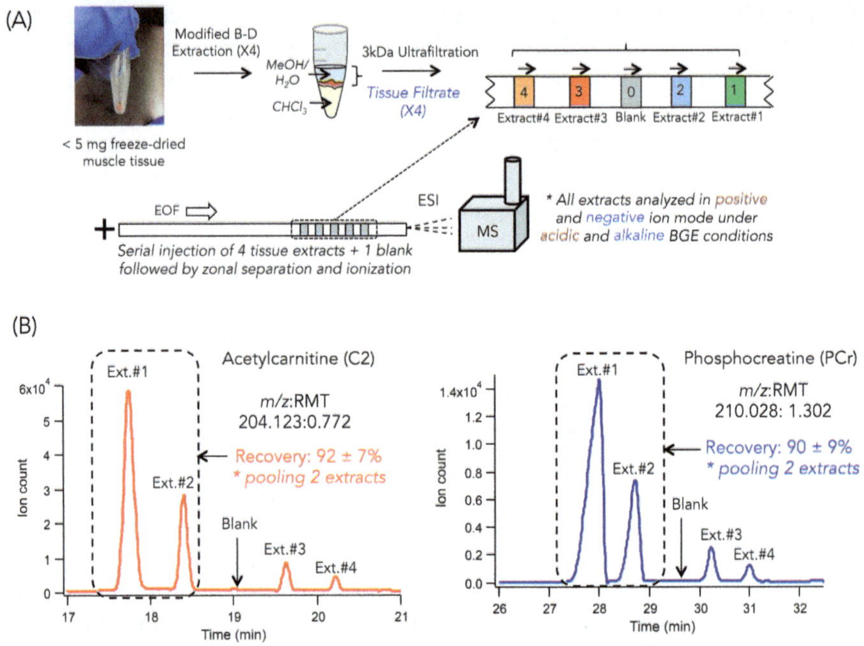

Figure 12.5 (A) Application of a 5-sample plug serial injection format in MSI-CE-MS as a way to optimize the recovery of cationic and anionic intracellular metabolites from a muscle tissue extract relative to a blank as annotated based on their characteristic *m*/*z*:RMT. (B) Overall, a quantitative recovery >90% was achieved by pooling together the first two aqueous methanol fractions of a lyophilized muscle tissue specimen based on a modified Bligh–Dyer method.

protein is sedimented at the bottom of the aqueous layer with the ether layer accessible on the top layer. In fact, such an approach has been proposed as a convenient way to enable a single "in-vial" extraction of lipids with subsequent proteomic analysis of human high density and low-density lipoprotein fractions when using LC-MS.[65] To date, CE-MS metabolomic studies have largely been assumed to be amenable to hydrophilic/ionic metabolites when using aqueous BGE systems, such as organic acids, nucleotides, amino acids and sugar phosphates. Indeed, there have been a few precedents of using NACE or NACE-MS for the routine analysis of ionic yet hydrophobic/surface-active metabolites, such as long-chain acylcarnitines,[66] phospholipids,[67] and non-esterified fatty acids;[68,69] however, these methods have not been extensively validated nor routinely used in CE-MS-based metabolomic studies, which rely on conventional reversed-phase LC-MS protocols.[32] Figure 12.6 highlights that high throughput screening of free fatty acids can be achieved when using MSI-NACE-MS with a non-aqueous buffer system in conjugation with standardized preparation of bare fused-silica capillaries to prevent deleterious polyimide swelling (Figure 12.1), including various long-chain polyunsaturated fatty acids (unpublished data). Moreover, this approach can

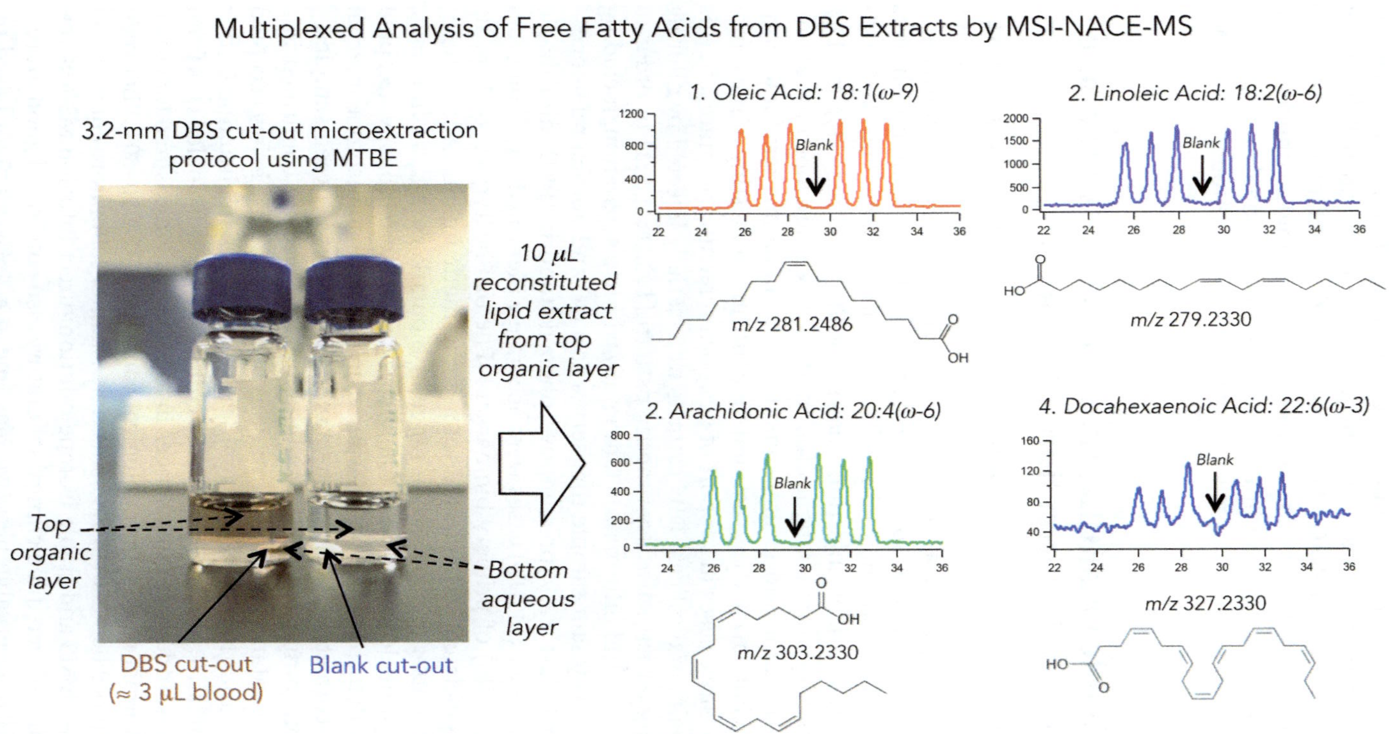

Figure 12.6 Development of a robust method for free (non-esterified) fatty acid analysis by MSI-NACE-MS when using a non-aqueous background electrolyte system from a single 3.2 mm diameter dried blood spot cut-out. Use of methyltertbutylether (MTBE)-based extraction allows for quantitative recovery of non-polar yet ionic long-chain polyunsaturated fatty acids from the top layer of extract without sample carry-over measured in the blank.

be coupled to a microscaled MTBE extraction format when using 3.2 mm diameter dried blood spot cut-outs (≈3 μL of whole blood) without sample carry-over in the blank extract. Current work in our laboratory is aimed at rigorously validating MSI-NACE-MS to enable high speed, precise, accurate yet robust analysis of fatty acids and other classes of lipids without complicated sample workup (*i.e.*, chemical derivatization) as required for large-scale population health studies that rely on conventional low throughput GC/MS methods.[70]

12.5 Reliable Quantification with Quality Assurance and Batch Correction Adjustment

A major advantage of CE-MS is that high efficiency separations are realized with effective desalting prior to ionization as required for analysis of highly saline biological samples with minimal ion suppression or enhancement effects. This feature allows for reliable quantification of ionic metabolites even when stable isotope internal standards are not available[71] unlike LC-MS and direct infusion/flow injection analysis MS/MS methods. Figure 12.7 depicts a serial injection configuration in MSI-CE-MS developed for unambiguous identification and accurate quantification of biomarkers for in-born errors of metabolism (IEM) as required for confirmatory testing of presumptive screen-positive cases in support of newborn screening programs. In this case, three pairs of duplicate DBS extracts were serially injected by diluting samples using a distinctive asymmetric pattern (1 : 2; 1 : 1; 2 : 1) in order to encode mass spectral information temporally within a separation, including a pooled healthy infant sample that serves as a QC and normal reference.[29] As shown in Figure 12.7, presumptive phenylketonuria (PKU) and maple syrup urine disease (MSUD) cases were accurately identified based on gross elevations (>4-fold) of primary biomarkers (Phe or Phe/Tyr; Val or Leu, Ile or *allo*-Ile) associated with affected yet asymptomatic infants. Additionally, absolute concentrations for a wide range of biomarkers (*e.g.*, amino acids, acylcarnitines, nucleosides) were found to exceed internationally accepted cut-off limits defined for each IEM when using a single non-isotope internal standard for correction of injection volume differences between samples.[29] However, a presumptive case of citrullinemia (CIT) was not confirmed in a screen-positive infant since concentrations of citrulline were below accepted cut-off levels, and there were no significant differences in citrulline in this case sample relative to two other disease controls, as well as pooled QCs from healthy infants. Indeed, minor changes in ion responses measured for citrulline reflect normal underlying biological variation (CV ≈ 50–60%) between newborn infants when using DBS collection. An inter-laboratory comparison between MSI-CE-MS and stable isotope dilution direct infusion-MS/MS performed within an accredited clinical laboratory also confirmed good mutual agreement with acceptable negative bias (mean = 46%, n = 320) based on retrospective analysis of DBS cut-out specimens due to variations in blood volume and storage.[29]

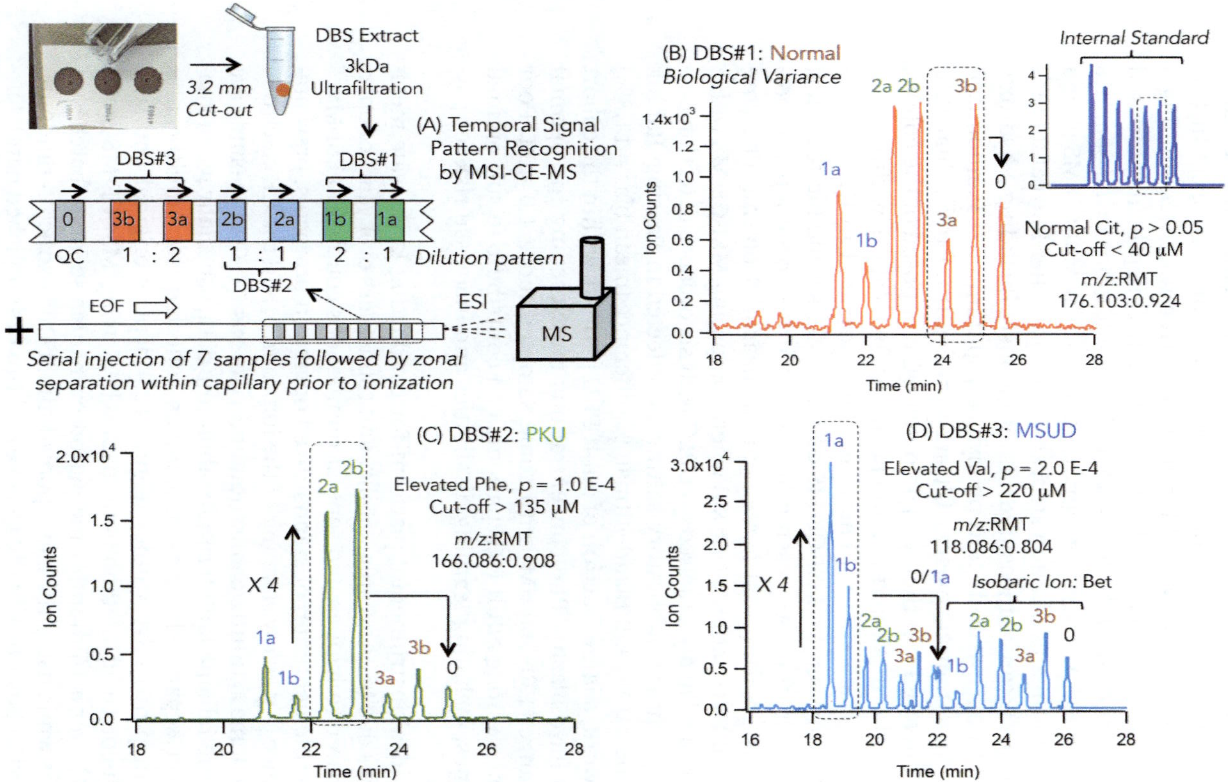

Figure 12.7 MSI-CE-MS in conjunction with temporal signal pattern recognition for unambiguous identification and accurate quantification of biomarkers associated with in-born errors of metabolism (IEM) from a dried blood spot (DBS). (A) Each run consists of a duplicate injection of a DBS extract from a presumptive IEM diluted in a distinctive pattern (1:2; 1:1; 2:1) together with a pooled QC (*i.e.*, healthy infant) as a reference. (B) No evidence of citrullinemia as shown by a lack of gross elevations in citrulline in three pairs of samples relative to the QC, unlike (C) phenylalanine and (D) valine for confirmed cases of phenylketonuria and maple syrup urine disease, whose concentrations exceed accepted cut-off limits. Reproduced with permission from ref. 29. Copyright (2017) American Chemical Society.

Similarly, Figure 12.8 demonstrates that routine therapeutic monitoring of patients with confirmed PKU can be realized by MSI-CE-MS with quality assurance (unpublished data), which is important to ensure proper dietary compliance (*i.e.*, protein restriction, amino acid supplements) to regulate circulating Phe levels within recommended therapeutic ranges.[72] In this case, all plasma specimens were diluted in deionized water (containing a recovery standard, 4-fluoro-phenylalanine) and subsequently ultrafiltrated (3 kDa MWCO) prior to MSI-CE-MS analysis. In addition, urine samples from a subset of PKU patients were analyzed directly by MSI-CE-MS after a simple centrifugation and dilution step, since several other Phe catabolites associated with PKU can be measured simultaneously, including *o*-hydroxyphenylacetate, phenylpyruvate and phenylacetylglutamine.[73] Randomized serial sample injections of plasma filtrate samples from individual PKU patients were analyzed together with a pooled QC composed of the cohort, which were placed at different injection positions for each run within a data workflow (Figure 12.4). Control charts for Phe measured in QC samples included with every run by MSI-CE-MS demonstrate acceptable technical variance (CV < 10%) for both plasma filtrate and urine specimens. Importantly, there was good mutual agreement in measured plasma Phe concentrations from PKU patients ($n = 84$) when comparing MSI-CE-MS with a validated amino acid analyzer based on UPLC with absorbance detection performed within a clinical laboratory setting as reflected in Passing–Bablok regression (slope = 0.78) and Bland–Altman % difference (mean bias = 14.8%) plots. In our work, a large fraction of adult PKU patients within the cohort was found to have plasma Phe concentrations far exceeding the optimal therapeutic range (120–360 μM) with some samples above 1.0 mM. Poor diet compliance among adult PKU patients can contribute to sub-optimal health outcomes, including learning disabilities and increased risk for psychiatric illness.[74]

NMR is an ideal instrumental platform for conducting large-scale metabolomic studies due to its greater stability and robustness over years of operation as compared to hyphenated separation methods coupled to ESI-MS.[75] In this case, implementation of standardized operation protocols, preventative maintenance (*e.g.*, daily cleaning of the ion source, and the CE electrode at the inlet *etc.*), mass tuning/calibration and rigorous QC/QA measures are critical to ensure reliable long-term metabolomic data, especially when analyzing data over separate batches of samples collected from multi-centre retrospective or prospective studies. Figure 12.9 depicts original and subsequent batch-corrected metabolomic data when using MSI-CE-MS (refer to Figure 12.7) over a 16 month time period over three separate batches of DBS specimens while using the same pooled infant DBS extract in randomly assigned injection positions as the QC in every run (unpublished data). Furthermore, during this same time interval, the QTOF-MS instrument used to acquire metabolomic data from screen-positive cystic fibrosis (CF) ($n = 94$) and screen-negative ($n = 34$) CF infants in full-scan operation under positive and negative ion modes was serviced several times, and required replacement

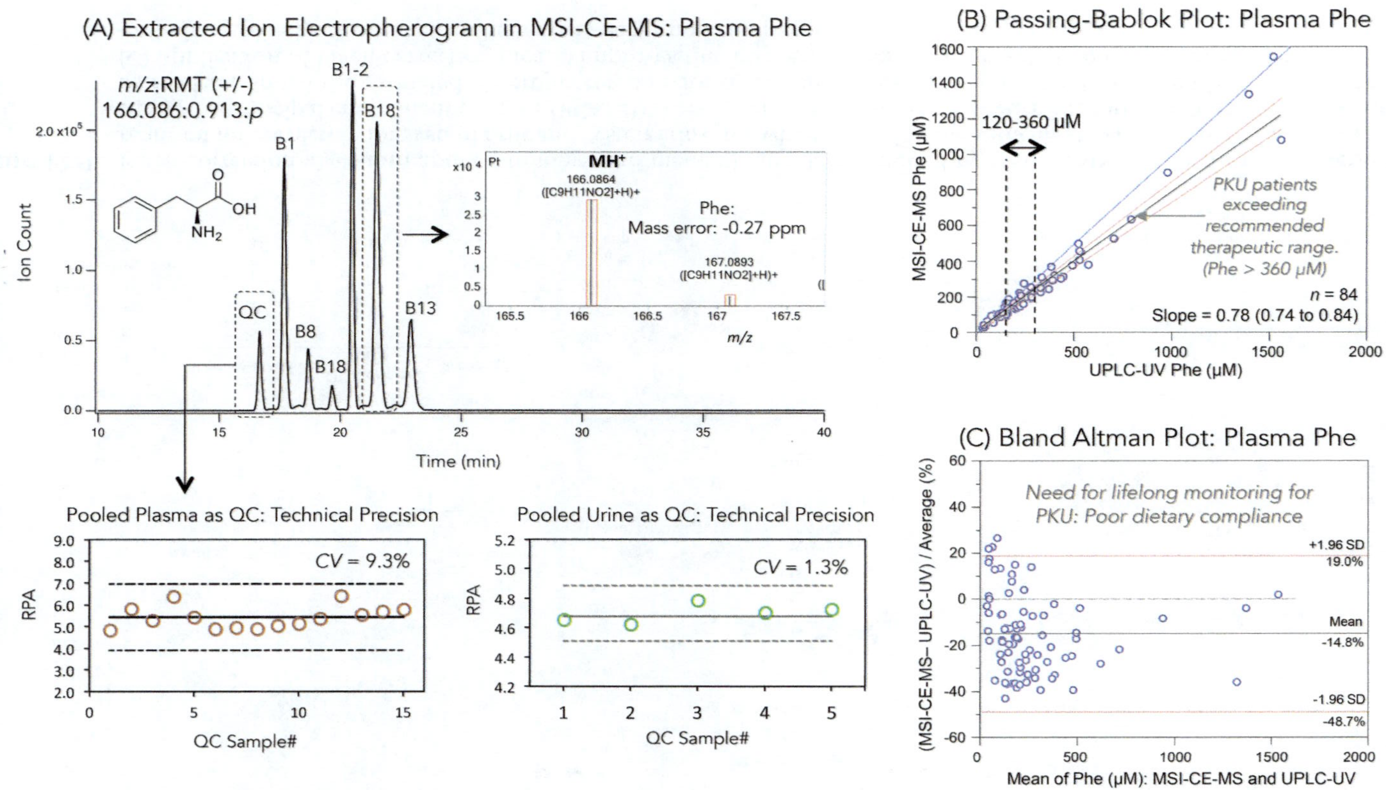

Figure 12.8 Inter-method comparison of MSI-CE-MS for accurate determination of plasma or urinary phenylalanine (Phe) as compared to UPLC as required for therapeutic monitoring of phenylketonuria (PKU) patients. (A) Representative extracted ion electropherogram for Phe from six PKU subjects together with a pooled QC in each run, including their respective control charts. Good mutual agreement in independently measured plasma Phe concentration levels was concluded when comparing MSI-CE-MS with UPLC-UV with a minor negative bias (14.8%) as reflected by (B) Passing–Bablok regression analysis and (C) Bland–Altman % difference plots.

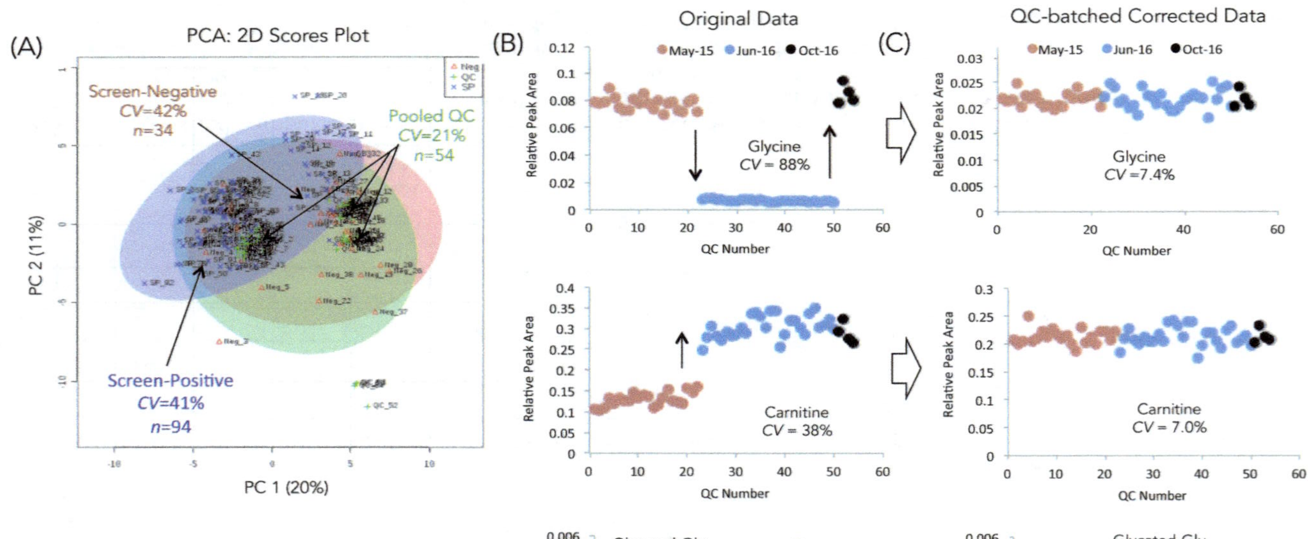

	Mean CV (%)		
	Uncorrected RPA	QC-Corrected RPA	% change
CF-positive	52	47	-11
Carriers	35	35	-1.6
Hypertrypsinogenic	37	37	-1.3
Screen-Negative	42	38	-11
QC	21	11	-66

Figure 12.9 Batch-correction algorithm applied to long-term metabolomic data acquired by MSI-CE-MS over an 18 month period as required for correcting for system drift and systematic error when analyzing the same quality control (QC) sample in every run. (A) A principal component analysis (PCA) 2D scores plot revealed three distinctive clustering patterns for QC samples as reflected in (B) the extracted electropherograms for three representative metabolites from dried blood spot extracts. (C) Application of batch-correction greatly improves the long-term precision of metabolomic data (D) without changes in underlying biological variance for better between-group comparisons.

of various electronic boards and the detector. A PCA 2D scores plot was first used as an unsupervised multivariate method to detect potential batch effects caused by system drift as reflected by a high median variance for technical replicates (CV = 21%) together with clustering of all QC samples (n = 54) into three distinct groups. Representative control charts for three metabolites measured from DBS extracts of pooled QC specimens revealed systematic changes in ion responses (normalized to internal standard) between batches of runs that were then subjected to a cluster-based batch normalization algorithm to increase overall data quality.[76] Indeed, within and between-batch variations in signal intensity in large-scale MS-based metabolomic studies contribute to greater noise and systematic bias, which lower statistical power to detect small treatment effects or metabolites of clinical significance.[77] Overall, batch-corrected data was found to significantly improve the technical precision (median CV = 11%) for QC samples by over 66% (notably for metabolites with large between-batch effects, such as glycine and carnitine) with minimal changes to the underlying data structure and intrinsic biological variance. MSI-CE-MS in conjunction with integrated QC/QA and data management strategies represents a powerful platform to conduct high throughput yet high quality metabolomic studies for large-scale epidemiological studies.

12.6 Applications of MSI-CE-MS: Biomarker Discovery in Clinical Metabolomic Studies

Multiplexed separations by MSI-CE-MS are ideally suited for the validation of the salutary benefits of lifestyle modifications relevant to human health when using well-designed time-course studies involving exercise and/or diet. For instance, between-subject variations to oral glucose challenge tests following adaptive changes to high intensity interval training were revealed based on time-dependent metabolomic changes within a single run, including plasma thiol redox and ornithine status.[78] Figure 12.10 demonstrates that time-resolved metabolomic studies by MSI-CE-MS can be designed to readily identify biomarkers associated with strenuous exercise and adaptive responses to 6 weeks of high intensity interval training (HIT) on an individual level.[61] In this case, time-dependent changes in plasma metabolome for a cohort of untrained/overweight women (*i.e.*, naïve status) revealed significant elevations in hypoxanthine (up to 116-fold) post-exercise (20 min) and following recovery (40 min) relative to baseline at rest (0 min), which is a biomarker of energetic stress derived from the purine degradation pathway due to irreversible ATP depletion within contracting muscle tissue.[61] Moreover, unambiguous treatment effects can be measured for each subject within a single run as reflected by differential attenuation of accumulation of hypoxanthine post-exercise, which was not dependent on body composition (*BMI*) and likely reflected baseline differences in cardiorespiratory fitness, as well as other factors. In all cases, a pooled QC sample was included as the fourth

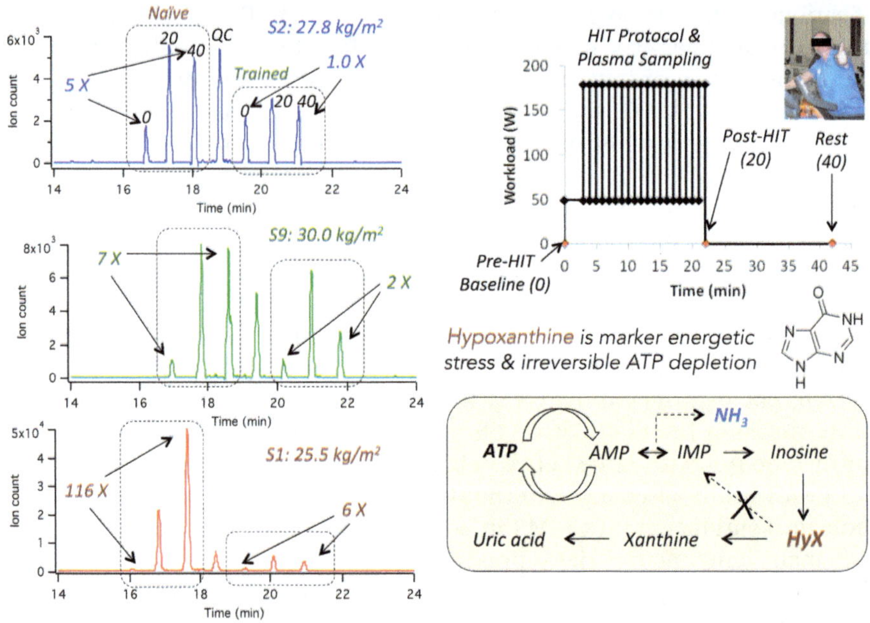

Figure 12.10 MSI-CE-MS for dynamic metabolomic studies of high intensity interval training (HIT) showing representative extracted ion electropherograms for plasma hypoxanthine (HyX) from three subjects collected at three time intervals (0 = baseline; 20 = post-exercise; 40 = recovery/rest) in their naïve and trained states. Both temporal changes to high intensity exercise and an individual's training responsiveness are reflected by a significant accumulation or attenuation in plasma HyX post-exercise, respectively. Reproduced with permission from ref. 62. Copyright 2015 WILEY-VCH Verlag GmbH & Co. KGaA, Weinheim.

sample injection segment in all runs, which also served as a reference to compare dynamic metabolomic changes to high intensity exercise for subjects in their naïve and trained states when performing standardized HIT regimes.

MS/MS technology has revolutionized expanded newborn screening programs by enabling high throughput screening of panels of biomarkers for rare yet treatable IEM with high sensitivity and specificity at incremental costs.[79] However, certain IEM still rely on classical enzyme kinetic assays, immunoassays and/or genetic mutation screens, such as a two-tiered algorithm based on immunoreactive trypsinogen and a *CFTR* mutation panel for population-based CF screening. Nevertheless, confirmatory testing of presumptive screen-positive CF cases is essential due to their low positive predictive value since over 85% of infants are subsequently determined as unaffected based on low sweat chloride (<29 mM). Although affected CF infants have elevated sweat chloride exceeding 60 mM, a sub-set of screen-positive CF cases have been shown to have intermediate sweat chloride with ambiguous diagnosis.[80] Our laboratory recently characterized the sweat metabolome when using MSI-CE-MS from screen-positive CF infants in order to better differentiate affected (n = 18) from unaffected (n = 50) infants beyond sweat chloride.[30] Figure 12.11 depicts a supervised orthogonal partial least squares-discriminate analysis (OPLS-DA) model with permutation testing to identify and rank sweat metabolites associated with CF disease status among asymptomatic infants at CF clinics from two regional pediatric hospitals. Unexpectedly, two exogenous metabolites were excreted at lower concentrations in the sweat of CF infants relative to screen-positive yet unaffected controls, which were subsequently identified as pilocarpic acid and monoethylhexylphthalate (MEHP). Both sweat-derived metabolites were demonstrated to be significantly depleted in CF infants after multi-hypothesis/Bonferroni correction with good discriminatory performance (area under the curve or AUC > 0.750) as reflected by receiver operating characteristic (ROC) curves. This work inferred that CF affected infants are prone to an enzyme deficiency in human paraoxanase, a non-specific arylesterase/lactonase that mediates lipid and xenobiotic metabolism, as well as regulating biofilm formation, and recurrent lung infections that are a hallmark of CF in older children/adults.[30] Thus, nontargeted metabolic phenotyping of sweat specimens can provide new mechanistic insights into the pathophysiology of CF at early stages of development during infancy, including novel therapeutic targets for treatment.

Given the worldwide opioid crisis and abuse/misuse of highly addictive prescription drugs, new methods are urgently needed for comprehensive surveillance of drugs of abuse beyond targeted approaches based on two-tiered screens that rely on immunoassays and confirmatory testing by LC-MS/MS.[81] Broad spectrum screening of drugs of abuse with high resolution full-scan data acquisition is also increasingly needed for identification of novel psychoactive substances and designer drugs in forensic toxicology.[82] Immunoassays are not only prone to cross-reactivity and false-positives in complex biological samples such as human urine, but they also suffer from batch variability and are not suitable for identification of emerging designer drugs when antibody reagents are unavailable. Figure 12.12 demonstrates that sample throughput can be further extended to 10 serial sample injection segments (<2–3 min per sample) when using a modified pressure-assisted MSI-CE-MS methodology, where most drugs of abuse (*e.g.*, amphetamines, opioids, stimulants) migrate within 20 min, such as cocaine and norfentanyl.[33] Moreover, inclusion of matching deuterated internal standards in diluted urine samples allows for unambiguous identification of drugs of abuse and their metabolites based on their co-migration, which complements accurate mass, most likely molecular formula and MS/MS with positive ion mode detection. Importantly, more than 50 drugs of abuse can be simultaneously analyzed with high specificity, adequate sensitivity at recommended screening cut-off concentration levels and minimal sample carry-over.[33] Moreover, MSI-CE-MS provides a robust platform for comprehensive drug screening with acceptable intermediate precision over three days of repeated analysis (n = 117) when using different capillaries and buffer/sheath liquid solutions as reflected by control charts for

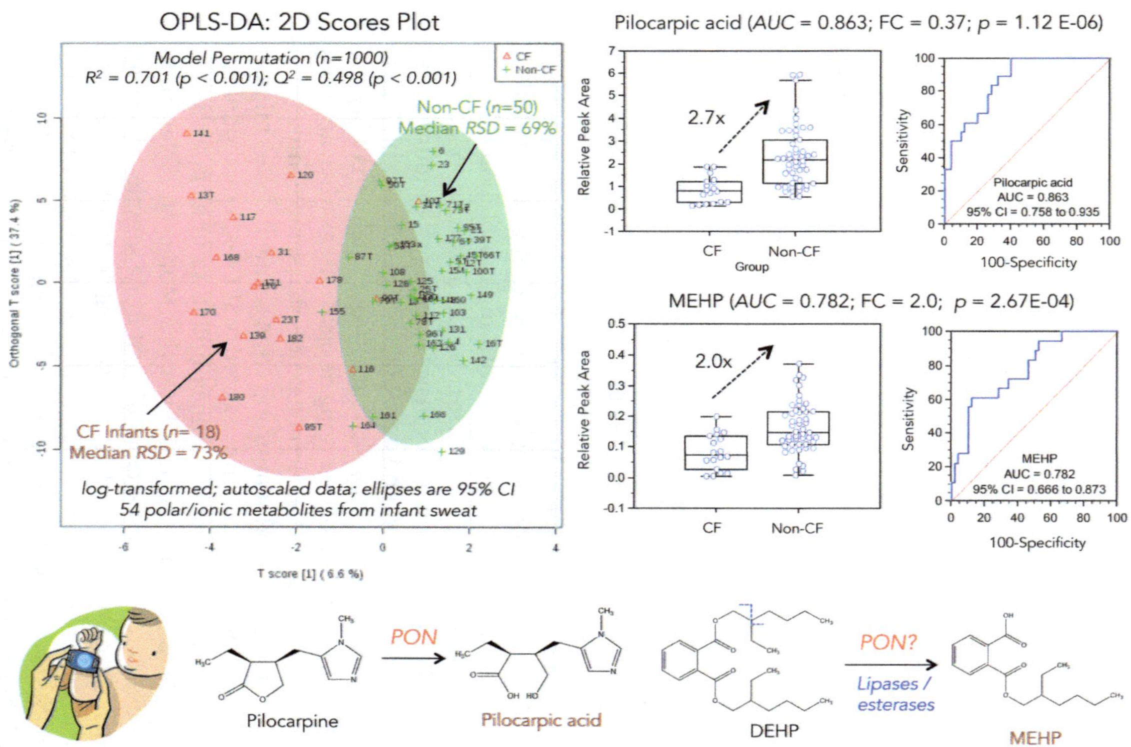

Figure 12.11 Characterization of the sweat metabolome by MSI-CE-MS from screen-positive cystic fibrosis (CF) infants when using orthogonal partial least squares-discriminant analysis (OPLS-DA) for differentiation of CF-affected from non-affected carriers (non-CF). Receiver operating characteristic (ROC) curves for two exogenous sweat metabolites as unanticipated biomarkers associated with CF disease status, which likely reflected an underlying deficiency in the arylesterase/lactonase, human paraoxanase. Reproduced with permission from ref. 30 (http://pubs.acs.org/doi/abs/10.1021/acscentsci.7b00299). Copyright 2017, American Chemical Society. Further permissions related to the material excerpted should be directed to the ACS.

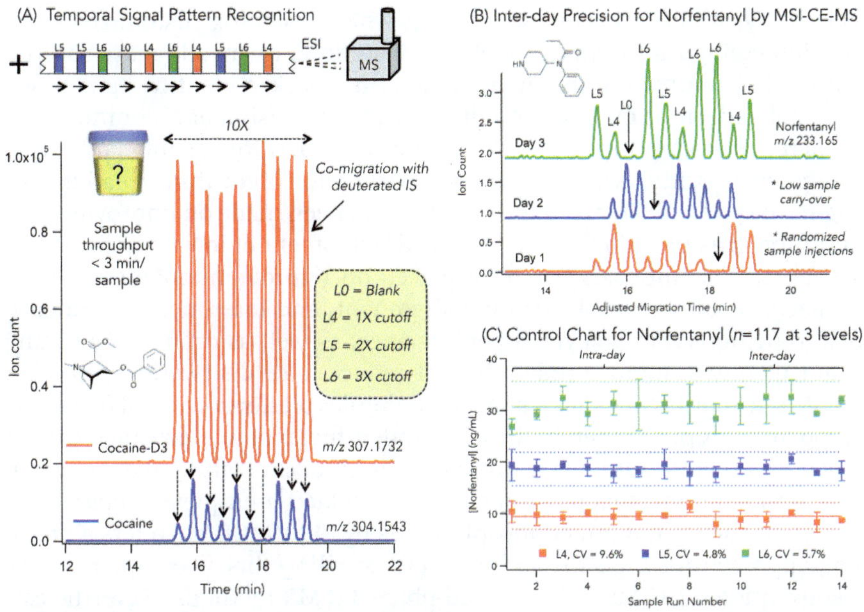

Figure 12.12 High throughput screening of a drug calibrator mixture in a synthetic urine matrix at three different concentration levels (L4: 1.0× cut-off, L5: 2.0× cut-off, L6: 3.0× cut-off) relative to the blank (L0) when using MSI-CE-MS with a 10 sample serial injection format, including (A) cocaine and its matching deuterated internal standard, cocaine-D3. (B) Good inter-day precision was achieved by MSI-CE-MS without significant sample carry-over for a broad array of drugs of abuse and their metabolites, including norfentanyl as reflected by (C) control charts acquired at three different concentration levels. Reproduced with permission from ref. 33 (http://pubs.acs.org/doi/10.1021/acs.analchem.7b03590). Copyright (2017) American Chemical Society. Further permissions related to the material excerpted should be directed to the ACS.

urinary norfentanyl at three different concentration levels. Current work will extensively validate MSI-CE-MS for broad spectrum drug surveillance of psychiatric patients as compared to conventional immunoassays and LC-MS/MS, which is needed for ensuring drug compliance, and/or potential drug substitution, including prescription or illicit drugs not included within established/targeted drug panels.

12.7 Conclusions

This chapter provides an overview of new advances in multiplexed separations based on MSI-CE-MS, which enhance sample throughout by an order of magnitude as compared to conventional "single injection" analyses without added infrastructure costs or modifications. Moreover, this approach also greatly enhances overall data quality when designing customized serial injection configurations and novel data workflows to encode mass spectral information temporally for unambiguous biomarker identification based on signal pattern recognition. Additionally, the inclusion of randomized QC or reference samples within each separation allows for reliable large-scale metabolomic studies when using batch correction algorithms to adjust for long-term system drift. Although progress has been made on improving concentration sensitivity in CE-MS with the development of low flow/sheathless interfaces, less emphasis has been placed on rigorous method validation, open-access software tools for time alignment, and inter-method comparisons to demonstrate analytical performance, as well as proficiency testing and/or quality assurance practices that reduce false discoveries. Although CE-MS is primarily considered as a metabolomics platform optimal for resolving polar/hydrophilic metabolites, expanded metabolome coverage can be extended to include ionic yet hydrophobic lipids of clinical significance when using optimal non-aqueous buffer systems. In fact, multiplexed separations of free fatty acids and intact phospholipids by MSI-NACE-MS offers a promising approach for high throughput lipid profiling that is complementary to conventional GC/MS and reversed-phase LC-MS methods. Nevertheless, increased transparency and better method validation together with well-designed multi-centre round-robin studies are still needed to better showcase method robustness for routine metabolite profiling by MSI-CE-MS that is applicable to a wider network of potential end-users at regional metabolomic facilities, as well as accredited laboratories in clinical and forensic toxicology. In summary, MSI-CE-MS offers a highly versatile platform for overcoming major bottlenecks in discovery-based metabolomics, as well as allowing for population-based metabolite screening with superior cost effectiveness, productivity and quality assurance as compared to conventional separation methods.

Acknowledgements

P. B. M. acknowledges funding from the Natural Sciences and Engineering Research Council of Canada, Cystic Fibrosis Canada, the Canada Foundation for Innovation and Genome Canada. The author also acknowledges support from the Faculty of Science and Health Sciences at McMaster University.

References

1. D. A. Beger, W. Dunn, M. A. Schmidt, S. S. Gross, J. A. Kirwan, M. Cascante, L. Brennan, D. S. Wishart, M. Oresic, T. Hankemeier, D. I. Broadhurst, A. N. Lane, K. Suhre, G. Kastenmüller, S. J. Sumner, I. Thiele, O. Fiehn and R. Kaddurah-Daouk, *Metabolomics*, 2016, **12**, 149.
2. A. M. Fogelman, *Circ. Res.*, 2015, **116**, 396.

3. C. H. Johnson, J. Ivanisevic and G. Siuzdak, *Nat. Rev. Mol. Cell Biol.*, 2016, **17**, 451.
4. R. Langley, E. Tsalik, J. van Velkinburgh, S. Glickman and B. Rice, *et al.*, *Sci. Transl. Med.*, 2013, **5**, 195.
5. L. M. Raamsdonk, B. Teusink, D. Broadhurst, N. Zhang, A. Hayes, M. C. Walsh, J. A. Berden, K. M. Brindle, D. B. Kell, J. J. Rowland, H. V. Westerhoff, K. van Dam and S. G. Oliver, *Nat. Biotechnol.*, 2001, **19**, 45.
6. G. J. Patti, O. Yanes and G. Siuzdak, *Nat. Rev. Mol. Cell Biol.*, 2012, **13**, 263.
7. S. Bouatra, F. Aziat, R. Mandal, A. C. Guo and M. R. Wilson, *et al.*, *PLoS One*, 2013, **8**, e73076.
8. J. Xia, D. I. Broadhurst, M. Wilson and D. S. Wishart, *Metabolomics*, 2013, **9**, 280.
9. N. L. Kuehnbaum and P. Britz-McKibbin, *Chem. Rev.*, 2013, **113**, 2437.
10. J. Xia, I. V. Sinelnikov, B. Han and D. S. Wishart, *Nucleic Acids Res.*, 2015, **43**, W251.
11. J. E. Peironcely, M. Rojas-Chertó, A. Tas, R. Vreeken, T. Reijmers, L. Coulier and T. Hankemeier, *Anal. Chem.*, 2013, **85**, 3576–3883.
12. J. A. Kirwan, R. J. M. Weber, D. I. Broadhurst and M. R. Viant, *Sci. Data*, 2014, **1**, 140012.
13. T. Fuhrer and N. Zamboni, *Curr. Opin. Biotechnol.*, 2015, **31**, 73.
14. J. C. May, C. R. Goodwin and J. A. McLean, *Curr. Opin. Biotechnol.*, 2015, **31**, 117.
15. D. Mung and L. Li, *Anal. Chem.*, 2017, **89**, 4435.
16. M. A. Kamleh, T. M. D. Ebbels, K. Spagou, P. Masson and E. J. Want, *Anal. Chem.*, 2012, **84**, 2670.
17. P. A. Vorkas, G. Isaac, M. A. Anwar, A. D. Davies, E. J. Want, J. K. Nicholson and E. Holmes, *Anal. Chem.*, 2015, **87**, 4184.
18. J. Haggarty and K. E. V. Burgess, *Curr. Opin. Biotechnol.*, 2017, **43**, 77.
19. Y. Ma, N. Tanaka, A. Vaniya, T. Kind and O. Fiehn, *J. Agric. Food Chem.*, 2016, **64**, 505.
20. K. Heinig and F. Bucheli, *J. Chromatogr. B: Anal. Technol. Biomed. Life Sci.*, 2003, **795**, 337.
21. W. Lu, M. F. Clasquin, E. Melamud, D. Amador-Noguez, A. A. Caudy and J. D. Rabinowitz, *Anal. Chem.*, 2010, **82**, 3212.
22. A. García, J. Godzien, Á. López-Gonzálves and C. Barbas, *Bioanalysis*, 2017, **9**, 99.
23. D. Q. Tang, L. Zou, X. X. Yin and C. N. Ong, *Mass Spectrom. Rev.*, 2016, **35**, 574.
24. M. C. Gulersonmez, S. Lock, T. Hankemeier and R. Ramataur, *Electrophoresis*, 2016, **37**, 1007.
25. L. Sun, G. Zhu, Z. Zhang, S. Mou and N. J. Dovichi, *J. Proteome Res.*, 2015, **14**, 2312.
26. J. M. Risely and D. D. Y. Chen, *Electrophoresis*, 2017, **38**, 1644.
27. N. L. Kuehnbaum, A. Kormendi and P. Britz-McKibbin, *Anal. Chem.*, 2013, **85**, 10664.
28. E. P. Diamandis, *BMC Med.*, 2012, **10**, 87.

29. A. DiBattista, N. McIntosh, M. Lamoureux, O. Y. Al-Dirbashi, P. Chakraborty and P. Britz-McKibbin, *Anal. Chem.*, 2017, **89**, 8112.
30. A. Nori de Macedo, S. Mathiaparanam, L. Brick, K. Keenan, T. Gonska, L. Pedder, S. Hill and P. Britz-McKibbin, *ACS Cent. Sci.*, 2017, **3**, 904.
31. J. Cheng, L. Wang, W. Liu and D. D. Y. Chen, *Anal. Chem.*, 2017, **89**, 1411.
32. K. Fukai, S. Harada, M. Iida, A. Kurihara and K. Kuwabara, *et al.*, *PLoS One*, 2016, **11**, e0164877.
33. A. DiBattista, D. Rampersaud, H. Lee, M. Kim and P. Britz-McKibbin, *Anal. Chem.*, 2017, **89**, 11853.
34. N. J. Dovichi and J. Zhang, *Angew. Chem., Int. Ed.*, 2000, **39**, 4463.
35. F. Cotton, F. Wolff and B. Gulbis, *Methods Mol. Biol.*, 2013, **984**, 227.
36. C. D. Whitmore and L. A. Gennaro, *Electrophoresis*, 2012, **33**, 1550.
37. L. A. Gennaro, O. Salas-Solano and S. Ma, *Anal. Biochem.*, 2006, **355**, 249.
38. B. J. Williams, C. J. Cameron, R. Workman, C. D. Broeckling, L. W. Sumner and J. T. Smith, *Electrophoresis*, 2007, **28**, 1371.
39. A. Nori de Macedo, J. Macri, P. L. Hudecki, M. Saoi, M. J. McQueen and P. Britz-McKibbin, *J. Appl. Lab. Med.*, 2017, **2**, 649.
40. S. Sangkitporn, S. K. Sangkitporn, S. Tanjatham, B. Suwannakan, S. Rithapirom, C. Yodtup, A. Yowang and S. Duangruang, *Southeast Asian J. Trop. Med. Public Health*, 2011, **42**, 1224.
41. C. Wenz, C. Barbas, A. López-Gonzálvez, A. Garcia and F. Benavente, *et al.*, *Sep. Sci.*, 2015, **38**, 3262.
42. T. Soga and D. N. Heiger, *Anal. Chem.*, 2000, **72**, 1236.
43. T. Soga, Y. Ueno, H. Naraoka, Y. Ohashi, M. Tomita and T. Nishioka, *Anal. Chem.*, 2002, **74**, 2233.
44. T. Soga, Y. Ueno, H. Naraoka, K. Matsuda, M. Tomita and T. Nishioka, *Anal. Chem.*, 2002, **74**, 6224.
45. J. M. Büscher, D. Czernik, J. C. Ewald, U. Sauer and N. Zamboni, *Anal. Chem.*, 2009, **81**, 2135.
46. T. Soga, K. Igarashi, C. Ito, K. Mizobuchi, H. P. Zimmermann and M. Tomita, *Anal. Chem.*, 2009, **81**, 6165.
47. J. P. A. Ioannidis, *PLoS Med.*, 2005, **2**, e124.
48. P. M. Nowak, M. Wozniakiewicz, M. Gladysz, M. Janus and P. Koscielnak, *Anal. Bioanal. Chem.*, 2017, **409**, 4388.
49. M. Sugimoto, M. Kawakami, M. Robert, T. Soga and M. Tomita, *Curr. Bioinf.*, 2012, **7**, 96.
50. P. M. Nowak, M. Wozniakiewicz and P. Koscielnak, *Anal. Chem.*, 2017, **89**, 3630.
51. W. Zhang, T. Hankemeier and R. Ramataur, *Curr. Opin. Biotechnol.*, 2017, **43**, 1.
52. E. Kenndler, *J. Chromatogr. A*, 2014, **1335**, 16.
53. M. Yamamoto, R. Ly, B. Gill, Y. J. Zhu, J. Moran-Mirabal and P. Britz-McKibbin, *Anal. Chem.*, 2016, **88**, 10710.
54. S. Zhao, X. Luo and L. Li, *Anal. Chem.*, 2016, **88**, 10617.

55. E. J. Want, P. Masson, F. Michopoulos, I. D. Wilson and G. Theodoridis, *et al.*, *Nat. Protocol*, 2013, **8**, 17.
56. T. Soga, M. Sugimoto, M. Honma, M. Mori and K. Igarashi, *et al.*, *J. Hepatol.*, 2011, **55**, 896.
57. A. Hirayama, K. Igarashi, M. Tomita and T. Soga, *J. Chromatogr. A*, 2014, **1369**, 161.
58. F. Matsuda, *Mass Spectrom.*, 2016, **5**, S0052.
59. N. G. Mahieu and G. J. Patti, *Anal. Chem.*, 2017, **89**, 10397.
60. A. Jankevics, M. E. Merlo, M. de Vries, E. Takano and R. Breitling, *Metabolomics*, 2012, **8**, 29.
61. A. Koulman, G. A. Lane, S. J. Harrison and D. A. Volmer, *Anal. Bioanal. Chem.*, 2009, **394**, 663.
62. N. L. Kuehnbaum, A. Kormendi, A. DiBattista, K. P. Lam, J. B. Gillen, M. J. Gibala and P. Britz-McKibbin, *Electrophoresis*, 2015, **36**, 2226.
63. R. D. A. M. Alves, A. D. Dane, A. Harms, K. Strassburg and R. M. Seifar, *et al.*, *Metabolomics*, 2015, **11**, 271.
64. D. Y. Lee, T. Kind, Y. R. Yoon, O. Fiehn and K. H. Liu, *Anal. Bioanal. Chem.*, 2014, **406**, 7275.
65. J. Godzien, M. Ciborowski, E. G. Armitage, I. Jorge and E. Camafeita, *et al.*, *J. Proteome Res.*, 2016, **15**, 1762.
66. K. R. Chalcraft and P. Britz-McKibbin, *Anal. Chem.*, 2009, **81**, 307.
67. K. Raith, R. Wolf, J. Wagner and R. H. H. Neubert, *J. Chromatogr. A*, 1998, **802**, 185.
68. J. H. Kim, S. J. Kim, S. Lee, J. K. Rhee, S. Y. Lee and Y. C. Na, *Anal. Chim. Acta*, 2017, **984**, 223.
69. M. A. de Oliveira, B. L. Porto, I. D. Faria, P. L. de Oliviera, P. M. de Castro Barra, R. de Jesus Coelho Castro and R. T. Sato, *Molecules*, 2014, **19**, 14094.
70. S. A. Abdelmagid, S. E. Clarke, D. E. Nielsen, A. Badawi, A. El-Sohenmy, D. M. Mutch and D. W. K. Ma, *PLoS One*, 2015, **10**, e0116195.
71. K. R. Chalcraft, R. Lee, C. Mills and P. Britz-McKibbin, *Anal. Chem.*, 2009, **81**, 2506.
72. M. Giovanninni, E. Verduci, E. Salvatici, S. Paci and E. Riva, *Nutr. Metab.*, 2012, **9**, 7.
73. H. Blasco, C. Veyrat-Durebex, M. Bertrand, F. Patin and F. Labarthe, *et al.*, *JIMD Rep.*, 2017, **32**, 69.
74. K. Balaji, M. Narayan, S. Bradbury and D. Michael, *Prog. Neurol. Psychiatry*, 2016, 7.
75. L. M. Smith, A. D. Maher, E. J. Want, P. Elliott, J. Stamler, G. E. Hawkes, E. Holmes, J. C. Lindon and J. K. Nicholson, *Anal. Chem.*, 2009, **81**, 4847.
76. C. Brunius, L. Shi and R. Landberg, *Metabolomics*, 2016, **12**, 173.
77. J. A. Kirwan, D. I. Broadhurst, R. L. Davidson and M. R. Viant, *Anal. Bioanal. Chem.*, 2013, **405**, 5147.
78. N. L. Kuehnbaum, J. B. Gillen, M. J. Gibala and P. Britz-McKibbin, *Sci. Rep.*, 2014, **4**, 6166.

79. D. H. Chace and T. A. Kalas, *Clin. Biochem.*, 2005, **38**, 296.
80. F. Vermeulen, P. Lebecque, K. De Boeck and T. Leal, *J. Cystic Fibrosis*, 2017, **16**, 30.
81. A. Saitman, H.-D. Park and R. L. Fitzgerald, *J. Anal. Toxicol.*, 2014, **38**, 387.
82. C. Montesano, G. Vannutelli, A. Gregori, L. Ripani, D. Compagnone, R. Curini and M. Sergi, *J. Anal. Toxicol.*, 2016, **40**, 519.

Subject Index